Modern Cryptography

William Easttom

Modern Cryptography

Applied Mathematics for Encryption and
Information Security

 Springer

William Easttom
Georgetown University
Plano, TX, USA

ISBN 978-3-030-63117-8   ISBN 978-3-030-63115-4   (eBook)
https://doi.org/10.1007/978-3-030-63115-4

This Springer imprint is published by the registered company Springer Nature Switzerland AG
The registered company address is: Gewerbestrasse 11, 6330 Cham, Switzerland

Introduction

Cryptography is an important but difficult topic. A great many professions require some level of use of cryptography. This includes programmers, network administrators, and cybersecurity professionals. However, such people often are not provided adequate training in cryptography. There are a number of excellent cryptography books available, but most assume some level of mathematical sophistication on the part of the reader. This renders them inaccessible to a substantial number of potential readers.

The entire purpose of this book is to bridge this gap. If you are a mathematician or cryptographer, this book is not intended for you. In fact, you might feel frustrated that certain mathematical details are not fully explored. As one glaring example, there are no mathematical proofs in this book. This book is aimed at the person who wants or needs a better understanding of cryptography, but may not have a background in number theory, linear algebra, and other similar topics. When needed, just enough math is provided for you to follow along, and no more detail than is necessary. As a consequence, this book is meant to be accessible to a broad audience. However, that also means that should you master every topic in this book, you would not be a cryptographer. You will, however, be an intelligent consumer of cryptography. You will know the questions to ask vendors, and be able to understand and evaluate their responses.

There are a great many practical questions this book will prepare you to analyze. Should you use elliptic curve or RSA? If you wish to use RSA, what is the appropriate key size? What makes a good cryptographic hashing algorithm? Why is an HMAC better than a hash? Why should you always use CBC mode with symmetric ciphers? What impact will quantum computing have on the current cryptography? These are just a few of the questions you will gain answers to in this book.

Furthermore, this book should provide you a solid foundation should you later wish to delve into more mathematically rigorous cryptography books. Having studied this book, you will be able to venture into deeper waters. There are some excellent books from Springer that go deeper into the mathematics. *Understanding Cryptography* by Paar and Pelsl is one such book. For readers who really want to dive into the math, *An Introduction to Mathematical Cryptography* by Hoffstein, Pipher, and Silverman is an excellent choice.

Contents

About the Author

Dr. William Easttom is the author of 30 books, including several on computer security, forensics, and cryptography. He has also authored scientific papers on many topics including cryptography topics such as s-box design, lattice-based cryptography, and cryptographic backdoors. He is an inventor with 22 computer science patents, several related to steganography. He holds a Doctor of Science in cyber security (Dissertation topic was "A Comparative Study of Lattice Based Algorithms for Post Quantum Computing"), a Ph.D. in Nanotechnology (Dissertation topic was "The Effects of Complexity on Carbon Nanotube Failures"), and three master's degrees (one in applied computer science, one in education, and one in systems engineering). He is a senior member of both the IEEE and the ACM. He is also a Distinguished Speaker of the ACM and a Distinguished Visitor of the IEEE. He is an adjunct lecturer at the Georgetown University teaching several graduate courses including cryptography and cybersecurity.

Chapter 1
History of Cryptography to the 1800s

Abstract Cryptography is an integral part of information security. Thus, the need to understand cryptography is substantial. The study of cryptography can be daunting, particularly for readers who do not have a rigorous mathematical background. It is often useful to begin with studying ancient ciphers. While these ciphers lack sufficient security to be useful today, they are effective at teaching the fundamental concepts of cryptography. These ancient ciphers still have a plaintext, an algorithm, and a key, used to produce the ciphertext. For this reason, ancient ciphers are an appropriate starting point for the study of cryptography.

Introduction

Cryptography is not a new endeavor. Certainly, the way in which we accomplish cryptography is different in the computer age than it was in the past. However, the essential goal of cryptography is the same. The desire to send and receive secure communications is almost as old as written communication itself. For centuries, it almost exclusively involved military secrets or political intrigue. Generals needed to communicate about troop movements to ensure that if a message was intercepted, it wouldn't be intelligible to the enemy. Political intrigue, such as a palace coup, required clandestine communications. Prior to the advent of the computer age, there was little need for cryptography outside of military and political applications. There were, of course, hobbyists who used cryptography for their own intellectual edification. We will encounter some of those hobbyists in this and the next chapter.

Modern cryptography certainly still includes military communications, as well as political communications, but it has expanded into more mundane areas as well. Online banking and shopping, for example, have made cryptography a part of most people's daily lives, whether they are aware of it or not. Many people also choose to encrypt their computer hard drives or files. Others encrypt their e-mail transmissions. Today, cryptography permeates our lives. Most people use cryptography with little or no awareness of how it works. And for many people that lack of detailed knowledge is sufficient. But many professionals require a deeper understanding of cryptography. Cybersecurity professionals, network administrators, and cyber security personnel would benefit from a better understanding. For these professionals, knowing more about cryptography will at least allow them to make better decisions regarding the implementation of cryptography. Deciding which symmetric algorithm to utilize for encrypting sensitive information or which key exchange algorithm is most resistant to man-in-the middle attacks requires a bit more knowledge of cryptography.

In This Chapter We Will Cover

- Single substitution ciphers
- Multi-alphabet substitution
- Devices
- Transposition ciphers

Why Study Cryptography?

It is an unfortunate fact that most people have almost no knowledge of cryptography. Even within the discipline of computer science, and more specifically the profession of cyber security, a lack of cryptographic understanding plagues the industry. Most cybersecurity professionals have only the most basic understanding of cryptography. For many in cybersecurity, their knowledge of cryptography does not extend beyond the few basic concepts that appear on common cybersecurity certification tests such as CompTIA Security+, or ISC2 CISSP. Many feel that this level of knowledge makes them well informed about cryptography and are not even aware of how much they do not know. Some would even argue that a deeper knowledge of cryptography is unnecessary. It can be assumed that since you are reading this book, you feel a need to deepen and broaden your knowledge of cryptography, and there are clearly practical reasons to do so, particularly for those in the cybersecurity profession.

By understanding cryptology, you can select the most appropriate cryptographic implementations to suit your needs. Even if you have no desire to be a cryptographer, you still have to choose which tool to use to encrypt a hard drive, for example. Should you use the Data Encryption Standard (DES)? Triple DES (3DES)? Blowfish? The Advanced Encryption Standard (AES)? If you use AES, then what key size do you use, and why? Why is CBC mode preferable to ECB mode? If you are interested in message integrity for e-mail, should you use the Secure Hash Algorithm (SHA-2)? Perhaps a message authentication code (MAC) or hash message authentication code (HMAC)? Which will provide the most appropriate solution to your particular problem, and why? Many people are not even aware of the differences in algorithms.

In addition, knowing about cryptology helps you understand issues that occur when cryptographic incidents broadly impact cybersecurity. A good example occurred in 2013, when *The New York Times* reported that among the documents released by National Security Administration, subcontractor Edward Snowden was evidence that the NSA had placed a cryptographic backdoor in the random number generator known as *Dual_EC_DRBG (Elliptic Curve Deterministic Random Bit Generator)*. This news story generated a flood of activity in the cybersecurity community. But what is a cryptographic backdoor? What does this mean for privacy and security? Does this mean that the NSA could read anyone's e-mail as if it were published on a billboard? And more importantly, why did this surprise the cybersecurity community, when the cryptography community had been speculating about this possibility since 2005? This story illustrates a substantial gap between the cryptography community and the cybersecurity community.

We will be exploring all of these issues, as well as random number generators and even quantum computing, later in this book. For the time being, it is sufficient for you to realize that you cannot answer any questions about this particular news story without having some knowledge of cryptography. This story, in fact, was not news to the cryptographic community. As early as 2005, papers had been published that suggested the possibility of a backdoor in this random number generator. Well-known and respected cryptographer Bruce Schneier, for example, blogged about this issue in 2006. Had the security community possessed a deeper knowledge of cryptography, this backdoor would have been a nonstory.

Another example comes from the aforementioned cybersecurity certifications. Many such certification exams ask about wireless security. Particularly about WPA2, and now WPA3. Many of you reading this text have undoubtedly passed such certifications. Based on that you may well be able to assert that WPA2 utilizes The Advanced Encryption Standard (AES) using the Counter Mode-Cipher Block Chaining (CBC)-Message Authentication Code (MAC) Protocol (CCMP). But do you know what CBC is, or why it is so useful. In this book, you will learn about these issues and many more.

I could continue with other reasons—very practical reasons—why learning cryptography is essential, and you will see some of those reasons in later chapters. Cryptography is not merely a mathematical endeavor to be engaged in by a select few mathematicians and cryptologists. In this chapter, you will begin your journey into the world of cryptography by learning some essential terms and then exploring some historical ciphers. It must be clear, however, that this book will not make you a cryptographer. That does require much more extensive mathematical knowledge. However, it will provide you with sufficient understanding to make good cryptographic choices and to ask effective questions about cryptographic solutions.

What Is Cryptography?

Before you can begin studying cryptography, you need to know what exactly cryptography is. A number of people seem to have some misconceptions about what *cryptography* and related terms actually mean. The Merriam-Webster online dictionary defines cryptography as follows "1) secret writing 2) the enciphering and deciphering of messages in secret code or cipher; also: the computerized encoding and decoding of information." This definition does not seem overly helpful and may not provide you with much insight into the topic. Columbia University provides a slightly better definition in its "Introduction to Cryptography" course: to "process data into unintelligible form, reversibly, without data loss—typically digitally."

The Columbia University definition adds the important element of not losing information, certainly a critical component in secure communications. However, I cannot help but think that the definition could be a bit clearer on the issue of exactly what cryptography is. So, allow me to try my hand at defining cryptography:

> Cryptography is the study of how to alter a message so that someone intercepting it cannot read it without the appropriate algorithm and key.

This definition is certainly not radically different from that of either Merriam-Webster or Columbia University. However, I think it is a good, concise definition, and one we will use throughout this book.

Note that *cryptography* and *cryptology* are not synonyms, though many people mistakenly use the terms as if they were. In fact, many textbooks utilize the terms interchangeably. However, I will define and differentiate these two terms, as well as some other common terms you will need throughout your study of cryptography. These terms are used in both ancient and modern cryptography.

- **Cipher** A synonym for the algorithm used in transforming plaintext to ciphertext.
- **Ciphertext** The coded or encrypted message. If your encryption is sufficiently strong, your ciphertext should be secure.
- **Cryptanalysis** Also known as code breaking; the study of principles and methods of deciphering ciphertext without knowing the key. This is more difficult than movies or television would indicate, as you will see in Chap. 17.
- **Cryptography** The study of how to alter a message so that someone intercepting it cannot read it without the appropriate algorithm and key.
- **Cryptology** Although some people, including more than a few cybersecurity books, use the terms *cryptography* and *cryptology* interchangeably, that is inaccurate. Cryptology is more comprehensive and includes both cryptography and cryptanalysis.
- **Decipher (decrypt)** *Decipher* and *decrypt* are synonyms. Both terms mean to convert the ciphertext to plaintext.
- **Encipher (encrypt)** *Encipher* and *encrypt* are synonyms. Both words mean to convert the plaintext into ciphertext.
- **Key** The information, usually some sort of number, used with the algorithm to encrypt or decrypt the message. Think of the key as the fuel the algorithm requires in order to function.
- **Key space** The total number of possible keys that could be used. For example, DES uses a 56-bit key; thus, the total number of possible keys, or the key space, is 2^{56}.
- **Plaintext** The original message—the information you want to secure.

These are some of the most basic terms that permeate the study of cryptology and cryptography. In any discipline, it is important that you know, understand, and use the correct vocabulary of that field of study. These terms are essential for your understanding.

If you suppose that you cannot study cryptography without a good understanding of mathematics, to some extent you are correct. Modern methods of cryptography, particularly asymmetric cryptography, depend on mathematics. We will examine those algorithms later in this book, along with the mathematics you need to understand modern cryptography. It is often easier for students first to grasp the concepts of cryptography within the context of simpler historical ciphers, however. These ciphers don't require any substantive mathematics at all, but they do use the same concepts you will encounter later in this book. It is also good to have an historical perspective on any topic before you delve deeper into it. In this chapter, we will examine a history of cryptography, looking at specific ciphers that have been used from the earliest days of cryptography to the 1800s.

Let us begin our study of historical cryptography by examining the most common historical ciphers. These are fascinating to study and illustrate the fundamental concepts you need in order to understand cryptography. Each one will demonstrate an algorithm, plaintext, ciphertext, and a key. The implementations, however, are far simpler than those of modern methods and make it relatively easy for you to master these ancient methods. Keep in mind that these ciphers are totally inadequate for modern security methods. They would be cracked extremely quickly with a modern computer, and many can be analyzed and cracked with a pen, paper, and the application of relatively simple cryptanalysis legerdemain.

Substitution Ciphers

The first ciphers in recorded history are *substitution ciphers*. With this method, each letter of plaintext is substituted for some letter of ciphertext according to some algorithm. There are two types of substitution ciphers: single-alphabet (or mono-alphabet) and multi-alphabet (or poly-alphabet). In a single-alphabet substitution cipher, a given letter of plaintext is always substituted for the corresponding letter of ciphertext. For example, an *a* in the plaintext would always be a *k* in the ciphertext. Multi-alphabet substitution uses multiple substitutions, so that, for example, an *a* in the plaintext is sometimes a *k* and sometimes a *j* in the ciphertext. You will see examples of both in this section.

The Caesar Cipher

One of the most widely known historical encryption methods is the *Caesar cipher*. According to the Roman historian Gaius Suetonius Tranquillus (c. 70–130 CE),

Julius Caesar used this cipher to encrypt military messages, shifting all letters of the plaintext three places to the right (d'Agapeyeff 2016). So, for example, the message

```
Attack at dawn
```

becomes

```
Dwwdfn dw gdzq
```

As you can see, the *a* in the plaintext is shifted to the right three letters to become a *d* in the ciphertext. Then the *t* in the plaintext is shifted three letters to the right to become a *w* in the ciphertext. This process continues for all the letters in the plaintext. In our example, none of the shifts went beyond the letter *z*. What would happen if we shifted the letter *y* to the right three? The process would wrap around the alphabet, starting back at letter *a*. Thus, the letter *y* would be shifted to a letter *b* in the ciphertext.

Although Caesar was reputed to have used a shift of three to the right, any shifting pattern will work with this method, shifting either to the right or left by any number of spaces. Because this is a quite simple method to understand, it is an appropriate place to begin our study of encryption. It is, however, extremely easy to crack. You see, any language has a certain letter and word *frequency*, meaning that some letters are used more frequently than others. In the English language, the most common single-letter word is *a*, followed closely by the word *I*. The most common three-letter word is *the,* followed closely by the word *and*. Those two facts alone could help you decrypt a Caesar cipher. However, you can apply additional rules. For example, in the English language, the most common two letter sequences are *oo* and *ee*.[4] Examining the frequency of letter and letter combination occurrences is called *frequency analysis.*

It is claimed that other Caesars, such as Augustus, used variations of the Caesar cipher, such as 1 shift to the right. It should be obvious that any shift, left or right, of more than 26 (at least in English) would simply loop around the alphabet. So, a shift to the right of 27 is really just a shift of 1.

Although the Caesar cipher is certainly not appropriate for modern cryptographic needs, it does contain all the fundamental concepts needed for a cryptography algorithm. First, we have the plaintext message—in our current example, *Attack at dawn*. Then we have an algorithm—shift every letter. And then a key, in this case *+3*, or three to the right (−*3* would be three to the left). And finally, we have ciphertext, *Dwwdfn dw gdzq*. This is, essentially, the same structure used by all modern symmetric algorithms. The only differences between the Caesar cipher and modern symmetric ciphers are the complexity of the algorithm and the size of the key.

The size of the key brings us to one significant problem with the Caesar cipher— its small key space. Recall that key space is the total number of possible keys. Because there are only 26 letters in the English alphabet, the key space is *26* (that is, *+−26*). It would be relatively easy for a person working with pen and paper to check all possible keys, and it would be ridiculously trivial for a computer to do so. In the cybersecurity world, a malicious person who checks all possible keys to decipher an encrypted message is conducting what it called a *brute-force attack*. The smaller the key space, the easier a brute-force attack will be. Compare the Caesar cipher, with a key space of 26, to AES 128-bit, with a key space of 2^{128}, or about 3.4×10^{38}. Clearly, the larger key space makes a cipher more resistant to brute-force attacks. Note, however, that simply having a long key is not sufficient for security. You will learn more about this when we discuss cryptanalysis in Chap. 17.

Mathematical Notation of the Caesar Cipher

With the various ancient ciphers, we will be using, the math is trivial. However, it is a good idea for you to become accustomed to mathematical notation, at least with those algorithms where such notation is appropriate. It is common to use a capital letter P to represent plaintext and a capital letter C to represent ciphertext. We can also use a capital letter K to represent the key. This gives us the following mathematical description of a Caesar cipher:

$$C \equiv P + K \ (\mathrm{mod}\ 26)$$

Here we see a symbol some readers may not be acquainted with, the $\equiv$. This is not a misprint of the $=$ sign, rather it is the symbol for congruence. Do not be overly concerned about the $\equiv$ 26. We will explore modulus operations and congruence in detail in Chap. 4. For now, I just use the modulus operation to denote dividing by a given number (in this case, 26, because there are 26 letters in the alphabet) and listing only the remainder. That is not a rigorous mathematical explanation, but it will suffice for now.

Decryption can also be represented via mathematical symbols:

$$P \equiv C - K \ (\mathrm{mod}\ 26)$$

The mathematical representation of Caesar's method of shifting three to the right is

$$C \equiv P + 3 \ (\mathrm{mod}\ 26)$$

According to the book *The Lives of the Caesars*, written by Suetonius, Julius Caesar used this cipher extensively:

There are also the letters of his to Cicero, as well as to his intimates on private affairs, and in the latter, if he had anything confidential to say, he wrote it in cipher, that is by so changing the order of the letters of the alphabet, that not a word could be made out. If anyone wishes to decipher these, and get at their meaning, he must substitute the fourth letter of the alphabet, namely D, for A, and so with the others.[5]

If the plaintext is the 24th letter of the alphabet (which is the letter X), then the ciphertext is $(24 + 3)/26$, listing only the remainder. Thus, $27/26 = 1$, or the letter A.

We cannot know how effective the Caesar cipher was at concealing messages. However, at the time of Julius Caesar, illiteracy was common, and cryptography was not widely known. So, what may seem a trivial, even frivolous, cipher today may well have been effective enough more than 2000 years ago.

The Caesar cipher is probably the most widely known substitution cipher, but it is not the only one. All substitution ciphers operate in a similar fashion: by substituting each letter in the plaintext for some letter in the ciphertext, with a one-to-one relationship between the plaintext and ciphertext. Let's look at a few other substitution ciphers.

Atbash Cipher

Hebrew scribes copying the biblical book of Jeremiah used the *Atbash substitution cipher*. Applying the Atbash cipher is fairly simple: just reverse the order of the letters of the alphabet. This is, by modern standards, a very primitive cipher that is easy to break. For example, in English, *a* becomes *z*, *b* becomes *y*, *c* becomes *x*, and so on. Of course, the Hebrews used the Hebrew alphabet, with *aleph* being the first letter and *tav* the last letter. However, I will use English examples to demonstrate the cipher:

```
Attack at dawn
```

   becomes

```
Zggzxp zg wzdm
```

As you can see, the A (the first letter in the alphabet) is switched with Z (the last letter), and the t is the 19th letter (or 7th from the end) and gets swapped with g, the 7th letter from the beginning. This process is continued until the entire message is enciphered.

To decrypt the message, you simply reverse the process so that z becomes a, b becomes y, and so on. This is obviously a simple cipher and is not used in modern times. However, like the Caesar cipher example, it illustrates the basic concept of cryptography—to perform some permutation on the plaintext to render it difficult to read by those who don't have the key to "unscramble" the ciphertext. The Atbash cipher, like the Caesar cipher, is a single-substitution cipher (each letter in the plaintext has a direct, one-to-one relationship with each letter in the ciphertext). The same letter and word frequency issues that can be used to crack the Caesar cipher can be used to crack the Atbash cipher.

Affine Ciphers

Affine ciphers are any single-substitution alphabet ciphers (also called *mono-alphabet substitution*) in which each letter in the alphabet is mapped to some numeric value, permuted with some relatively simple mathematical function, and then converted back to a letter. For example, using the Caesar cipher, each letter is converted to a number, shifted by some amount, and then converted back to a letter.

The basic formula for any affine cipher is

$$ax + b \ (\text{mod } m)$$

M is the size of the alphabet—so in English that would be 26. The *x* represents the plaintext letter's numeric equivalent, and the *b* is the amount to shift. The letter *a* is some multiple—in the case of the Caesar cipher, *a* is 1. So, the Caesar cipher would be

$$1x + 3 \ (\text{mod } 26)$$

What has been presented thus far is rather simplified. To actually use an affine cipher, you need to pick the value a so that it is coprime with m. We will explore coprime in more detail later in this book. However, for now simply understand that two numbers are coprime if they have no common factors. For example, the number 8 has the factors 2 and 4. The number 9 has the factor 3. Thus, 8 and 9 have no common factors are coprime. If you don't select *a* and *m* that are coprime, it may not be possible to decrypt the message.

Continuing with a simplified example (ignoring the need for coprime *a* and *m*), you could obviously use any shift amount you want, as well as any multiplier. The *ax* value could be 1*x*, as with Caesar, or it could be 2*x*, 3*x*, or any other value. For example, let's create a simple Affine cipher:

$$2x + 4 \ (\text{mod } 26)$$

To encrypt the phrase *Attack at dawn*, we first convert each letter to a number, then multiply that number by 2 and add 4. So, A is *1*, 2 multiplied by 1 is still 2, add 4, gives us 6 mod 26 yielding 6, or *F*.

Then we have *t*, which is 20, and 2 multiplied by 20 is 40, add 4, which gives us 44, and 44 mod 26 yields 18, or *r*. Ultimately, we get this:

```
Attack at dawn
Frrfj0 fr lfxf
```

Notice that the letter *k* did not convert to a letter; instead, a 0 (zero) appears. *K* is the 11th letter of the alphabet, and 2*x* + 4, where *x* = 11, equals 26. And 26 mod 26 is 0.

This is one example of an affine cipher, and there are quite a few others. As you have just seen, you can easily create one of your own. You would want to limit your selection of *a* to values that produce only integer results, rather than decimals. A value of 1.3*x*, for example, would lead to decimal values, which could not easily be converted to letters. We know that *1* is *a* and *2* is *b*, but what letter is *1.3*?

All affine ciphers have the same weaknesses as any single-substitution cipher. They all preserve the letter and word frequencies found in the underlying language and are thus susceptible to frequency analysis. In fact, no matter how complex you make the permutation, any single-substitution cipher is going to be vulnerable to frequency analysis.

ROT 13

ROT 13 is a trivial single-substitution cipher that also happens to be an affine cipher. *ROT* is short for *rotate*: each letter is rotated to the right by 13. So, the affine representation of the ROT 13 (in English) is

$$1x + 13 \pmod{26}$$

Since the Latin alphabet has 26 letters, simply applying ROT 13 a second time will decrypt the message. As you can probably guess, this is not at all secure by modern standards. However, it is actually used in some situations. For example, some of the keys in the Microsoft Windows Registry are encrypted using ROT 13. In this case, the reasoning is likely to be that, first and foremost, you need access to the system before you can explore the Windows Registry, and second, most people are not well versed in the Windows Registry and would have difficulty finding specific items there even if they were not encrypted at all, so ROT 13 may be secure enough for this scenario. I am not necessarily in agreement with that outlook, but it is a fact that the Windows Registry uses ROT 13.

It has also been reported that in the late 1990s, Netscape Communicator used ROT 13 to store e-mail passwords. ROT 13 has actually become somewhat of a joke in the cryptology community. For example, cryptologists will jokingly refer to "ROT 26," which would effectively be no encryption at all. Another common joke is to refer to "triple ROT 13." Just a brief reflection should demonstrate to you that the second application of ROT 13 returns to the original plaintext, and the third application of ROT 13 is just the same as the first.

Homophonic Substitution

Over time, the flaws in single-substitution ciphers became more apparent. *Homophonic substitution* was one of the earlier attempts to make substitution ciphers more robust by masking the letter frequencies, as plaintext letters were mapped to more than one ciphertext symbol, and usually the higher frequency plaintext letters were given more ciphertext equivalents. For example, *a* might map to either *x* or *y*. This had the effect of disrupting frequencies, making analysis more difficult. It was also possible to use invented symbols in the ciphertext and to have a variety of mappings. For example, *a* maps to *x*, but *z* maps to ¥. The symbol ¥ is one I simply created for this example.

There are variations of this cipher, and one of the most notable versions is called the *nomenclator cipher*, which used a codebook with a table of homophonic substitutions. Originally, the codebook used only the names of people, thus the term nomenclator. So, for example, Mr. Smith might be *XX* and Mr. Jones would be *XYZ*. Eventually, nomenclators were created that used a variety of words rather than just names. The codes could be random letters, such as those already described, or code words. Thus, Mr. Jones might be enciphered as *poodle* and Mr. Smith enciphered as *catfish*. Such codebooks with nomenclator substitutions where quite popular in espionage for a number of years. The advantage of a nomenclator is that it does not provide any frequencies to analyze. However, should the codebook become compromised, all messages encoded with it will also be compromised.

The Great Cipher

The *Great Cipher* is one famous nomenclator used by the French government until the early 1800s. This cipher was invented by the Rossignol family, a French family with several generations of cryptographers, all of whom served the French court. The first, a 26-year-old Rossignol mathematician, served under Louis XIII, creating secure codes.

The Great Cipher used 587 different numbers that stood for syllables (note that there were variations on this theme, some with a different number of codes). To help prevent frequency analysis, the ciphertext would include nulls, or numbers that meant nothing. There were also traps or codes that indicated the recipient should ignore the previous coded message.

Copiale Cipher

This is an interesting homophonic cipher. It was a 105-page, 75,000-character, handwritten manuscript that went unbroken for many years. The Copiale cipher used a complex substitution code that used symbols and letters for both texts and spaces. The document is believed to date from the 1700s from a secret society named the "high enlightened occultist order of Wolfenbüttel." The cipher included abstract symbols, Greek letters, and Roman letters. It was finally cracked in 2011 with the help of computers.

Polybius Cipher

The *Polybius cipher* (also known as the Polybius square) was invented by the Greek historian Polybius (c. 200–118 BCE). Obviously, his work used the Greek alphabet, but we will use it with English here. As shown in the following grid, in the Polybius cipher, each letter is represented by two numbers (Mollin 2000). Those two numbers being the x and y coordinates of that letter on the grid. For example, *A* is 1 1, *T* is 4 4, *C* is 1 3, and *K* is 2 5. Thus, to encrypt the word *attack*, you would use 114444111325. You can see this in Fig. 1.1.

Despite the use of two numbers to represent a single letter, this is a substitution cipher and still maintains the letter and word frequencies found in the underlying language of the plaintext. If you used the standard Polybius square, which is a widely known cipher, it would be easily cracked, even without any frequency analysis. If you wanted to use a different encoding for letters in the square, that would require that the sending and receiving parties share the particular Polybius square in advance, so that they could send and read messages.

It is interesting to note that the historian Polybius actually established this cipher as a means of sending codes via torches. Messengers standing on hilltops could hold up torches to represent letters, and thus send messages. Establishing a series of such messengers on hilltops, each relaying the message to the next, allowed communications over a significant distance, much faster than any messenger on foot or horseback could travel.

Ancient Cryptography in Modern Wars

Here is a very interesting story that does not necessarily fit with the timeline of this chapter (pre-twentieth century), but it does concern the Polybius square. The Polybius square was used by prisoners of war in Vietnam, who communicated via tapping on a wall to signal letters. Therefore, for example, four taps, a pause, and then two

Fig. 1.1 Polybius square

	1	2	3	4	5
1	A	B	C	D	E
2	F	G	H	I/J	K
3	L	M	N	O	P
4	Q	R	S	T	U
5	V	W	X	Y	Z

taps would be the letter *R*. When used in this fashion, it is referred to as a *tap code*. This cipher was introduced into the POW camps in 1965 by Captain Carlyle Harris, Lieutenant Phillip Butler, Lieutenant Robert Peel, and Lieutenant Commander Robert Shumaker, all imprisoned at the Hoa Lo prisoner of war camp. It is reported that Harris recalled being introduced to the Polybius square by an instructor during his training. He then applied the Polybius square to a tap code so that he and his fellow prisoners could communicate. This technique was taught to new prisoners and became widespread in the POW camp. Vice Admiral James Stockdale wrote about using the tap code, stating, "Our tapping ceased to be just an exchange of letters and words; it became conversation. Elation, sadness, humor, sarcasm, excitement, depression—all came through."[8] This is a poignant example of cryptography being applied to very practical purposes.

Null Cipher

The *null cipher* is a very old cipher—in fact, by today's standards, it might be considered more steganography than cipher (you'll read about steganography in Chap. 16). Essentially, the message is hidden in unrelated text. So, in a message such as

We are having breakfast at noon at the cafe, would that be okay?

the sender and recipient have prearranged to use some pattern, taking certain letters from the message. So, for example, the numbers

3 20 22 27 32 48

would signify the letters in the sentence and provide the message

```
attack
```

The pattern can be complex or simple—such as always using the second letter of each word or any other pattern. In addition, punctuation and spaces could be counted as characters (our example ignored punctuation and spaces).

Multi-Alphabet Substitution

As you know, any single-alphabet substitution cipher is susceptible to frequency analysis. The most obvious way to improve such ciphers would be to find some mechanism whereby the frequency of letters and words could be disrupted. Eventually, a slight improvement on the single-substitution cipher was developed, called *multi-alphabet substitution*. In this scheme, you select multiple numbers by which to shift letters (that is, multiple substitution alphabets). For example, if you select three substitution alphabets (+1, +2, and + ×), then

Attack at dawn

becomes

Bvwben bv gbxo

In this example, the first letter was shifted forward by one, so *A* became *B*; the second letter was shifted forward by two, so *t* became *v*; the third letter was shifted forward by three, so in this case *t* became *w*. Then you start over with one shift forward. It should be abundantly clear that the use of multiple alphabets changes letter and word frequency. The first letter *t* became *v*, but the second letter *t* became *w*. This disrupts the letter and word frequency of the underlying plaintext. The more substitution alphabets that are utilized, the more disruption there will be to the letter and word frequency of the plaintext. This disruption of the letter and word frequency overcomes the weaknesses of traditional single-substitution ciphers. There are a variety of methods for making a multi-alphabet substitution cipher. We will examine a few of the most common multi-alphabet ciphers in the following sections.

Tabula Recta

Tabula recta is one of the earliest major multi-alphabet substitution ciphers. It was invented in the sixteenth century by Johannes Trithemius. A tabula recta is a square table of alphabets made by shifting the previous alphabet to the right, as shown in Fig. 1.2.

This essentially creates 26 different Caesar ciphers. Trithemius described this in his book *Polygraphia*, which is presumed to be the first published book on cryptology. To encrypt a message using this cipher, you substitute the plaintext letter for the letter that appears beneath it in the table. Basically, the first letter of the plaintext (denoting the row) is matched with the first letter of the keyword (denoting the column), and the intersection of the two forms the ciphertext. This is repeated with each letter. When the end of the keyword is reached, you start over at the beginning of the keyword. Trithemius used a fixed keyword, so although this did change the frequency distributions found in single-substitution ciphers, it still had a significant flaw when compared to later developments such as Vigenère.

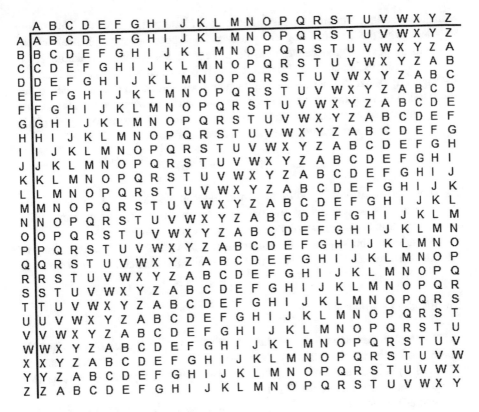

Fig. 1.2 Tabula recta

Vigenère

Perhaps the most widely known multi-alphabet cipher is the *Vigenère cipher*. This cipher was first described in 1553 by Giovan Battista Bellaso, though it is misattributed to nineteenth century cryptographer Blaise de Vigenère (Singh 2000). It is a method of encrypting alphabetic text by using a series of different mono-alphabet ciphers selected based on the letters of a keyword. Bellaso also added the concept of using any keyword, thereby making the choice of substitution alphabets difficult to calculate. Essentially, the Vigenère cipher uses the tabula recta with a keyword. So, let us assume you have the word *book*, and you wish to encrypt it. You have a keyword for encryption, that keyword is *dog*. You would like up the first letter of your plaintext *b* on the left-hand side of the tabula recta, with the first letter or your keyword *d* on the top. The first letter of your ciphertext is then *e*. Then you take the second letter of your plaintext *o* and line it up with the second letter of the keyword, also *o*, producing the second letter of your cipher text *c*. The next *o* in book, will line up with the *g* in dog, producing *u*. Now that you have reached the end of your keyword, you start over at *d*. So, the *k* in book is lined up with the d in dog, producing the last letter of your ciphertext, which is n. Thus, using Vigenère, with the keyword dog, the plaintext book becomes the cipher text *ecun*.

For many years, Vigenère was considered very strong—even unbreakable. However, in the nineteenth century, Friedrich Kasiski published a technique for breaking the Vigenère cipher. We will revisit that when we discuss cryptanalysis later in this book. It is important that you get accustomed to mathematical notation. Here, using P for plaintext, C for ciphertext, and K for key, we can view Vigenère very similarly to Caesar, with one important difference: the value K changes.

$$Ci = Pi + Ki \pmod{26}$$

The i denotes the current key with the current letter of plaintext and the current letter of ciphertext. Note that many sources use M (for message) rather than P (for plaintext) in this notation. Let us assume the word you wish to.

A variation of the Vigenère, the *running key cipher*, simply uses a long string of random characters as the key, which makes it even more difficult to decipher.

The Beaufort Cipher

The *Beaufort cipher* also uses a tabula recta to encipher the plaintext. A keyword is preselected by the involved parties. This cipher was created by Sir Francis Beaufort (1774–1857) and is very similar to the Vigenère cipher. A typical tabula recta was shown earlier in this chapter in Fig. 1.1.

When using the Beaufort cipher, you select a keyword, except unlike Vigenère, you locate the plaintext in the top row, move down until you find the matching letter of the keyword, and then choose the letter farthest to the left in the row as the ciphertext.

For example, using the tabula recta in Fig. 1.1, and the keyword *falcon*, you would encrypt the message *Attack at dawn* in the following manner:

1. Find the letter A on the top row.
2. Go straight down that column until you find the letter F (the first letter of the keyword).
3. Use the letter in the far-left column as the ciphertext letter. In this case, that would be F.
4. Repeat this, except this time use the next letter of the keyword, a. Locate the second letter of the plaintext t in the top row.
5. Move down that column until you find an a, and then we select the letter on the far left of that row, which would be h.
6. When you reach the last letter of the keyword, you start over at the first letter of the keyword, so that.

```
Attack at dawn
```

becomes

```
Fhscmdfhicsa
```

Devices

In modern times, devices are almost always used with cryptography. For example, computers are used to encrypt e-mail, web traffic, and so on. In ancient times, there were also ciphers based on the use of specific devices to encrypt and decrypt messages.

Scytale Cipher

The *Scytale cipher* is one such ancient cypher. Often mispronounced (it actually rhymes with "Italy"), this cipher used a cylinder with a strip of parchment wrapped around it. If you had the correct diameter cylinder, then when the parchment was wrapped around it, the message could be read (Dooley 2018). You can see the concept shown in Fig. 1.3.

If you did not have the correct size of cylinder, however, or if you simply found the parchment and no cylinder, the message would appear to be a random string of letters. This method was first used by the Spartans and later throughout Greece. The earliest mention of Scytale was by the Greek poet Archilochus in the seventh century BC. However, the first mention of how it actually worked was by Plutarch in the first century BCE, in his work *The Parallel Lives*:

> The dispatch-scroll is of the following character. When the ephors send out an admiral or a general, they make two round pieces of wood exactly alike in length and thickness, so that each corresponds to the other in its dimensions, and keep one themselves, while they give the other to their envoy. These pieces of wood they call "scytale." Whenever, then, they wish to send some secret and important message, they make a scroll of parchment long and narrow, like a leather strap, and wind it round their "scytale," leaving no vacant space thereon, but covering its surface all round with the parchment. After doing this, they write what they wish on the parchment, just as it lies wrapped about the "scytale;" and when they have written their message, they take the parchment off, and send it, without the piece of wood, to the

Fig. 1.3 Scytale

commander. He, when he has received it, cannot other get any meaning of it—since the letters have no connection, but are disarranged—unless he takes his own "scytale" and winds the strip of parchment about it, so that, when its spiral course is restored perfectly, and that which follows is joined to that which precedes, he reads around the staff, and so discovers the continuity of the message. And the parchment, like the staff, is called "scytale," as the thing measured bears the name of the measure.[9]

Alberti Cipher Disk

The Alberti *cipher disk*, created by Leon Battista Alberti, is an example of a multi-alphabet substitution. Alberti wrote about this cipher in 1467 in his book *De Cifris*. It consists of two disks attached in the center with a common pin. Each disk had 24 equal cells. The larger, outer disk, called the *stabilis*, displayed an uppercase Latin alphabet used for the plaintext. The smaller, inner disk, called the *mobilis*, displayed a lowercase alphabet for the ciphertext.

To encrypt a message, a letter on the inner disk was lined up with a letter on the outer disk as a key. If you knew what letter to line up with, you would know which key to use. This has the effect of offering multiple substitution alphabets. You can see an example of the cipher disk, with the English alphabet, in Fig. 1.4.

In Alberti's original cipher disk, he used the Latin alphabet. So, the outer disk had the Latin alphabet minus a few English letters, as well as numbers 1 through 4 for use with a codebook that had phrases and words assigned four-digit values.

The Jefferson Disk

The *c*, which was called a "wheel cipher" by its inventor, Thomas Jefferson, is a rather complex device, at least for its time. Invented in 1795, the disk is a set of wheels or disks, each displaying the 26 letters of the English alphabet. The disks are all on a central axle and can be rotated about the axle. The order of the disks is the

Fig. 1.4 Cipher disk

Fig. 1.5 Jefferson disk

key, and both sender and receiver had to order their disks according to the key. An example of the Jefferson disk is shown in Fig. 1.5.

When using the Jefferson disk, the sender would rotate the letters on the disks until the message was spelled out in a single row. The sender would then copy down any row of text on the disks other than the one that contained the plaintext message. That enciphered message would then be sent to the recipient. The recipient then arranged the disk letters according to the predefined order and then rotated the disk until the message was displayed.

It should be noted that this device was independently invented by Étienne Bazeries (1846–1931), a French cryptographer, although Jefferson improved on the disk in his version. Bazeries was known for being a very skilled cryptographer and cryptanalysis. After he broke several transposition systems used by the French military, the French government hired him to work for the Ministry of Foreign Affairs. During World War I, he worked on breaking German ciphers. Stories such as this are not uncommon in cryptography. Two different parties may independently invent the same or remarkably similar ciphers. This often occurs from time to time in modern times, when at least some work in cryptography is classified by various governments. You will see other examples of this in later chapters on modern ciphers.

Book Ciphers

Book ciphers have probably been around for as long as books have been available. Essentially, the sender and receiver agree to use a particular book as its basis. The simplest implementation is to send coordinates for words. So, for example, *3 3 10* means "go to page 3, line 3, tenth word." In this way, the sender can specify words with coordinates and write out entire sentences. There are numerous variations of this cipher. For example, you could combine book ciphers with Vigenère and use the book coordinates to denote the keyword for Vigenère.

Beale Ciphers

In 1885, a pamphlet was published describing treasure buried in the 1820s by one Thomas J. Beale in Virginia. The *Beale ciphers* are three ciphertexts that allegedly give the location, contents, and names of the owners of the buried treasure. The first

Beale cipher, which has not been solved, provides the location. The second cipher provides details of the contents of the treasure and has been solved. The second cipher was a book cipher that used the US Declaration of Independence as the book. Each number in the cipher represents a word in the document. There is a great deal of controversy regarding the Beal ciphers, which we will explore in more detail in Chap. 2. They are presented here simply as an example of a book cipher.

Dorabella Cipher

The Dorabella Cipher is not a book, but rather a letter. It was composed by Edward Elgar to Dora Penny, and sent in July of 1897. Ms. Penny never delivered the message, and it remains an unexplained cipher. It consists of 87 characters spread over three lines. There have been proposed solutions, but no one has been able to verify their proposed solution.

Babington Plot Ciphers

In 1586, there was a plot to assassinate Queen Elizabeth I. The reason for the plot was to depose Queen Elizabeth who was a protestant and replace her with Mary Queen of Scots who was Roman Catholic. Queen Mary was currently imprisoned. Anthony Babington was one of the conspirators and communicated with encrypted messages with Queen Mary. The details of the ciphers are less important than the illustration of the role of ciphers in political intrigue.

Transposition Ciphers

So far, we have looked at ciphers in which some sort of substitution is performed. However, this is not the only way to encrypt a message. It is also possible to transpose parts of a message. Transposition ciphers provide yet another avenue for encryption.

Reverse Order

The simplest implementation of a transposition cipher is to reverse the plaintext. In this way

```
Attack at dawn
```

becomes

```
Nwadtakcatta
```

Obviously, this is not a particularly difficult cipher to break, but it demonstrates a simple transposition.

Rail Fence Cipher

The *rail fence cipher* may be the most widely known transposition cipher. You encrypt the message by alternating each letter on a different row. So

```
Attack at dawn
```

is written like this:

```
A  t  c  a  d  w
  t  a  k  t  a  n
```

Next you write down the text on both lines, reading from left to right, as you normally would, thus producing.

```
Atcadwtaktan
```

To decrypt the message, the recipient must write it out on rows:

```
A  t  c  a  d  w
  t  a  k  t  a  n
```

Then the recipient reconstructs the original message. Most texts use two rows as examples, but any number of rows can be used.

Geometric Shape Cipher

In the *geometric shape cipher*, the sender writes out the plaintext in rows and then maps a path (that is, a shape) through it to create the ciphertext. So, if the plaintext is.

```
Attack the beach at sunrise
```

Fig. 1.6 Geometric shape
cipher

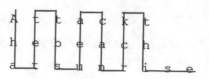

this message would be written in rows like this:

```
At  a c k t
h e b e a c h
a t s u n r i s  e
```

Then the sender chooses some path through the message to create the ciphertext, perhaps start at bottom right and go up and down the columns as shown in Fig. 1.6.

Using the path depicted in Fig. 1.6, the ciphertext reads

```
esihtkcrnacaeusottetaha
```

For this example, I used a very simple geometric path through the plaintext, but you could use other, more complex patterns as well. This method is sometimes called a route cipher, as it is encrypted using a specific route through the plaintext.

Columnar Cipher

The *columnar cipher* is an intriguing type of transposition cipher. In this cipher, the text you want to encrypt is written in rows usually of a specific length and determined by some keyword. For example, if the keyword is *falcon*, which is six characters long, you would write out your messages in rows of six characters each. So, you would write out

```
Attack
thebea
chatsu
nriseq
```

Notice the added *q* at the end. That was added because the last row is only five characters long. In a regular columnar cipher, you pad the last row so that all rows are of equal length.

If you leave the blank spaces intact, this would be an irregular columnar cipher, and the order of columns would be based on the letters in the keyword as they appear in the alphabet. So, if the keyword is *falcon*, the order is 3 1 4 2 6 5 as *f* is the third lowest letter in the alphabet, *a* is the lowest, *l* is the fourth lowest, and so on. So, if we apply 3 1 4 2 6 5 to encrypt the message, we first write out the letters down column

3, then column 1, then column 4, then column 2, then column 6, and then column 5. So, the message.

```
attack
thebea
chatsu
nriseq
```

is encrypted like so:

```
teaiatcnabtskauqcese
```

Many variations of the columnar cipher, such as the Myskowski variation, have been created over the years, each adding some subtle twist to the concept.

Myskowski Variation

When using a columnar cipher, shat happens if the keyword includes the same letter twice? Normally, you treat the second occurrence as if it were the next letter. For example, if *babe* is the keyword, the second *b* is treated as if it were a *c*, so the order would be 2 1 3 4.

In 1902, Emile Myskowski proposed a variation that did something different. The repeated letters were numbered identically, so *babe* would be 2 1 2 3. Any plaintext columns that had unique numbers (in this case 1 and 3) would be transcribed downward as usual. However, the recurring numbers (in this case 2) would be transcribed left to right.

Combinations

One of the first thoughts that may occur when you're first learning cryptography is to combine two or more of the classic ciphers, such as those covered in this chapter. For example, you might use a Vigenère cipher first, and then put the message through a columnar transposition or a rail fence cipher. Combining some substitution cipher with a transposition cipher would increase the difficulty a human would have in breaking the cipher. You can think of this in mathematical terms as a function of a function:

$$f(g(x))$$

Where *g* is the first cipher, *x* is the plaintext, and *f* is the second cipher. And you could apply them in any order—first a substitution and then a transposition, or vice versa.

75628 28591 62916 48164 91748 58464 74748 28483 81638 18174
74826 26475 83828 49175 74658 37575 75936 36565 81638 17585
75756 46282 92857 46382 75748 38165 81848 56485 64858 56382
72628 36281 81728 16463 75828 16483 63828 58163 63630 47481
91918 46385 84656 48565 62946 26285 91859 17491 72756 46575
71658 36264 74818 28462 82649 18193 65626 48484 91838 57491
81657 27483 83858 28364 62726 26562 83759 27263 82827 27283
82858 47582 81837 28462 82837 58164 75748 58162 92000

Fig. 1.7   D'Agapeyeff cipher

When you're exploring this train of thought, be aware that if you simply apply two mono-alphabet substitution ciphers, you have not improved secrecy at all. The ciphertext will still preserve the same letter and word frequencies. In fact, the best improvement will come from combining transposition and substitution ciphers. As you will see beginning in Chap. 6, modern block ciphers combine substitution and transposition, albeit in a more complex fashion. Don't think, however, that such innovations will lead to ciphers that are sufficient for modern security needs. Performing such combinations is an intriguing intellectual exercise and will hone your cryptography knowledge, but these methods would not provide much security against modern computerized cryptanalysis.

D'Agapeyeff Cipher

For the more adventuresome reader, this cipher may capture your attention. It is, as of yet unbroken. So, I cannot tell you specifically how it works. I can only give you a bit of history, then present to you the unbroken ciphertext, should you choose to undertake this rather herculean challenge.

In 1939, cryptography Alexander D'Agapeyeff authored the first edition of the book *Codes and Ciphers*. In that book, he offered the following ciphertext, shown in Fig. 1.7, as a challenge. In later editions of the book, this challenge was omitted.

Conclusions

In this chapter, you have been exposed to a variety of historical ciphers. You were shown single-substitution ciphers such as Caesar and Atbash and multi-alphabet ciphers such as Vigenère. You learned about the weaknesses of mono-alphabet substitution ciphers and how multi-alphabet methods attempt to overcome those issues. You were introduced to a variety of transposition ciphers, including the rail fence and columnar ciphers. This chapter also introduced you to devices such as Scytale and the Jefferson disk. It is important that you get very comfortable with these ciphers before proceeding on.

You were also introduced to some basic mathematical notation to symbolize some of the ciphers in this chapter, as well as some general cryptographic terminology such as *ciphertext* and *key space*. That notation and those terms should be very familiar to you because they will help form the basis for modern symmetric ciphers you'll read about, beginning in Chap. 4.

Test Your Knowledge

A few questions are provided here to aid you in testing your knowledge before you proceed.

1. What is the most obvious weakness in a mono-alphabet cipher?

   A. They preserve word frequency.
   B. They can be cracked with modern computers.
   C. They are actually quite strong.
   D. They don't use complex mathematics.

2. The total number of possible keys for a given cipher is referred to as the

   _____.

   A. key group
   B. key domain
   C. key space
   D. key range

3. Which of the following methods used a cylinder with text wrapped around it?

   A. Vigenère cipher
   B. Jefferson disk
   C. Cipher disk
   D. Scytale

4. What is an affine cipher?

   A. Any cipher of the form $ax + b \; (\equiv m)$
   B. Only single-substitution ciphers
   C. Any single-substitution cipher
   D. A multi-alphabet cipher

5. What are the key features of homophonic substitution?

   A. Multiple substitution alphabets are used.
   B. A single plaintext letter may have several ciphertext representations.
   C. The ciphertext is phonically similar to the plaintext.
   D. It combines substitution with transposition.

References

d'Agapeyeff, A. (2016). Codes and ciphers-A history of cryptography. Read Books Ltd.
Dooley, J. F. (2018). History of cryptography and cryptanalysis: Codes, Ciphers, and their algorithms. Springer..
Mollin, R. A. (2000). An introduction to cryptography. CRC Press.
Singh, S. (2000). The code book: the science of secrecy from ancient Egypt to quantum cryptography. Anchor.

Chapter 2
History of Cryptography from the 1800s

Abstract When studying cryptography, it is often advantageous for the novice to start with analyzing ancient ciphers. These tend to be easier to understand. This chapter will cover cryptography from the 1800s up until the modern computer age. Coupled with Chap. 1, this provides a broad-based coverage of pre-computer cryptography. This will also provide a foundation in cryptographic concepts that will facilitate learning modern cryptography.

Introduction

In Chap. 1, you were introduced to relatively simple ciphers. These ciphers were widely used for many centuries. However, beginning in the 1800s, cryptography entered an age of growth. More complex ciphers were developed. These might not be considered particularly complex by modern standards, but they were certainly more complex than the earlier ciphers. In this chapter, we look at the end of the pre-computer age of cryptography and continue on into the twentieth century. If you have any difficulty with a cipher, I strongly urge you to take a pencil and paper, and experiment with it until you are comfortable.

While these ciphers are more complex than those in Chap. 1, they are still not adequate for the modern computer age. It would be an egregious mistake to attempt to implement one of these, perhaps via computer code, and then presume your data is secure. Modern computers can utilize brute force methods. And in fact, non-computerized cryptanalysis methods have been developed to attack many of these ciphers. We examine those methods in Chap. 17, when we explore cryptanalysis in general.

In This Chapter We Will Cover

- Playfair cipher
- Two-square and four-square ciphers
- Hill cipher
- ADFGVX cipher
- Bifid and trifid ciphers
- Gronsfeld cipher
- Vernam cipher
- Enigma
- Historical figures in cryptography

Playfair

The Playfair cipher was devised in 1854 by Charles Wheatstone; however, it was popularized by Lord Playfair, thus it bears his name. It is sometimes called the Wheatstone-Playfair cipher or the Playfair square (Rahim and Ikhwan 2016). While this cipher is clearly less complex than the modern ciphers we shall explore in later chapters, it was first rejected by the British government as being too complex. They later changed their mind, and this cipher was actually used by the British military in World War I and to some extent in World War II. This cipher works by encrypting pairs of letters, also called digraphs, at a time.

Charles Wheatstone lived from February 6, 1802, to October 19, 1875. He was an accomplished inventor and scientists. Among his various contributions to science was the Wheatstone bridge. This is an electrical circuit that is used to measure electrical resistance. The original bridge was invented by Samuel Hunter Christie, but Wheatstone improved it. Wheatstone also made a number of contributions to our knowledge and application of electricity.

The Playfair cipher depends on a 5×5 table that contains a keyword or key phrase. To use the Playfair cipher, one need only memorize that keyword and four rules. To use the cipher, first break the plaintext message into digraphs. So that "Attack at Dawn" becomes "At ta ck at da wn." If the final digraph is incomplete (just a single letter), you can pad it with a letter z.

Any square of 5×5 letters can be used. You first fill in the keyword, then start, in order, adding in letters that did not appear in the keyword. I/J are combined. You can see this in the table below. In this example, the keyword is "falcon."

F	A	L	C	O
N	B	D	E	G
H	I/J	K	M	P
Q	R	S	T	U
V	W	X	Y	Z

Since the 5 × 5 matrix is created by starting with a keyword, then filling in letters that did not appear in the keyword, the matrix will be different when different keywords are used. The next step is to take the plaintext, in this case "attack at dawn," and divide it into digraphs. If there are any duplicate letters, replace the second letter with an x. For example, "letter" would be "letxer." Playfair does not account for numbers or punctuation marks, so you will need to remove any punctuation from the plaintext and spell out any numbers.

Next, you take the plaintext, in our case

At ta ck at da wn

And find the pairs of letters in the table. Look to the rectangle formed by those letters. In our example, the first letters are AT, thus forming the rectangle shown below, with A in the upper left-hand corner and T in the lower right-hand corner.

A	L	C
B	D	E
I/J	K	M
R	S	T

Then you will take the opposite ends of the rectangle to create the ciphertext. A is in the upper left hand, so you replace it with whatever appears in the upper right hand, which in this case is the letter C. T is in the lower right-hand corner, so you replace it with whatever letter is in the lower left hand, which in this case is the letter R. Thus, AT gets enciphered as CR.

The next digraph is TA and will form the same rectangle. However, since T is the first letter of the plaintext, R will be the first letter of the ciphertext, yielding RC. Next, we have CK. Those letters form the rectangle shown here.

A	L	C
B	D	E
I/J	K	M

C becomes L, and K becomes M, so we have LM. If you continue this process through to the end of the plaintext, you will have "Attack at dawn" encrypted to yield.

CRRCLMCRBLVB

Yes, this cipher does take some practice to master. However, as previously noted, it was eventually used by the British military. It should be noted that the cipher disrupts the letter and word frequencies found in single substitution ciphers, merely by encrypting digraphs instead of single letters. It was also used by the government of New Zealand as late as the early 1940s for non-military communications.

Two-Square Cipher

The two-square cipher is also often referred to as a double Playfair, for reasons that will soon become apparent. Like the Playfair, the two-square cipher operates on digraphs. There are two variations of the two-square cipher, the horizontal and the vertical. These refer to how the squares are arranged. In the horizontal method, they are side by side, in the vertical the squares are arranged one on top of the other. As you might suppose, each square requires its own keyword, so the two-square cipher is based on using two keywords, as well as two matrices.

Let us consider an example of a vertical two-square cipher with the keywords "falcon" and "osprey" shown below.

F	A	L	C	O
N	B	D	E	G
H	I/J	K	M	P
Q	R	S	T	U
V	W	X	Y	Z

O	S	P	R	E
Y	A	B	C	D
F	G	H	I/J	K
L	M	N	Q	T
U	V	W	X	Z

Just as with the Playfair, you divide the plaintext into digraphs. For example, "attack at dawn" becomes:

At ta ck at da wn

Now you will form a rectangle, just like with Playfair, but the rectangle will bridge the two matrices. You find the first letter of plaintext in the top matrix, and the second letter in the bottom matrix. In this case, A in the top matrix and T in the bottom forms the rectangle shown here:

A	L	C	O
B	D	E	G
I/J	K	M	P
R	S	T	U
W	X	Y	Z

S	P	R	E
A	B	C	D
G	H	I/J	K
M	N	Q	T

The letter A is in the upper left-hand corner, so you replace it with the letter in the upper right-hand corner, which in this case is the letter O. The letter T is in the bottom right-hand corner, so you replace it with the letter in the bottom left-hand corner, in this case M, giving you the ciphertext of

OM

If you continue this process through the entire plaintext, you will finally generate the ciphertext shown here:

OM RC OI OM BB XM

There were also specialized rules should the two letters line up (i.e., be in the exact same column). In that case, the ciphertext is the same as the plaintext. If you are using the horizontal two-square cipher, then the same rule applies if the two plaintext characters are in the same row.

Four-Square Cipher

As you can probably surmise, the four-square cipher takes the Playfair model and expands it to use four squares. This cipher was invented by Felix Marie Delastelle. Delastelle was a Frenchman who lived from 1840 until 1902 and invented several ciphers including the bifid, trifid, and four square (Aishwarya et al. 2014).

Delastelle contributions to cryptography during the 1800s extend well beyond the four-square cipher. He also developed the bifid and trifid ciphers, which will be covered a bit later in this chapter. He finished work on a book on cryptography entitled *ité Élémentaire de Cryptographie*, just 11 months before he died in April 1902. His book was published 3 months after his death. What is most remarkable was that Delastelle was an amateur cryptographer with no formal training in math. Given the state of cryptography in the 1800s, this was possible, though uncommon. However, with modern cryptography, it is simply impossible to do any substantive work in cryptography without a strong mathematics background.

The four-square works by first generating four squares of 5×5 letters, often either the Q is omitted, or I/J are combined in order to account for the 26 letters in the alphabet. Notice that in the matrices shown below, in some cases I have omitted the Q, in others combined I and j. Also note that the letters may or may not be in alphabetical order. It should also be noted that the two upper case matrices are going to be ciphertext, the two lower case matrices are for plaintext. This should also explain why it is the two upper case matrices that are not in alphabetical order.

```
a b c d e        E D A R P
f g h i j        L B C X F
k l m n o        G H I J K
p r s t u        N O M S T
v w x y z        U V W Y Z

D E Y W O        a b c d e
R K A B C        f g h i/j k
F G H I J        l m n o p
L S N P M        q r s t u
T U V X Z        v w x y z
```

You then split the message into digraphs (two letter pairs) so that "Attack at dawn" becomes:

at ta ck at da wn

Now the next part may seem a bit complex, so follow along carefully. First, you find the first letter in the upper left plaintext matrix. In this case, it is the first row. Then find the second letter of the digraph in the lower right-hand plaintext matrix. It is in the fourth column. Below these two letters are in bold and underlined.

a b c d e	E D A R P
f g h i j	L B C X F
k l m n o	G H I J K
p r s t u	N O M S T
v w x y z	U V W Y Z

D E Y W O	a b c d e
R K A B C	f g h i/j k
F G H I J	l m n o p
L S N P M	q r s t u
T U V X Z	v w x y z

Because we are working with a digraph, we need to encrypt both the *a* and the *t*. To find the first letter of the ciphertext, you use the row of the first letter of plaintext, in this case one, and the column of the second letter of ciphertext, in this case 4. Now find that letter in the upper right-hand ciphertext matrix. Row 1, letter 4 is R. Then to find the second letter of the ciphertext, you take the row of the second letter of plaintext, in this case four, and the column of the first letter, in this case 1, and find that letter in the lower left-hand ciphertext matrix. In this case, that gives the letter L. You can see this in the matrices below, where the relevant letters are in bold and underlined.

a b c d e	E D A **R** P
f g h i j	L B C X F
k l m n o	G H I J K
p r s t u	N O M S T
v w x y z	U V W Y Z

D E Y **W** O	a b c d e
R K A B C	f g h i/j k
F G H I J	l m n o p
L S N P M	q r s t u
T U V X Z	v w x y z

So, at in plaintext becomes RL in ciphertext. This process is continued throughout the entire plaintext phrase. Yielding

RL NW PA RL EW WG

Hill Cipher

The Hill cipher was invented by Lester A. Hill in 1929. It is a bit more complex than other cipher we have studied thus far. It is based in linear algebra and uses matrix mathematics. We will be exploring those topics in much more detail in later chapters. To encrypt a message, you break the plaintext into a block of n letters. Each letter being represented by a number (a = 1, b = 2, c = 3, etc.) it should be noted that the number assignment is not critical. Some implementations start with a = 0. That block of numbers representing the block of plaintext forms a matrix that is multiplied by some invertible $n \times n$ matrix, mod 26. If you do not have some math background, this may seem rather intimidating, not to worry, it will all be made clear.

The matrix used is the key for this cipher. It should be selected randomly and be mod 26 (26 for the alphabet used in English, other alphabets would require a different mod). For our discussions on this cipher, we will use the following matrix as a key:

$$\begin{bmatrix} 4 & 5 & 10 \\ 3 & 8 & 19 \\ 21 & 5 & 14 \end{bmatrix}$$

Bear in mind that while I am showing you a 3×3 matrix, any size can be used as long as it is a square. For those readers not familiar with matrix math, I will show you just enough to allow you to understand the Hill cipher. Chapters 4 and 5 will give you a better introduction to mathematics for cryptography. To understand multiplying a matrix by a vector, examine this example using letters.

$$\begin{bmatrix} a & b & c \\ d & e & f \\ g & h & I \end{bmatrix} \begin{bmatrix} x \\ y \\ z \end{bmatrix} = \begin{bmatrix} ax & by & cz \\ dx & ey & fz \\ gx & hy & iz \end{bmatrix}$$

You take each row of the matrix and multiply it times each item in the vector. That provides a resulting matrix that is the product of the original matrix multiplied by the vector. In the case of the Hill cipher, the vector is simply numbers representing a block of text. So "Attack at dawn" is

1 20 20 1 3 11 1204 1 23 14

Please note that this conversion assumes 1 is a. Some sources assume 0 is a. Now we break this down into three-character blocks giving

1 20 20
1 3 11
1 20 4
1 23 14

And multiply each vector times the key. This is shown in Fig. 2.1.
The resulting vector (304, 542, 401) has to be mod 26 giving

18 22 11

Then you convert these numbers back to letters, giving ATT plaintext encrypts to RWK (if we assume a = 1). Now repeat this process for each block of letters until the

$$\begin{bmatrix} 4 & 5 & 10 \\ 3 & 8 & 19 \\ 21 & 5 & 14 \end{bmatrix} \times \begin{bmatrix} 1 \\ 20 \\ 20 \end{bmatrix} = \begin{bmatrix} 4\,(4 \times 1)\ 100\,(5 \times 10)\ 200\,(10 \times 20) \\ 3\,(3 \times 1)\ 160\,(4 \times 20)\ 380\,(19 \times 20) \\ 21\,(21 \times 1)\ 100\,(5 \times 20)\ 280\,(14 \times 20) \end{bmatrix} = \begin{matrix} 304 \\ 543 \\ 401 \end{matrix}$$

Fig. 2.1 Hill cipher

entire message is encrypted. The ciphertext will be Rwkybh ne uykt. To decrypt a message, you need to convert the letters of the ciphertext back to numbers, then taking blocks of ciphertext, multiply them by the inverse of the key matrix.

ADFGVX

The ADFGVX cipher is an improvement to the previous ADFGX cipher. ADFGVX was invented by Colonel Fritz Nebel in 1918 and was used by the German Army during World War I. It is essentially a modified Polybius square combined with a columnar transposition. You may recall in Chap. 1, a discussion of combining ciphers in order to create a new cipher. It is interesting to note that ADFGVX was designed to work well when transmitted by Morse code.

Let us start by examining the ADFGX cipher, then we will examine the improvement. The cipher works by first creating a modified Polybius square. Rather than number coordinates, the letters A, D, F, G, and X are used (the letter V is not used to create the modified Polybius square).

	A	D	F	G	X
A	B	T	A	L	P
D	D	H	O	Z	K
F	Q	F	V	S	N
G	G	I/J	C	U	X
X	M	R	E	W	Y

Notice that the letters are not in order in the Polybius square. This is a common variation to help make the cipher more secure. The next step is to encrypt the message using this modified Polybius square. Let us assume the message is "attack at dawn." As with the traditional Polybius square, each letter is represented by its coordinates, in this case by two letters, giving us:

AF AD AD AF GF DX AF AD DA AF XG FX

Next, the message is written out under in columns under some key word. Let us continue using "falcon" as our keyword.

```
F A L   C O N
A F A D A D
A F G F D X
A F A D D A
A F X G F X
```

Now the columns are sorted by alphabetical order of the keyword. So, we have

```
A C F L N O
F D A A D A
F F A G X D
F D A A A D
F G A X X F
```

Then each column is written down in sequence as the ciphertext. That gives us

FFFF DFDG AAAA AGAX DXAX ADDF

Obviously one can use any size transposition key word one wishes. In practice, longer keywords were frequently used. In 1918, the ADFGX cipher was expanded to add the letter V making a 6 × 6 cipher, which allowed for all 26 letters (no need to combine I and J) and allowed digits 0 to 9. Other than that, the ADFGVX works exactly the same as the ADFGX cipher.

	A	D	F	G	V	X
A	B	T	2	L	P	H
D	D	7	O	Z	1	K
F	3	F	V	S	4	B
G	0	I/J	C	U	X	8
V	G	A	0	J	Q	5
X	M	R	E	W	Y	9

Compared to various classical ciphers, and prior to the computer age, this was a very strong cipher and difficult to break. A French army officer, Lieutenant Georges Painvin, worked on analyzing this cipher and was eventually able to break the cipher. However, his method required large amounts of ciphertext to cryptanalyze. It was a forerunner to modern ciphertext-only attacks that we will examine in Chap. 17.

Bifid

The Bifid cipher is another cipher that combines the Polybius square with a transposition cipher. It was invented by Felix Delastelle in 1901. The first step is, of course, to create a Polybius square, as shown here:

	1	2	3	4	5
1	B	G	W	K	Z
2	P	D	S	N	Q
3	I/J	A	L	X	E
4	O	F	C	U	M
5	T	H	Y	V	R

As you can see, the letters are not necessarily put in alphabetical order. The coordinates are written down, as with the traditional Polybius cipher, except the coordinates are written vertically forming rows. Thus, the phrase "Attack at dawn" is written:

```
3  5  5  3  4  1  3  5  2  3  1  2
2  1  1  2  3  4  2  1  2  2  3  4
```

Then read out in rows, much like a rail fence cipher, producing the following stream of numbers:

3 5 5 3 4 1 3 5 2 3 1 2 2 1 1 2 3 4 2 1 2 2 3 4

Next, those numbers are put into pairs, and converted back to letters, yielding:

EYOESGPGXPDX

To decrypt a message, the process is reversed.

The Gronsfeld Cipher

The Gronsfeld cipher is just a variant of the Vigenere cipher we discussed in Chap. 1. It uses 10 different alphabets, corresponding to the digits 0–9. It was widely used throughout Europe (Dooley 2018). The numbers can be picked at random, which is more secure but harder to remember, or the numbers might be of some significance to the sender and recipient.

```
  ABCDEFGHIJKLMNOPQRSTUVWXYZ
0 ABCDEFGHIJKLMNOPQRSTUVWXYZ
1 BCDEFGHIJKLMNOPQRSTUVWXYZA
2 CDEFGHIJKLMNOPQRSTUVWXYZAB
3 DEFGHIJKLMNOPQRSTUVWXYZABC
4 EFGHIJKLMNOPQRSTUVWXYZABCD
5 FGHIJKLMNOPQRSTUVWXYZABCDE
6 GHIJKLMNOPQRSTUVWXYZABCDEF
7 HIJKLMNOPQRSTUVWXYZABCDEFG
8 IJKLMNOPQRSTUVWXYZABCDEFGH
9 JKLMNOPQRSTUVWXYZABCDEFGHI
```

You would select each letter of plaintext and substitute it for the ciphertext corresponding to the appropriate alphabet. Thus, for the first letter, use alphabet 0, then for the next letter use alphabet 1, then use alphabet 2, etc. Once you use alphabet 9, you start over with alphabet 0. So, to encrypt the plaintext "attack at dawn," we would first find the first letter, in this case A and in this case the first alphabet (alphabet 0) yields A as the ciphertext. Then you find the second letter of plaintext T and the second alphabet yields U. Then you find the second T, and the third alphabet (alphabet 2) yields V. Continuing this process, we arrive at the ciphertext of:

AUVDGP GZ LJWO

As an interesting historical note, in 1892 the French government arrested several anarchists who were utilizing the Gronsfeld cipher for communications.

The Vernam Cipher

The Vernam cipher is a type of one-time pad (Mollin 2000). The concept behind a one-time pad is that the plaintext is somehow altered by a random string of data so that the resulting ciphertext is truly random. Gilbert Vernam (April 3, 1890, to February 7, 1960) proposed a stream cipher that would be used with teleprinters. It would combine character by character a prepared key that was stored on a paper tape, with the characters of the plaintext to produce the ciphertext. The recipient would again apply the key to get back the plaintext.

In 1919, Vernam patented his idea (U.S. Patent 1,310,719). In Vernam's method, he used the binary XOR (Exclusive OR) operation applied to the bits of the message. We will be examining binary operations including XOR, in more detail in Chap. 4. To truly be a one-time pad, by modern standards, a cipher needs two properties. The first is suggested by the name: the key is only used once. After a message is enciphered with a particular key, that key is never used again. This makes the one-time pad quite secure, but also very impractical for ongoing communications such as one encounters in e-commerce. The second property is that the key be as long as the message. That prevents any patterns from emerging in the ciphertext. It should be noted that Vernam also patented three other cryptographic inventions: U.S. Patent 1,416,765; U.S. Patent 1,584,749; and U.S. Patent 1,613,686.

One-time pads are still used for communications today, but only for the most sensitive communications. The keys must be stored in a secure location, such as a safe, and used only once for very critical messages. The keys for modern one-time pads are simply strings of random numbers sufficiently large enough to account for whatever message might be sent.

Edgar Allen Poe

While Edgar Allen Poe is most known for his literary works, he also was an accomplished amateur cryptographer. In 1841, Poe wrote an essay entitled "A Few Words on Secret Writing," it was published in *Graham's Magazine*. In his story, "The Gold-Bug" cryptography plays a central role in the story. While Poe is not known to have done any research in cryptography, nor created any ciphers, he did play a key role in increasing the public interest in the field of cryptography. William Friedman, a cryptographer who lived from September 24, 1891, to November 12, 1969, credited his interest in cryptography to Poe's story *The Gold-Bug*. Friedman was the director of research for the Army Signals Intelligence Service in the 1930s. He is also credited with coining the term "cryptanalysis."

In December 1839, Poe published a challenge in the "Alexander's Weekly Messenger." In that challenge he stated he could solve any single substitution cipher than readers could submit. His challenge stated.

> It would be by no means a labor lost to show how great a degree of rigid method enters into enigma-guessing. This may sound oddly; but it is not more strange than the well-known fact that rules really exist, by means of which it is easy to decipher any species of hieroglyphical writing -- that is to say writing where, in place of alphabetical letters, any kind of marks are made use of at random. For example, in place of A put % or any other arbitrary character --in place of B, a *, etc., etc. Let an entire alphabet be made in this manner, and then let this alphabet be used in any piece of writing. This writing can be read by means of a proper method. Let this be put to the test. Let anyone address us a letter in this way, and we pledge ourselves to read it forthwith--however unusual or arbitrary may be the characters employed.

From December 1839 to May 1840, he was able to successfully solve all of the ciphers that readers submitted. There was one submitted that he could not break.

However, various analysts have determined that this cipher was simply a set of random characters and not a legitimate example of ciphertext.

Cryptography Comes of Age

The various ciphers used in the 1800s through the first part of the twentieth century were certainly more complex than the ciphers of more ancient times. However, they were still ciphering that could be handled manually, with pencil and paper. The twentieth century moved more toward encryption related to devices.

Enigma

Contrary to popular misconceptions, the Enigma is not a single machine, but rather a family of machines. The first version was invented by German engineer Arthur Scherbius toward the end of World War I. It was also used by several different militaries, not just the Nazi Germans. Some military texts encrypted using a version of Enigma were broken by Polish cryptanalysts: MarrianRejewsky, Jerzy Rozycki, and Henry Zygalski. The three basically reverse engineered a working Enigma machine. The team then developed tools for breaking Enigma ciphers, including one tool named the cryptologic bomb.

The core of the Enigma machine was the rotors. These were disks that were arranged in a circle with 26 letters on them. The rotors where lined up. Essentially each rotor represented a different single substitution cipher. You can think of the Enigma as a sort of mechanical poly-alphabet cipher. The operator of the Enigma machine would be given a message in plaintext, then type that message into Enigma. For each letter that was typed in, Enigma would provide a different ciphertext based on a different substitution alphabet. The recipient would type in the ciphertext, getting out the plaintext, provided both Enigma machines had the same rotor settings. Fig. 2.2 is a picture of an Enigma machine.

Enigma Variations

There were several Enigma models including:

Enigma A, the first public Enigma
Enigma B
Enigma C
Enigma B, used by United Kingdom, Japan, Sweden, and others
Navy Cipher D used by the Italian Navy
Funkschlüssel C, used by the German navy beginning in 1926

Fig. 2.2 An Enigma machine

Enigma G used by the German Army

Wehrmacht Enigma I, a modification to the Enigma G. Used extensively by the German Military

M3, an improved Enigma introduced in 1930 for the German military

There have been systems either derived from Enigma, or similar in concept. These include the Japanese system codenamed GREEN by American cryptographers, the SIGABA system, NEMA, and others.

Alan Turing

One cannot tell the story of Enigma without at least mentioning Alan Turing. He is famous for leading the team that was able to crack the Enigma version used by the German Navy during World War II, a story that has been immortalized in the movie *The Imitation Game*. During World War II, Turing worked for the Government Code and Cipher School that operated out of Bletchley Park. Bletchley Park is 50 miles northwest of London. The mansion and 58 acres where purchased by MI6 (British Intelligence) for use by the Code and Cipher School. A number of prominent people worked at Bletchley Park including Derek Taunt (a mathematician), Max Newman (a mathematician), and Hugh Alexander (a chess champion).

Turing's impact on the field of cryptography is profound. He led the British team responsible for attempting to break the German naval Enigma. He worked to improve the pre-war Polish machine that would find Enigma settings. He was successful and MI6 was able to decipher German naval messages. This work was considered highly classified. They feared that if the Germans realized that the Enigma had been compromised, they would change how they communicated. Therefore, intelligence gathered via Turing's team was kept a closely guarded secret and when it needed to be shared with others (such as military commanders), a cover story was given to account for the source of the information. Many sources claim that Turing's work at Bletchley Park shortened the war by at least 2 years, saving millions of lives.

The work at Bletchley Park was kept in strict secrecy, but in the decades since the war, many details have been declassified and are now available to the public. Turing's team was by no means the only team working at Bletchley Park. There were various "huts" or buildings used for various purposes. As was already mentioned, Hut 8 was where Turing and his team worked on the Naval Enigma. In Hut 6, work was ongoing on the Army and Air Force Enigmas. Hut 7 concentrated on Japanese codes and intelligence. Other Huts were used for intelligence work, support work (such as communications and engineering), there was even Hut 2 which was a recreational hut where the workers at Bletchley could enjoy beer, tea, and other recreation.

The work at Hut 8 centered on use of what is called a *bombe*. That is an electromechanical device that helps to determine what settings the Enigma machine was using. The initial design used at Bletchley Park was created by Alan Turing and later refined by others such as Gordon Welchman, a British mathematician.

The work at Bletchley was kept very secret, and much of the equipment and documents were destroyed at the end of the war. Even close relatives did not know what the people at Bletchley were working on. Today it is home to a museum with extensive displays of cryptography history, including Enigma machines.

SIGABA

SIGABA was an encryption machine used by the United States military from World War II through the 1950s (Lee 2003). The machine was specifically developed to overcome weaknesses found in other rotor systems, such as Enigma. The device was a joint effort of the US Army and Navy. One can find detailed specification on this machine today, but some of the operating principles remained classified until the year 2000. SIGABA was patented as patent 6,175,625. That patent was filed on 15 December 1944 but remained classified until 2001. There is a picture of a SIGABA machine in Fig. 2.3.

There were several similar devices used during the mid-twentieth century. The British Air Ministry used a device named *Mercury* as late as the early 1960s.

Fig. 2.3 SIGABA

Mercury was designed by E.W. Smith and F. Rudd. The machine was in wide use in the British military by 1950. Rotor-based machines were in wide use. In addition to SIGABA and Mercury, there was the M-325 (U.S. Patent 2877, 565) used by the United States Foreign Service in the late 1940s.

The French government made use of a rotor machine in the 1950s, named HX-63. It was designed by a Swiss company named Crypto AG who specialized in secure communications equipment. In the 1940s, the U.S. Army used a device known as M-228 (also known as SIGCUM) to encrypt communications. This device was also designed by William Friedman who would go on to become director of research for the Army Signals Intelligence Service.

Lorenz Cipher

The Lorenz cipher is actually a group of three machines (SZ40, SZ42A, and SZ42B) that were used by the German Army during World War II (Davies 1995). These machines were rotor stream cipher machines. They were attached to standard teleprinters to encrypt and decrypt the message sent via the teleprinter.

These machines were attachments added to teleprinters to encrypt the data automatically. The machine generated a stream of pseudorandom characters. These characters were the key that was combined with the plaintext using the exclusive or operation to create ciphertext. We will be discussing the exclusive or operations in more detail in Chap. 3.

Navajo Code Talkers

While this does not represent a true cipher, it is a topic that is important in the history of cryptography. Phillip Johnston was a World War I veteran and had been raised on a Navajo Reservation. He was one of the few non-Navajo people who spoke the language. At the beginning of World War II, he proposed to the United States Marine Corps the use of Navajo. His reasoning was that very few non-Navajo knew the language, and being an unwritten language, it would be difficult for someone to learn it.

The new system was tested, and it was proved that Navajo men could encode, transmit, and decode a three-line message in 20 s or less. This was critical because the machines that were being used at the time required 30 s for the same steps to be accomplished. And in the heat of battle time is of the essence.

In both World War I and World War II, Navajo code talkers were used for secure communications by the US military, specifically by the U.S. Marines in World War I. However, there were earlier uses of other indigenous languages. Such as Cherokee and Choctaw code talkers in World War I. This was the most publicized use of an obscure language to facilitate secure communication, but not the only such use. During the Sino-Vietnamese war, a brief border war fought between the People's Republic of China and the Socialist Republic of Vietnam in early 1979, China used Wenzhounese-speakers for coded messages. Wenzhounese is a language spoken in the Wenzhou province of China.

During the Arab-Israeli war of 1973, Egypt utilized Nubian speakers to transmit sensitive communications. At one time, Nubian dialects were spoken in much of Sudan, but in more recent times, this language is limited to small numbers of people in the Nile valley and parts of Darfur.

VIC Cipher

This is an interesting cipher, in that it was done completely by pencil and paper, but used in the 1950s. Soviet spy Reino Häyhänen uses this cipher. It is considered an extension of the Nihilist ciphers which were manual symmetric ciphers used by Russian Nihilists in the 1980's when organizing against the tsars.

The algorithm uses three inputs: a short phrase, a date (in six-digit format), and a personal 1- to 2-digit number. There was also a 5-digit key group that was embedded in the ciphertext at some location that was not publicly known. The four inputs (phrase, date, personal number, and key group) were combined to form a 50-digit block of pseudo random numbers. This was used to create keys for a straddling checkerboard and two columnar transpositions. The plaintext was first encrypted with the straddling checkerboard, then was put through transpositions with the columnar transposition ciphers. A straddling checkerboard takes plaintext and converts into digits followed by fractionation (dividing each plaintext symbol into multiple ciphertext symbols) and data compression. Then the key group was inserted into the ciphertext at a location determined by the personal number.

IFF Systems

The Identify Friend or Foe system for aircraft depends on cryptography. Britain was the first to work out an IFF system. The most primitive of these systems simply sent a pre-arranged signal at specific intervals. With the advent of radar, it was important to identify what aircraft the radar system had detected. The IFF Mark I was first used in 1939. It was an active transponder that would receive a query (called an interrogation) from the radar system. The query consisted of a distinctive pulse using a specific frequency. The transponder would then respond with a signal that used a steadily increasing amplitude, thus identifying the aircraft as friendly.

By 1940, the Mark III was in use by Western Allies and continued to be used throughout World War II. The Mark III expanded the types of communications that could be accomplished, including a coded Mayday response. In Germany, IFF systems were also being developed. The first widely implemented system was the FuG 25a. Before a plane took off, two mechanical keys were inserted, each of 10 bits. These provided the keys to encode the IFF transmissions. British scientists, however, were able to build a device that would trigger a response from any FuG 25a system within range, thus revealing the position of German planes flying at night.

Since World War II, the IFF systems have been used for a variety of purposes. The operating in four primary modes:

Mode 1 is not secure and simply used to track position of aircraft and ships.
Mode 2 is used for landings on aircraft carriers.
Mode 3 is a standard system used by commercial (i.e., non-military) aircraft to
   communicate their position. This is used around the world for air traffic control.
Mode 4 is encrypted and thus secure.

In modern times, secure IFF systems are a part of military air defense operations. Cryptography is critical in ensuring these systems are secure and reliable. In some cases, the cryptographic key used is changed daily. These systems represent one, very practical application of cryptography. The primary goal is to ensure that friendly aircraft are not shot down.

The NSA—The Early Years

It is impossible to discuss the history of cryptography without some discussion of the history of the United States National Security Agency. Today, they are a large organization and are often reported to be the single largest employer of mathematicians, anywhere in the world. The history of cryptography in the latter half of the twentieth century, and beyond, is closely intertwined with the history of the NSA.

While the NSA formally was founded in 1952, there were many pre-cursors to it. As early as 1919, the U.S. Department of State created the Cipher Bureau, often simply called the "Black Chamber." The Black Chamber operated in an office in Manhattan, and its main purpose was to crack the communications of foreign governments. They persuaded Western Union to allow them to monitor telegraphs transmitted by Western Union Customers. The group had significant initial successes but was shut down in 1929 by the Secretary of State. He felt that spying was not a gentlemanly or honorable activity.

In 1924, the U.S. Navy formed its Radio Intelligence office with the purpose of developing intelligence from monitoring radio communications. By 1942, the US Army renamed its Signal Intelligence Service, as Signal Security Service. At this time, the various military branches had their own initiatives on communications, radio intelligence, and cryptography, and cooperation was at a minimum.

In 1949, the various military agencies coordinated cryptology activities with a new, centralized organization named the Armed Forces Security Agency. This agency was part of the Department of Defense, rather than a specific branch of the military. In 1951, President Harry Truman setup a panel to investigate the shortcomings of the AFSA. Among those shortcomings was the failure to predict the outbreak of the Korean War. From this investigation came the National Security Agency. President Truman issued a classified directive entitled "Communications Intelligence Activities" that, among other things, established the NSA.

For much of its early history, the existence of the NSA was not acknowledged. This led to those who did know, jokingly referring to the NSA as "No Such Agency." Obviously, the history of any intelligence agency is not completely public. But let's examine some highlights that are.

After World War II, Soviet encryption was unbreakable and thus posed a significant issue for U.S. Intelligence agencies. This fact, coupled with the discovery of Soviet Spies in various western governments, lead to a renewed emphasis on signals intelligence (SIGINT) and cryptanalysis.

The NSA had two primary roles. The first being to be able to monitor and decipher the communications of other nations. This would enable the gathering of important intelligence. The second being the protection of United States communications from other nations eavesdropping. This led the NSA to develop a standard now known as TEMPEST, an acronym for Transient Electromagnetic Pulse Emanation Standard. This standard applies to both equipment used and to deployment and configuration of communications equipment.

During the cold war, the NSA grew and had some significant successes. As one prominent example, in 1964 the NSA intercepted and decrypted communications regarding China's first nuclear weapon test. There were many others, some are still classified today. In recent years, the *Washington Times* reported that NSA programs have foiled terrorist plots in over 20 different countries. We will see the NSA again in later chapters, particularly when we study modern cryptographic ciphers such as DES and AES in Chaps. 6 and 7, then when we discuss cryptographic backdoors in Chap. 18.

Conclusions

In this chapter, you were introduced to more complex ciphers than in Chap. 1. Some of these algorithms used matrices in order to encrypt and decrypt text. There were many variations of the Polybius square you learned about in Chap. 1. The Hill cipher introduced you to the basics of matrix algebra, which you will see more of in Chaps. 4 and 5. You also learned about contributions to cryptography made by historical figures such as Edgar Allen Poe and Allen Turing.

We also explored some significant developments in the history of cryptography including the famous Navajo Code Talkers and IFF systems that are critical to air traffic control and air defense. We ended the chapter with a discussion of the history of the National Security Administration and its impact on the history of cryptography. This information combined with Chap. 1 provides foundational knowledge of the history of cryptography prior to the computer age.

Test Your Knowledge

A few questions provided here to aid you in testing your knowledge before your proceed.

1. The _____is based on using a matrix as the key and grouping letters in m letter groups.
2. Which of the following is a variation of Vigenere using 10 alphabets, one for each decimal numeral 0 through 9?

   (a) Playfair
   (b) Hill
   (c) Bifid
   (d) Gronsfeld

3. What was Edgar Allen Poe's main contribution to cryptography?

   (a) Popularizing cryptography
   (b) Creating the Hill cipher
   (c) Creating SIGABA
   (d) Breaking the Hill cipher

4. What did Alan Turing work on?

   (a) Breaking the German Naval Enigma
   (b) Breaking the German Army Enigma
   (c) Breaking all Enigma variations
   (d) Breaking the SIGABA system

5. The _____ cipher works by modifying a Polybius square. Rather than number coordinates, letters are used.

References

Aishwarya, J., Palanisamy, V., & Kanagaram, K. (2014). An Extended Version of Four-Square Cipher using 10 X 10 Matrixes. International Journal of Computer Applications, 97(21).

Davies, D. W. (1995). The Lorenz cipher machine SZ42. Cryptologia, 19(1), 39–61.

Dooley, J. F. (2018). History of cryptography and cryptanalysis: Codes, Ciphers, and their algorithms. Springer.

Lee, M. (2003). Cryptanalysis of the SIGABA (Doctoral dissertation, University of California, Santa Barbara).

Mollin, R. A. (2000). An introduction to cryptography. CRC Press.

Rahim, R., & Ikhwan, A. (2016). Cryptography technique with modular multiplication block cipher and playfair cipher. Int. J. Sci. Res. Sci. Technol, 2(6), 71–78.

Chapter 3
Basic Information Theory

Abstract Information theory is a foundational topic for modern cryptography. Without a basic working knowledge of information theory, it can be extremely difficult to understand modern cryptography. This chapter provides the reader with the essential concepts of information theory. This includes an introduction to Claude Shannon's work, discussion of key topics such as diffusion and Hamming weight, as well as basic equations.

Introduction

Information theory may seem like an obscure topic, but it is very important to modern cryptography. More importantly, it is not overly difficult to grasp the essentials. And all you need to progress through this book are the essentials. In this chapter, we are going to explore the fundamentals of this topic. Of course, entire books have been written on the topic, so in this chapter, we will give you just an introduction, but enough for you to understand the cryptography we will be discussing in subsequent chapters. For some readers, this will be a review. We will also explore the essentials of what a mathematical theory is as well as what a scientific theory is. This chapter also provides an introduction to basic binary mathematics, which should be a review for most readers.

While information theory is relevant to cryptography, that is not the only application of information theory. At its core, information theory is about quantifying information. This will involve understanding how to code information, information compression, information transmission, and information security (i.e., cryptography).

In This Chapter We Will Cover

Information theory
Claude Shannon's theorems
Information entropy
Confusing & diffusion
Hamming distance & Hamming weight
Scientific and mathematical theories
Binary math

The Information Age

It is often stated that we live in an information age. This would clearly make information theory pertinent not only to an understanding of cryptography, but to an understanding of modern society. First, we must be clear on what is meant by the phrase "information age." To a great extent, the information age and the digital age go hand in hand. Some might argue the degree of overlap; however, it is definitely the case that without modern computers, the information age would be significantly stifled. While information theory began before the advent of modern digital computers, the two topics are still inextricably intertwined.

From one perspective, the information age is marked by information itself becoming a primary commodity. Clearly, information has always been of value. But it was historically just a means to a more concrete end. For example, even prehistoric people needed information, such as were to locate elk or other game. However, that information was just peripheral to the tangible commodity of food. In our example, the food was the goal; that was the actual commodity. The information age is marked by information itself being widely considered a commodity.

If you reflect on this just briefly, I think you will concur that in modern times information itself is often viewed as a product. For example, you purchased this book you now hold in your hands. Certainly, the paper and ink used to make this book was not worth the price paid. It is the information encoded on the pages that you paid for. In fact, you may have an electronic copy and not actually have purchased any pages and ink at all. If you are reading this book as part of a class, then you paid tuition for the class. The commodity you purchased was the information transmitted to you by the professor or instructor (and of course augmented by the information in this book!). So, clearly information as a commodity can exist separately from computer technology. The efficient and effective transmission and storage of information, however, requires computer technology.

Yet another perspective on the information age is the proliferation of information. Just a few decades ago, news meant a daily paper, or perhaps a 30-min evening news broadcast. Now news is 24 hours a day on several cable channels and on various internet sites. In my own childhood, research meant going to the local library and consulting a limited number of books that were, hopefully, not more than 10 years out of date. Now, at the click of a mouse button, you have access to scholarly journals, research websites, almanacs, dictionaries, encyclopedias, an avalanche of information. Thus, one could view the information age as the age in which most people have ready access to a wide range of information.

Younger readers who have grown up with the internet, cell phones, and generally being immersed in a sea of instant information may not fully comprehend how much information has exploded. It is important to understand the explosion of information in order to fully appreciate the need for information theory. To provide you some perspective on just how much information is being transmitted and consumed in our modern civilization, consider the following facts. As early as 2003, experts estimated that humanity had accumulated a little over 12 exabytes of data over the entire course of human history. Modern media such as magnetic storage, print, and film had produced 5 exabytes in just the first 2 years of the twenty-first century. In 2009, researchers claim that in a single year Americans consumed over 3 zettabytes of information. As of 2019, the World Wide Web is said to have had 4.7 zettabytes, or 4700 exabytes of data. That is just the internet, not including offline storage, internal network servers, and similar data. If you were to try to put all that data on standard external drives, such as 4 terabyte external drives that are quite common as of this writing, it would take 109,253,230,592 such drives.

Such large numbers can be difficult to truly comprehend. Most readers understand kilobytes, megabytes, and gigabytes, but may not be as familiar with larger sizes. Here, you can see the various sizes of data:

Kilobyte 1000 bytes
Megabyte 1,000,000 bytes
Gigabyte 1,000,000,000 bytes
Terabyte 1,000,000,000,000 bytes
Petabyte 1,000,000,000,000,000 bytes
Exabyte 1,000,000,000,000,000,000 bytes
Zettabyte 1,000,000,000,000,000,000,000 bytes
Yottabyte 1,000,000,000,000,000,000,000,000 bytes

These incredible scales of data can be daunting to grasp but should provide you an idea as to how much information is in our modern society. It should also clearly demonstrate that whether you measure that by the amount of information we access, or the fact that we value information itself as a commodity, we are in the information age. And this age has come upon us quite rapidly. Consider hard drives for computers. In 1993, I owned an IBM XT with a 20-megabyte hard drive. While working on this second edition of this book, in the year 2020, I purchased a 10-terabyte external drive for well under $200. That is an increase in capacity of several orders of magnitude. And it seems this increase will continue for at least the foreseeable future.

Claude Shannon

It is impossible to seriously examine information theory without discussing Claude Shannon. Claude Shannon is often called the father of information theory (Gray, 2011). He was a mathematician and engineer who lived from April 30, 1916 until February 24, 2001. He did a great deal of fundamental work on electrical applications of mathematics, and work on cryptanalysis. His research interests included using Boolean algebra (we will discuss various algebras at length in Chap. 5) and binary math (which we will introduce you to later in this chapter) in conjunction with electrical relays. This use of electrical switches working with binary numbers and Boolean algebra is the basis of digital computers.

During World War II, Shannon worked for Bell Labs on defense applications. Part of his work involved cryptography and cryptanalysis. It should be no surprise that his most famous work, information theory, has been applied to modern developments in cryptography. It should be noted that in 1943, Shannon became acquainted with Alan Turing, whom we discussed in Chap. 2. Turing was in the United States to work with the US Navy's cryptanalysis efforts, sharing with the United States some of the methods that the British had developed.

Information theory was introduced by Claude Shannon in 1948 with the publication of his article "A Mathematical Theory of Communication" (Guizzo, 2003). Shannon was interested in information, specifically in relation to signal processing operations. Information theory now encompasses the entire field of quantifying, storing, transmitting, and securing information. Shannon's landmark paper was eventually expanded into a book. The book was entitled "The Mathematical Theory of Communication" and was co-authored with Warren Weaver and published in 1963.

In his original paper, Shannon laid out some basic concepts that might seem very elementary today, particularly for those readers with an engineering or mathematics background. At the time, however, no one had ever attempted to quantify information nor the process of communicating information. The relevant concepts he outlined are given below with a brief explanation of their significance to cryptography:

- An information source that produces a message. This is perhaps the most elementary concept Shannon developed. There must be some source that produces a given message. In reference to cryptography, that source takes plaintext and applies some cipher to create ciphertext.
- A transmitter that operates on the message to create a signal which can be sent through a channel. A great deal of Shannon's work was about the transmitter and channel. These are essentially the mechanisms that send a message, in our case an encrypted message, to its destination.
- A channel, which is the medium over which the signal, carrying the information that composes the message, is sent. Modern cryptographic communications often take place over the internet. However, encrypted radio and voice transmissions are often used. In Chap. 2, you were introduced to IFF (Identify Friend or Foe) systems.

- A receiver, which transforms the signal back into the message intended for delivery. For our purposes, the receiver will also decrypt the message, producing plaintext from the ciphertext that is received.
- A destination, which can be a person or a machine, for whom or which the message is intended. This is relatively straightforward. As you might suspect, sometimes the receiver and destination are one and the same.

Information entropy is a concept that was very new with Shannon's paper and one we delve into in a separate section of this chapter.

In addition to these concepts, Shannon also developed some general theorems that are important to communicating information. Some of this information is more general and applies primarily to electronic communications but does have relevance to cryptography.

Theorem 1: Shannon's Source Coding Theorem

This theorem can be stated simply as follows: It is impossible to compress the data such that the code rate is less than the Shannon entropy of the source, without it being virtually certain that information will be lost (Yamano, 2002). In information theory, entropy is a measure of the uncertainty associated with a random variable. We will explore this in detail later in this chapter. What this theorem is stating is that if you compress data in a way that the rate of coding is less than the information content, then it is quite likely that you will lose some information. It is frequently the case that messages are both encrypted and compressed. Compression is used to reduce transmission time and bandwidth. Shannon's coding theorem is important when compressing data.

Theorem 2: Noisy Channel Theorem

This theorem can be simply stated as follows: For any given degree of noise contamination of a communication channel, it is possible to communicate discrete data (digital information) nearly error-free up to a computable maximum rate through the channel. This theorem addresses the issue of noise on a given channel. Whether it be a radio signal, or a network signal traveling through a twisted pair cable, there is usually noise involved in signal transmission. This theorem essentially states that even if there is noise, you can communicate digital information. However, there is some maximum rate at which you can compute information. That rate is computable and is related to how much noise there is in the channel.

Concepts

There are some key concepts of information theory that are absolutely pivotal for your study of cryptography. You have been given a brief overview of information theory, and those concepts have some relevance to cryptography. However, in this section, you will learn core concepts that are essential to cryptography.

Information Entropy

Shannon thought this was a critical topic in information theory. His landmark paper had an entire section entitled "Choice, Uncertainty And Entropy" that explored this topic in a very rigorous fashion. We need to not explore all the nuances of Shannon's ideas here, just enough for you to move forward in your study of cryptography.

Entropy has a different meaning in information theory than it does in physics. Entropy, in the context of information, is a way of measuring information content. There are two difficulties people have in mastering this concept, and I would like to address both. The first difficulty lies in confusing information entropy with thermodynamic entropy that you probably encountered in elementary physics courses. In such courses, entropy is often described as the measure of disorder in a system. This is usually followed by a discussion of the second law of thermodynamics which states that in a closed system entropy tends to increase. In other words, without the addition of energy, a closed system will become more disordered over time. Before we can proceed to discuss information entropy, you need to firmly grasp that information entropy and thermodynamic entropy are simply not the same things. So, you can take the entropy definition you received in freshman physics courses and put it out of your mind, at least for the time being.

The second problem you might have with understanding information theory is that so many references explain the concept in different ways, some of which can seem contradictory. In this section, I demonstrate for you some of the common ways that information entropy is often described so that you can have a complete understanding of these seemingly disparate explanations.

In information theory, entropy is the amount of information in a given message (Yeung, 2012). This is simple, easy to understand, and, as you will see, it is essentially synonymous with other definitions you may encounter. One example of an alternative explanation of information entropy is that it is sometimes described as the number of bits required to communicate information. This definition is simply stating that if you wish to communicate a message, that message contains information, if you represent the information in binary format, how many bits of information is contained. It is entirely possible that a message might contain some redundancy, or even data you already have (which, by definition, would not be information), thus the number of bits required to communicate information could be less than the total bits in the message. This is actually the basis for lossless data compression. Lossless data compression seeks to remove redundancy in a message and thus compress the message. The first step is to determine the minimum number of bits to communicate the information in the message. Or put another way, to calculate the information entropy.

Another example of a definition that might sound peculiar to you is that many texts describe entropy as the measure of uncertainty in a message (Hu et al. 2010). This may seem puzzling to some readers. I have defined information entropy as "the measure of uncertainty in a message" and "the number of bits required to communicate information" How can both of these be true? Actually, they are both saying the same thing. Let us examine the definition that is most likely causing you some consternation: entropy as a measure of uncertainty. It might help you to think of it in the following manner: only uncertainty can provide information. For example, if I tell you that right now you are reading this book, this does not provide you with any new information. You already knew that, and there was absolutely zero uncertainty regarding that issue. However, the content you are about to read in remaining chapters is uncertain. You don't know what you will encounter, at least not exactly. There is, therefore, some level of uncertainty, and thus information content. Put even more directly: uncertainty is information. If you are already certain about a given fact, then no information is communicated to you. New information clearly requires there was uncertainty that the information you received cleared up. Thus, the measure of uncertainty in a message is the measure of information in a message.

Now, we need to move to a somewhat more mathematical expression of this concept. Let us assume X is some random variable. The amount of entropy measured in bits that X is associated with is usually denoted by $H(X)$. Next, we will further assume that X is the answer to a simple yes or no question. If you already know the answer, then there is no information. It is the uncertainty of the value of X that provides information. Put mathematically, if you already know the outcome, then $H(X) = 0$. If you do not know the outcome, then you will get 1 bit of information from this variable X. That bit could be a 1 (a yes) or a 0 (a no), so $H(X) = 1$.

Hopefully, you have a good grasp of the concept of information entropy, let's try to make it more formal and include the concept of probability. What is the information content of a variable X? First, we know X has some number of possible values, we will call that number n. With a binary variable that is one single bit, then $n = 2$ (values 0 and 1). However, if you have more than one bit, then you have $n > 2$. For example, 2 bits has $n = 2^2$. Let us consider each possible value of X to be i. Of course, all possible values of X might not have the same probability. Some might be more likely than others. For example, if X is a variable giving the height of your next-door neighbor in inches, a value of 69 (5 feet 9 inches) is a lot more probable than a value of 89 (7 feet 5 inches). So, consider the probability of each given i to be P_i. Now, we can write an equation that will provide the entropy (i.e., the information contained) in variable X, taking into account the probabilities of individual values of X. That equation is shown in Fig. 3.1.

The symbol p_i denotes the probability of the occurrence of the ith category of symbol in the string being examined. Symbol H denotes the Shannon entropy. The value is given in bits, thus $\log_2$. This formula allows one to calculate the information

Fig. 3.1 The information
entropy of X

$$H(X) = -\sum_{i=1}^{n} p_i \log_2 p_i$$

Fig. 3.2 Joint entropy

$$H(X, Y) = -\sum_{x,y} p(x, y) \log p(x, y)$$

Fig. 3.3 Conditional
entropy

$$H(X|Y) = -\sum_{x,y} p(x, y) \log p(x|y)$$

$$0 \leq H(X) \leq \log_2 n$$

Fig. 3.4 Range of information contained in X

entropy (i.e., Shannon entropy) in a given message. As stated earlier in this book, we won't be examining proofs in this text. In fact, for our purposes, you can simply accept the math as given. There are a number of excellent mathematics books and courses that delve into the proofs for these topics.

For two variables or two messages that are independent, their *joint entropy* is the sum of the two entropies. That formula is shown in Fig. 3.2.

If the two variables are not independent, but rather variable Y depends on variable X, then instead of joint entropy you have *conditional entropy*. That formula is shown in Fig. 3.3.

The equation shown in Fig. 3.1 provides a way of calculating the information contained in (i.e., the information entropy) some variable X, assuming you know the probabilities of each possible value of X. If you do not know those probabilities, you can still get at least a range for the information contained in X. That is done with the equation shown in Fig. 3.4.

This equation assumes that there is some information, some level of uncertainty, thus we set the lower range at greater than zero. Hopefully, this gives you a clear understanding of the concept of information entropy. The basic concept of information entropy, often called Shannon entropy, is essential to cryptography, and you will encounter it later in this book. Joint entropy and differential entropy are more advanced topics and are introduced here for those readers who wish to delve deeper, beyond the scope of this book.

Quantifying Information

Understanding information entropy is essential to understanding how we quantify information. If you have any doubts about your understanding of entropy, please re-read the preceding section before proceeding. Quantifying information can be a difficult task. For those new to information theory, even understanding what is meant by "quantifying information" can be daunting. Usually, it is done in the context of binary information (bits). In fact, while one can find many different definitions for what information is, within the context of information theory we look at information as bits of data. This works nicely with computer science, since computers store data as bits. A bit can only have one of two values, 0 or 1. This can be equivalent to yes or no. Of course, one can have several bits together containing several yes or no values, thus accumulating more information.

Once we have defined information as a single bit of data, the next issue becomes quantifying that data. One method often used to explain the concept is to imagine how many guesses it would take to ascertain the contents of a sealed envelope. Let me walk you through this helpful thought experiment.

Suppose someone hands you an envelope that is sealed and contains some message. You are not allowed to open the envelope. Rather you have to ascertain the contents merely by asking a series of yes or no questions (binary information: 1 being yes, 0 being no). What is the smallest number of questions required to accurately identify the contents of the message within the sealed envelope?

Notice that we stated the scenario as how many questions, on average, will it take. Clearly, if you try the same scenario with different people posing the queries, even with the same message, you will get a different number of questions needed. The game could be repeated with different messages. We then enumerate messages and any given message is M_i and the probability of getting that particular message is P_i. This leaves us with the question of what exactly is the probability of guessing a specific message? The actual equation takes us back to Fig. 3.1. Computing the entropy of a given message is quantifying the information content of that message.

Confusion & Diffusion

The concept of confusion, as relates to cryptography, was outlined in Claude Shannon's 1949 paper "Communication Theory of Secrecy Systems". In general, this concept attempts to make the relationship between the statistical frequencies of the ciphertext and the actual key as complex as possible. Put another way, the relationship between the plaintext, ciphertext, and key should be complex enough that it is not easy to determine what that relationship is.

If an algorithm does not produce sufficient confusion, then one could simply examine a copy of plaintext and the associated ciphertext and determine what the key is. This would allow the attacker to decipher all other messages that are encrypted with that same key.

Diffusion literally means having changes to one character in the plaintext affect multiple characters in the ciphertext, unlike historical algorithms (Caesar Cipher, Atbash, Vigenere) where each plaintext character only affected one ciphertext character.

Shannon thought the related concepts of confusion and diffusion where both needed to create an effective cipher:

> "Two methods (other than recourse to ideal systems) suggest themselves for frustrating a statistical analysis. These we may call the methods of diffusion and confusion. In the method of diffusion, the statistical structure of M which leads to its redundancy is "dissipated" into long range statistics—i.e., into statistical structure involving long combinations of letters in the cryptogram. The effect here is that the enemy must intercept a tremendous amount of material to tie down this structure, since the structure is evident only in blocks of very small individual probability. Furthermore, even when he has sufficient material, the analytical work required is much greater since the redundancy has been diffused over a large number of individual statistics."

These two goals are achieved through a complex series of substitution and permutation. While not secure enough to withstand modern cryptanalysis, an example can be made using the historic ciphers you studied in Chaps. 1 and 2. Let us assume you have a simple Caesar cipher whereby you shift each letter three to the right. This will provide you a small degree of confusion, but no diffusion. Now assume you swap every three letters. This transposition will provide another small degree of confusion. Now let us apply a second substitution, this time 2 letters to the right. The two substitutions, separated by a transposition, provide minimal diffusion. Consider the following example:

Plaintext:	Attack at dawn
Step 1 (shift right 3)	dwwdfn dw gdzq
Step 2 (swap 3 letter blocks)	dfndww dz. qdw
Step 3 (shift right 2)	fhpfyy fb sfy

Let us try changing just one letter of plaintext (though it will make for a misspelled plaintext word). Change "attack at dawn" to "attack an dawn."

Plaintext:	Attack an dawn
Step 1 (shift right 3)	Dwwdfn dq gdzq
Step 2 (swap 3 letter blocks)	dfndww dz. qdq
Step 3 (shift right 2)	fhpfyy fb sfs

Now compare this ciphertext to the one originally produced. You can see only one letter has changed; it is the last letter. Instead of sfy you now have sfs. This provides only minimal confusion and still no diffusion! What is missing? Two things are missing. The first being that, at least by modern standards, this simply is not complex enough. It is certainly an improvement on the basic Caesar cipher, but still it is not enough. The second problem is that there is no mechanism to have a change in one character of plaintext change multiple characters in the ciphertext. In modern ciphers, operations are at the bit level, not at the character level. Furthermore, in modern ciphers, there are mechanisms to propagate a change, and we will see those beginning in Chap. 6. However, this example should give you the general idea of combining substitution and permutation.

Avalanche

This term means that a small change yields large effects in the output, like an avalanche. This is Horst Fiestel's variation on Claude Shannon's concept of diffusion. We will see Fiestel's ideas used in many of the block ciphers we explore in Chap. 6. Obviously, a high avalanche impact is desirable in any cryptographic algorithm. Ideally, a change in one bit in the plaintext would affect all the bits of the ciphertext. This would be complete avalanche, but that has not been achieved in any current algorithm.

Hamming Distance

The Hamming distance is, essentially, the number of characters that are different between two strings (Bookstein et al. 2002). This can be expressed mathematically as $h(x, y)$.

Hamming distance is used to measure the number of substitutions that would be required to turn one string into another. In modern cryptography, we usually deal with binary representations, rather than text. In that context, Hamming distance can be defined as the number of ones if you exclusive or (XOR) two strings. The concept of Hamming distance was developed by Richard Hamming who first described the concept in his paper "Error detecting and error correcting codes." The concept is used widely in telecommunications, information theory, and cryptography.

Hamming distance only works when the strings being compared are of the same length. One application is to compare plaintext to ciphertext to determine how much has changed. However, if there are two strings of different lengths to be compared another metric must be used. On such metric is the Levenshtein distance. This is a measurement of the number of single character edits required to change one word into another. Edits can include substitutions (as with Hamming distance) but can also include insertions and deletions. The Levenshtein distance was first described by Vladimir Levenshtein in 1965.

Hamming Weight

The concept of Hamming weight is closely related to Hamming distance. It is essentially comparing the string to a string of all zeros. Put more simply, it is how many 1s are in the binary representation of a message. Some sources call this the population count or pop count. There are actually many applications for Hamming weight both within cryptography and in other fields. For example, the number of modular multiplications required for some exponent e is computed by log2 e + hamming weight (e). You will see modular arithmetic in Chap. 4. Later in this book, the RSA algorithm is examined at length; you will find that the e value in RSA is chosen due to low Hamming weight.

Unlike Hamming distance, Richard Hamming was not the first to describe the concept of Hamming Weight. The concept was first introduced by Irving S. Reed, though he used a different name. Richard Hamming later described a practically identical concept to the one Reed had described, and this latter description was more widely known, thus the term Hamming weight.

Kerckhoffs's Principle/Shannon's Maxim

Kerckhoffs's principle is an important concept in cryptography. Auguste Kerckhoffs first articulated this in the 1800s, stating that the security of a cipher depends only on the secrecy of the key, not the secrecy of the algorithm. Claude Shannon rephrased this stating that "One ought to design systems under the assumption that the enemy will ultimately gain full familiarity with them." This is referred to as Shannon's maxim and states essentially the same thing Kerckhoffs's principle states.

Let me attempt to restate and expound this in terms you might find more verbose but hopefully easier to understand. Either Kerckhoffs's principle or Shannon's maxim states that the only thing that you must keep secret is the key. You don't need to keep the algorithm secret. In fact, this book which you are currently holding in your hand will, in subsequent chapters, provide you the intimate details of most modern algorithms. And that in no way compromises the security of those algorithms. As long as you keep your key secret, it does not matter that I know you are using AES 256 bit, or Serpent, or Blowfish, or any other algorithm you could think of.

I would add to Kerckhoffs's principle/Shannon's maxim something I will humbly call Easttom's corollary: "You should be very wary of any cryptographic algorithm that has not been published and thoroughly reviewed. Only after extensive peer review should you consider the use of any cryptographic algorithm". Consider that you have just created your own new algorithm. I mean no offence, but the most likely scenario is that you have mistaken ignorance for fresh perspective and your algorithm has some serious flaw. One can find many examples on the internet of amateur cryptographers believing they have invented some amazing new algorithm. Often their "discovery" is merely some permutation on polyalphabetic ciphers you studied in Chap. 1. But even if you are an accomplished cryptographic researcher, one with a proven track record, it is still possible you could make an error. To demonstrate this fact, consider Ron Rivest. You will hear a great deal about Dr. Rivest in coming chapters. His name is the R in RSA and he has been involved in other significant cryptographic developments. He submitted an algorithm to the

SHA-3 competition (which we will discuss in detail in Chap. 9); however, after several months of additional analysis a flaw was found, and he withdrew his algorithm from consideration. If such a cryptographic luminary as Ron Rivest can make a mistake, certainly you can?

The purpose of publishing an algorithm is so that experts in the field can review it, examine it, and analyze it. This is the heart of the peer-review process. Once an algorithm is published, then other cryptographers get to examine it. If it withstands a significant amount of time (usually years) of such analysis, then and only then should it be considered for use.

This issue has serious practical implications. In September 2011, I was interviewed by CNN Money reporter David Goldman regarding a company who claimed to have an "unbreakable code." The reporter also interviewed cryptography experts from Symantec and Kaspersky labs. All of us agreed that the claim was nonsense. The algorithm was being kept "secret" and not open to peer review. It also happens that the person producing the unbreakable code produced ciphertext challenging people to break it. The ciphertext produced failed every randomness test it was put to, indicating it was very weak.

There is of course one, very glaring, exception to publishing an algorithm. The US National Security Agency (NSA) organizes algorithms into two groups. Suite B are published. This includes algorithms such as Advanced Encryption Standard (AES). You can find the complete details of that algorithm in many sources, including Chap. 7 of this book. The second group is suite A algorithms, which are classified and not published. This seems to violate the spirit of Kerckhoffs's principle and peer review. However, keep in mind that the NSA is the single largest employer of mathematicians in the world. This means that they can subject an algorithm to thorough peer review, entirely via their own internal staff, all of whom have clearance to view classified material.

Scientific and Mathematical Theories

We have discussed information theory, but we have not discussed what constitutes a theory within the scientific and mathematical world. Mathematics and science are closely related, but the processes are slightly different. We will examine and compare both in this section. Many words have a different meaning in specific situations than they do in normal, day-to-day conversation. For example, the word "bug" usually refers to an insect, but in computer programming it refers to a flaw in a program. Even a simple and common word such as "love" means different things in different situations. For example, you may "love" your favorite TV show, "love" your spouse, and "love" chocolate ice cream, but you probably don't mean the same thing in each case. The word "love" has a different meaning in different contexts. So, when one is attempting to define a term, it is important to consider the context one is using it in.

This is even truer, when a word is being used in a specific technical or professional context. Various professions have very specific definitions for certain words. The legal community is a good example. Many words have very specific meanings within the law, meanings which might not exactly match their ordinary daily use.

It is also true that scientists (as well as mathematicians) have some very specific meanings for words they use. The term *theory* is such a word. This word has a very different meaning in a scientific context than it does in everyday language. In the day-to-day vernacular, a theory is often synonymous with a guess. For example, your favorite sports team is on a losing streak, you might have a "theory" about why they are losing. And in this case, by *theory* you probably mean just a guess. You may or may not have a shred of data to support your "theory." In fact, it may be little more than a "gut feeling."

In science, however, a theory is not a guess or "gut feeling," it is not even just an "educated guess." An educated and testable guess is called an hypothesis (Larson, 2004). The key part being that it is a testable guess. In fact, a guess that is untestable has no place in science at all. When you have a testable, educated, guess, you then have a hypothesis. Once you have tested that hypothesis you have a fact. The fact may be that the test results confirm or reject your hypothesis. Usually, you will repeat the test several times to make sure the results were not an error. But even an hypothesis is more than a wild guess or a hunch. It is an educated estimate that must be testable. If it is not testable, it is not even an hypothesis.

What Is a Mathematical Theory?

We will see, in our discussion of scientific theories, that theories are conceptual models. We will also see that a mathematical theory and a scientific theory are very closely related. Theories in math flow from axioms. An axiom is a single fact that is assumed to be true without proof. I think it should be obvious to all readers that one must have only the most basic facts taken on an a priori basis as axioms.

There are a number of interesting math theories. Graph theory studies mathematical structures that represent objects and the relationship between them. It does this by using vertices (the objects) and edges (the connections between objects). Set theory studies collections of objects or sets. It governs combining sets, overlapping sets, subsets, and related topics.

However, mathematics deals with proofs not experiments. So, the hypothesis, which we will see momentarily in the scientific process, is not strictly applicable to mathematics. That is why we begin with axioms. The statements are derived from those axioms, statements whose veracity can be objectively determined. This is sometimes taught using truth tables.

Individual mathematical theories address a particular system. These theories begin with a foundation that is simply the data describing what is known or what is assumed (i.e., axioms). New concepts are logically derived from this foundation and are (eventually) proven mathematically. Formal mathematics courses will discuss first-order theories, syntactically consistent theory, deductive theories, and other nuances of mathematical theory. However, this book is focused on applied mathematics, and more narrowly on one single narrow sub-topic in applied mathematics: cryptography. We won't be delving into the subtleties of mathematical theory, nor will you be exposed to proofs. For our purposes, we will simply apply mathematics that others have rigorously proven.

The Scientific Process

The scientific process is primarily about experimentation. Nothing is really "true" until some experiment says it is true. Usually, multiple experiments need to confirm that something is, indeed, true. The first step is to create an hypothesis, an educated guess that can be tested. You will then develop facts from that. Testing, experimentation, and forming hypothesis are all important parts of the scientific process. However, the pinnacle of the scientific process is the theory.

This was not always the case. In ancient times, it was far more common for philosophers to simply formulate compelling arguments for their ideas, without any experimentation. Ptolemy, in the second century C.E., was one of the early proponents of experimentation and what is today known as the scientific method. Roger Bacon, who lived in the thirteenth century C.E. is often considered the father of the modern scientific method. His experimentation in optics was a framework for the modern approach to science.

A Scientific Theory

You know that an hypothesis is a testable guess. You also know that after conducting a number of tests, you will have one or more facts. Eventually, one has a number of facts that require some explanation. The explanation of those facts is called a theory. Or put another way, "A theory is an explanation of a set of related observations or events based upon proven hypotheses and verified multiple times by detached groups of researchers. How does a scientific theory compare to a scientific law? In general, both a scientific theory and a scientific law are accepted to be true by the scientific community as a whole. Both are used to make predictions of events. Both are used to advance technology."

Now think about this definition for the word theory for just a moment: a "theory is an explanation." That is the key part of this definition. After you have accumulated data, you must have some sort of explanation. A string of facts with no connection, no explanation is of little use. This is not only true in science but in other fields as well. Think about how a detective works. Anyone can notice a string of facts. The detective's job is to put those facts together in a manner which is consistent with all the facts. This is very similar to what scientists do when trying to formulate a theory. Note that with both the scientist and the detective, the theory must match all the facts.

It is sometimes difficult for nonscientists to become accustomed to this use of the word theory. People often make the mistake of simply consulting a standard dictionary for a definition. Yet even a basic dictionary usually has multiple definitions. For example, the Merriam-Webster online dictionary lists many alternative definitions for the word theory. Some of those definitions are synonymous with guess. However, even this standard dictionary offers alternative definitions for the word theory. The ones applicable to sciences use of the word theory are the following:

> "1: the analysis of a set of facts in their relation to one another"
> "3: the general or abstract principles of a body of fact, a science, or an art <music theory>"
> "5: a plausible or scientifically acceptable general principle or body of principles offered to explain phenomena <wave theory of light>²"

As you can see, these three definitions are not synonymous with guess or gut feeling. An even better explanation of this was given by the Scientific American Magazine:

> "Many people learned in elementary school that a theory falls in the middle of a hierarchy of certainty—above a mere hypothesis but below a law. Scientists do not use the terms that way, however. According to the National Academy of Sciences (NAS), a scientific theory is "a well-substantiated explanation of some aspect of the natural world that can incorporate facts, laws, inferences, and tested hypotheses." No amount of validation changes a theory into a law, which is a descriptive generalization about nature. So, when scientists talk about the theory of evolution—or the atomic theory or the theory of relativity, for that matter— they are not expressing reservations about its truth."

The point to keep in mind is that a theory is an explanatory model. It explains the facts we know and gives us context for expanding our knowledge base. While scientific theories are based on experimentation and mathematical theories are based on proofs and axioms, the issue of being an explanatory model is the same for a scientific theory or a mathematical theory. Putting it in the context of information theory, it is a theory that gives a conceptual model of information and provides the framework to continue adding to our knowledge of information.

A Look at Successful Scientific Theories

Examining in general terms what a theory is, and what it is not, may still not be an adequate explanation for all readers. So, let's look at a concrete example. Let's look at an example of a very well-accepted theory, and how it relates to specific facts. For this purpose, we will consider gravity. Newton observed the fact things tend to fall. In a very general way, this is the essence of his law of gravity: gravity makes things fall downward. Of course, he expressed that law mathematically with this equation: $F = Gm_1m_2/d^2$.

In this equation, F is the force of gravity, G is a constant (the Gravitational Constant) which can be measured, m_1 and m_2 are the masses of the two objects (such as the earth and a falling apple), and d is the distance between them. This law of gravity can be measured and verified thousands of times, and in fact has been.

But this law, by itself, is incomplete. It does not provide an adequate explanation for why this occurs. We know there is gravity, we all feel its pull, but what is it? Why does gravity work according to this formula? Einstein's Theory of General Relativity was an explanation for why there is gravity. With General Relativity, Einstein proposed that matter causes time and space to curve. Much like a ball sitting on a stretched-out sheet causes a curving in the sheet. Objects that get close enough to the ball will tend to fall toward it, due to this curvature.

Clearly, Einstein's General Theory of Relativity explains all the observed facts. But there is still yet another component to make it a scientific theory. It must make some predictions. By predictions, I don't mean that it foretells the future. What I mean is that if this theory is valid, then one would expect certain things to also be true. If those predictions turn out not to be true, then Einstein's General Theory of Relativity must either be modified or rejected. Whether you choose to modify a theory or reject it depends on how well its predictions work out. If it is 90% accurate, then you would probably just look to adjust the theory, so it matched the facts. However, if it was only right less than half the time, you should probably consider simply rejecting the theory outright. Well what predictions do we get from the Theory of General Relativity? A few of those predictions are:

1. **That gravity can bend light**. This has been reconfirmed literally hundreds of times with more precision each time. More recently, in 2003, the Cassini Satellite once again confirmed this prediction. And keep in mind this 2003 test was just one more among many, all confirming that gravity can bend light.
2. **Light loses energy escaping from a gravitational field**. Because the energy of light is proportional to its frequency, a shift toward lower energy means a shift to lower frequency and longer wavelength. This causes a "red shift" or shift toward the red end of the visible spectrum. It was experimentally verified on earth using gamma-rays at the Jefferson Tower Physics Laboratories.
3. **Deflection of Light by Gravity:** General relativity predicts that light should be deflected or bent by gravity. This was first confirmed by observations and measurements of a solar eclipse in 1919. It has since been confirmed by hundreds of experiments over decades.

There are other predictions made by the General Theory of Relativity, and every single one has been confirmed by dozens, and in some cases hundreds of independent experiments. Note the word independent. These experiments were carried out by people with no particular vested interest in proving Einstein correct. And in a few cases, they really wanted to prove him wrong. Yet in every case the data confirmed Einstein's theory. You would be hard put to find any physicist who does not agree that the General Theory of Relativity is valid is true. Some may argue it is not yet complete, but none will argue that it is not true. Hundreds of experiments spanning many decades and literally thousands of scientists have confirmed every aspect of General Relativity. Now that this theory is well established, it can be used to guide future experiments and deeper understanding of gravity, as well as related topics.

Binary Math

Binary math is not actually a part of information theory, but it is critical information you will need in order to be able to understand cryptography in later chapters. Many readers will already be familiar with binary math, and this will be a review. However, if your background does not include binary math, then this section will teach you the essentials you will need to know.

First let us define what binary math is. It is actually rather simple. It is mathematics that uses a base two rather than the more familiar base ten. The only reason you find base ten (i.e., the decimal number system) to be more "natural" is because you have ten fingers and ten toes. Or at a minimum, most people have ten fingers and toes, so even if some accident or birth defect left you personally with a different number of fingers and toes, you understand that ten is normal for humans.

Once we introduce the concept of zero, then you have ten digits in the base ten system, 0, 1, 2, 3, 4, 5, 6, 7, 8, and 9. Keep in mind there is not really a number ten in the base ten system. There is a 1 in the tens place and a 0 in the one's place. The same thing is true with binary, or base two numbers. There is not actually a number two. There is a 1 in the two's place and a zero in the 1's place. This chart comparing the two might be helpful to you:

Number in base ten (decimal)	Number in base two (binary)
0	0
1	1
2	10
3	11
4	100
5	101
6	110
7	111
8	1000

Binary numbers are important because that is, ultimately, what computers understand. You can, of course, represent numbers in any system you wish. And a number system can be based on any integer value. Historically, there have been a few widely used. Hexadecimal (base 16) and octal (base 8) are sometimes used in computer science because they are both powers of 2 (2^4 and 2^3, respectively) and thus easily converted to and from binary. The Babylonians used a base 60 number system, which you still see today in representation of degrees in a geographic coordinate system. For example, 60 seconds makes a minute, 60 min makes a degree so one can designate direction by stating 45 degrees 15 min and 4 s.

The modern concept of binary numbers traces back to Gottfried Leibniz who wrote an article entitled "Explanation of the Binary Arithmetic". In that article, Leibniz noted that the Chinese I-Ching hexagrams corresponded to binary numbers ranging from 0 to 1111111 (or 0 to 63 in decimal numbers).

Converting

Since humans tend to be more comfortable with decimal numbers, you will frequently want to convert from binary to decimal and may need to make the inverse conversion as well. There are many methods: one common one takes advantage of the fact that the various "places" in a binary number are powers of two. So, consider for a moment a binary number 10111001 and break it down into the powers of two. If there is a 1 in a given place, then you have that value. If not, you have a zero. For example, if there is a 1 in the 2^7 place then you have 128, if not you have a zero.

Power of 2	2^7	2^6	2^5	2^4	2^3	2^2	2^1	2^0
Value of that place	128	64	32	16	8	4	2	1
Binary numeral	1	0	1	1	1	0	0	1
	128	0	32	16	8	0	0	1

Now just add the values you have: $128 + 32 + 16 + 8 + 1 = 185$.
This may not be the fastest method, but it works, and it is easy to understand.

Binary Operations

One can do all the mathematical operations with binary that one might do with decimal numbers. Fortunately, with cryptography you will be primarily concerned with three operations: AND, OR, and XOR. And these are relatively easy to understand. For all of our examples, we will consider two four-bit binary numbers: 1101 and 1100. We will examine how the AND, OR, and XOR operations work on these two numbers.

Binary AND

The binary AND operation combines two binary numbers, one bit at a time. It asks is there a 1 in the same place in both numbers. If there is, then the resulting number is a 1. If not, then the resulting number is a 0. You can see in Fig. 3.5 the comparison of the rightmost digit in two 4-digit binary numbers.

Fig. 3.5 Binary AND for rightmost digit

Fig. 3.6 Binary AND for
two numbers

```
  1 1 0 1
  1 1 0 0
 ┌─────────┐
 │ 1 1 0 0 │
 └─────────┘
```

Fig. 3.7 Binary OR
operation

```
  1 1 0 1
  1 1 0 0
 ┌─────────┐
 │ 1 1 0 1 │
 └─────────┘
```

Fig. 3.8 Binary XOR
operation

```
  1 1 0 1
  1 1 0 0
 ┌─────────┐
 │ 0 0 0 1 │
 └─────────┘
```

Since the first number has a 1 in this place, and the second has a 0, the resultant number is a zero. You continue this process from right to left. Remember if both numbers have a 1, then the resulting number is a 1. Otherwise the resulting number is a 0. You can see this in Fig. 3.6.

Binary OR

The binary OR operation is just a little different. It is asking the question of whether there is a 1 in the first number or the second number, or even in both numbers. So basically, if either number has a 1, then the result is a 1. The result will only be a zero if both numbers are a zero. You can see this in Fig. 3.7.

Essentially, the OR operation asks is there a 1 in the first number *and/or* the second number.

Binary XOR

The binary XOR is the one that really matters for cryptography. This is the operation that provides us with interesting results we can use as part of cryptographic algorithms. With the binary exclusive or (usually denoted XOR) we ask the question is there a 1 in the first number or second number, but not in both. In other words, we are exclusively asking the "OR" question, rather than asking the "AND/OR" question that the OR operation asks. You can see this in Fig. 3.8.

Fig. 3.9 Reversing the
binary XOR operation

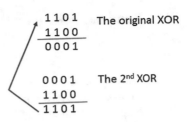

What makes this operation particularly interesting is that it is reversible. If you perform the XOR operation again, this time taking the result of the first operation and XORing it with the second number, you will get back the first number. You can see this in Fig. 3.9.

This is particularly useful in cryptographic algorithms. Our example used 4-bit numbers but extrapolate from that, considering longer numbers. Assume instead of 4 bits, the first number was a lengthy message, for example, 768 bits long. And the second number was a randomly generated 128-bit number; we use it as the key. You could then break the 768-bit plaintext into blocks of 128 bits each (that would produce 6 blocks). Then you XOR the 128-bit key with each 128-bit block of plaintext, producing a block of ciphertext. The recipient could then XOR each 128-bit block of ciphertext with the 128-bit key and get back the plaintext.

Before you become overly excited with this demonstration, this simple XOR process is not robust enough for real cryptography. However, you will see, beginning in Chap. 6, that the XOR operation is a part of most modern symmetric ciphers. It cannot be overly stressed, however, that it is only a part of modern ciphers. The XOR operation is not, in and of itself, sufficient.

Conclusions

In this chapter, you have been introduced to a number of concepts that will be critical for your understanding of modern ciphers, particularly symmetric ciphers we begin studying in Chap. 6. Information entropy is a critical concept that you must understand before proceeding. You were also introduced to confusion and diffusion, Hamming distance, and Hamming weight. These concepts are going to be frequently used in later chapters. There was also a significant coverage of what is a mathematical and scientific theory. Finally, the chapter ended with a basic coverage of binary math.

Test Your Knowledge

1. The difference in bits between two strings X and Y is called <u>Hamming Distance.</u>
2. If you take 1110 XOR 0101 the answer is _____.
3. A change in one bit of plaintext leading to changes in multiple bits of ciphertext is called _____.
4. The amount of information that a given message or variable contains is referred to as _____.
5. _____ refers to significant differences between plaintext, key, and ciphertext that make cryptanalysis more difficult.

References

Bookstein, A., Kulyukin, V. A., & Raita, T. (2002). Generalized hamming distance. Information Retrieval, 5(4), 353–375.

Gray, R. M. (2011). Entropy and information theory. Springer Science & Business Media.

Guizzo, E. M. (2003). The essential message: Claude Shannon and the making of information theory (Doctoral dissertation, Massachusetts Institute of Technology).

Hu, Q., Guo, M., Yu, D., & Liu, J. (2010). Information entropy for ordinal classification. Science China Information Sciences, 53(6), 1188–1200.

Larson, E. J. (2004). Evolution: The remarkable history of a scientific theory (Vol. 17). Random House Digital, Inc..

Yamano, T. (2002). Source coding theorem based on a nonadditive information content. Physica A: Statistical Mechanics and its Applications, 305(1–2), 190–195.

Yeung, R. W. (2012). A first course in information theory. Springer Science & Business Media.

Chapter 4
Essential Number Theory
and Discrete Math

Abstract Much of cryptography is predicated on a basic working knowledge of number theory. This is particularly true for asymmetric algorithms. It is impossible to really understand RSA, Diffie-Hellman, or elliptic curve without a solid working knowledge of number theory. Discrete mathematics also plays an important role in cryptography. Both of these topics will be explored in this chapter. Obviously, a single chapter cannot make you an expert at number theory or discrete mathematics. The goal of this chapter is to provide you just enough of both topics so that you may understand the algorithms presented subsequently in this book.

Introduction

Much of cryptography depends on at least a working knowledge of number theory. You will see elements of it in some symmetric ciphers; however, it will be most applicable when you study asymmetric cryptography in Chaps. 9 and 10. You will discover that all asymmetric algorithms are simply applications of some facet of number theory. This leads us to the question of what is number theory? Traditionally, number theory has been the study of positive, whole numbers. The study of prime numbers is a particularly important aspect of number theory. Wolfram Mathworld describes number theory as follows:

> Number theory is a vast and fascinating field of mathematics, sometimes called "higher arithmetic," consisting of the study of the properties of whole numbers. Primes and prime factorization are especially important in number theory, as are a number of functions such as the divisor function, Riemann zeta function, and totient function.

This current chapter is designed to provide the reader just enough number theory that the cryptographic algorithms discussed later will be comprehensible. There are a dizzying number of tomes on the topic of number theory. Each of the topics in this chapter could certainly be expanded upon. If you are someone with a rigorous mathematical background, you may feel that some topics are given a rather abbreviated treatment. In that, you would be correct. It should also be pointed out that there will be no mathematical proofs. In mathematics, proofs are foundational.

© The Author(s), under exclusive license to Springer Nature Switzerland AG 2021
W. Easttom, *Modern Cryptography*, https://doi.org/10.1007/978-3-030-63115-4_4

Anyone actually pursuing mathematics must learn about proofs and understand proofs. However, our goals in this text can be met without such proofs.

For those readers with a solid foundation in mathematics, this should be just a brief review. For those readers lacking in such a foundation, allow me to quote from the famed author Douglas Adams "Don't Panic". You will be able to understand the material in this chapter. Furthermore, this chapter will be sufficient for you to continue your study of cryptography. But keep in mind that if these topics are new to you, this chapter (along with Chap. 5) might be the most difficult chapters for you. You may wish to read the chapters more than once. I will also provide some study tips as we go through this chapter, directing you to some external sources should you feel you need a bit more coverage of a particular topic.

In This Chapter We Will Cover

Number Systems.
Prime Numbers.
Important Number Theory Questions.
Modulus Arithmetic.
Set Theory.
Logic.
Combinatorics.

Number Systems

The first step in examining number theory is to get a firm understanding of number systems. In this section, we will examine the various sets of numbers. You have probably encountered them at some point in your education and this is likely to be just a refresher. While any symbol can be used to denote number groups, certain symbols have become common. The following table summarizes those symbols.

Symbol	Description
N	N denotes the natural numbers. These are also sometimes called the counting numbers. They are 1, 2, 3, etc.
Z	Z denotes the integers. These are whole numbers $-1, 0, 1, 2$, etc. – the natural numbers combined with zero and the negative numbers
Q	Q denotes the rational numbers (ratios of integers). Any number that can be expressed as a ratio of two integers. Examples are 3/2, 17/4, 1/5
P	P denotes irrational numbers. Examples include $\sqrt{2}$
R	R denotes the real numbers. This includes the rational numbers as well irrational numbers
i	i denotes imaginary numbers. These are numbers whose square is a negative $\sqrt{-1} = 1i$

The preceding table summarizes number systems and their common symbol. Now, in the next few subsections, we can explore these systems in more detail.

Natural Numbers

The obvious place to begin is with the natural numbers. These are so called because they come naturally. That is to say that this is how children first learn to think of numbers. These are often also called counting numbers. Many sources count only the positive integers: 1, 2, 3, 4, ... Without including zero (Bolker 2012). Other sources include zero. In either case, these are the numbers that correspond to counting. If you look on your desk and count how many pencils you have, you can use natural numbers to accomplish this task.

Zero has a fascinating history. While to our modern minds, the concept of zero may seem obvious, it actually was a significant development in mathematics. While there are some indications that the Babylonians might have had some concept of zero, it usually attributed to India. By at least the ninth century C.E., they were carrying out calculations using zero. If you reflect on that for a moment, you will realize that zero as a number, rather than just a placeholder, was not known to ancient Egyptians and Greeks.

Integers

Eventually the natural numbers required expansion. While negative numbers may seem perfectly reasonable to you and I, they were unknown in ancient times. Negative numbers first appeared in a book from the Han Dynasty in China. In this book, the mathematician Liu Hui described basic rules for adding and subtracting negative numbers. In India, negative numbers first appeared in the fourth century C.E. and were routinely used to represent debts in financial matters by the seventh century C.E.

The knowledge of negative numbers spread to Arab nations, and by at least the tenth century C.E. Arab mathematicians were familiar with and using negative numbers. The famed book Al-jabr wa'l-muqabala written by Al-Khwarizmi, from which we derive the word algebra did not include any negative numbers at all.

Rational Numbers

Number systems evolve in response to a need produced by some mathematical problem. Negative numbers, discussed in the preceding section, were developed in response, subtracting a larger integer from a smaller integer (i.e., 3–5). Rational numbers were first noticed as the result of division. In fact, the two symbols one learns for division in modern primary and secondary school are "÷" and "/". Now that is not a mathematically rigorous definition, nor is it meant to be. A mathematical definition is: any number that can be expressed as the quotient of two integers (Hua 2012). However, like many early advances in mathematics, the formalization of rational numbers was driven by practical issues.

Irrational Numbers

Eventually, the division of numbers led to results that could not be expressed as the quotient of two integers. The classic example comes from geometry. If one tries to express the ratio of a circles circumference to its radius, the result is an infinite number. It is often approximated as 3.14159, but the decimals continue on and no pattern repeats. Irrational numbers are sometimes repeating, but they need not be. As long as a number is a real number that cannot be expressed as the quotient of two integers, it is classified as an irrational number.

Real Numbers

Real numbers are the superset of all rational numbers and all irrational numbers. It is likely that all the numbers you encounter on a regular basis are real numbers, unless of course you work in certain fields of mathematics or physics. For quite a long time, these where considered to be the only numbers that existed. The set of real numbers is infinite.

Complex Numbers

Imaginary numbers developed in response to a specific problem. The problem begins with the essential rules of multiplications. If you multiply a negative with a negative, you get a positive. For example, $-1 * -1 = 1$. This becomes a problem if one considers the square root of a negative number. Clearly, the square root of a positive number is also positive. $\sqrt{4} = 2$, $\sqrt{1} = 1$, etc. But what is the $\sqrt{-1}$? If you answer that it is -1, that won't work. $-1 * -1$ gives us positive 1. This conundrum led to the development of imaginary numbers. Imaginary numbers are defined as follows $i^2 = -1$ (or conversely $\sqrt{-1} = i$). Therefore, the square root of any negative number can be expressed as some integer multiplied by i. A real number combined with an imaginary number is referred to as a complex number.

Imaginary numbers work in much the same manner as real numbers do. If you see the expression $4i$, that denotes $4 * \sqrt{-1}$. It should also be noted that the name imaginary number is unfortunate. These numbers do indeed exist and are useful in a number of contexts. For example, such numbers are often used in quantum physics and quantum computing.

Complex numbers are simply real numbers and imaginary together in an expression. For example:

$$3 + 4i$$

This is an example of a complex number. There is the real number 4, combined with the imaginary number i to produce $4i$ or $4 * \sqrt{-1}$. Let us put this a bit more formally. A complex number is a polynomial with real coefficients and i for which $i^2 + 1 = 0$ is true. You can perform all the usual arithmetic operations with complex numbers that you have performed with real numbers (i.e., rational numbers, irrational numbers, integers, etc.). Let us consider a basic example:

$$(3 + 2i) + (1 + 1i)$$
$$= (3 + 1) + (2 + 1)i$$
$$= 4 + 3i$$

Here is another example, this time subtracting.

$$(1 + 2i) - (3 + 4i)$$
$$= (1 - 3) - (2i - 4i)$$
$$= 2 - 2i$$

As you can see, basic addition and subtraction with complex numbers is very easy. Multiplying complex numbers is also quite straight forward. We use the same method you probably learned in elementary algebra as a youth: FOIL or First – Outer – Inner – Last. This is shown in Fig. 4.1.

Transcendental Numbers

Some would argue that transcendental numbers are not necessary for introductory cryptography. And I would agree. However, you see such numbers with some

Fig. 4.1 FOIL method

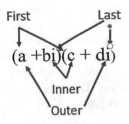

frequency, and it is useful to have some understanding of what they are. Transcendental numbers are numbers other than algebraic numbers. This of course begs the question of what is an algebraic number. An algebraic number that is a root of a non-zero polynomial. Algebraic numbers include both real numbers and complex numbers.

Two examples of transcendental numbers are π and e. However, there are actually infinitely many transcendentals (Fel'dman and Nesterenko 2013). One fact regarding real transcendental numbers is that they are irrational. However, the reverse is not true. Certainly not all irrational numbers are transcendental. A good example is $\sqrt{2}$, because that is actually a root of the polynomial equation $x^2 - 2 = 0$.

Euler's number is not only one of the more common examples of a transcendental number, but Leonard Euler was one of the first to provide a definition of transcendental numbers. He did so in the eighteenth century. It was Joseph Liouvill in 1844 who first provided the existence of transcendental numbers. In 1873, Charles Hermite proved that e was indeed a transcendental number. The famous mathematician David Hilbert proposed a number of problems. Hilbert's seventh problem is stated as: If a is an algebraic number that is not zero or one, and b is an irrational algebraic number, is a^b necessarily transcendental?

Prime Numbers

Prime numbers have a particular significance in cryptography. As we will see in later chapters, the RSA algorithm would simply not exist without an understanding and use of prime numbers. A prime number is defined as a natural number greater than 1 that has no positive divisors other than itself and 1. Note that the number 1 itself is not considered prime. There are a number of facts, conjectures, and theorems in number theory related to prime numbers.

The fundamental theorem of arithmetic states that every positive integer can be written uniquely as the product of primes, where the prime factors are written in order of increasing size. The practical implication of this is that any integer can be factored into its prime factors. This is probably a task you had to do in primary school, albeit with relatively small integers. If you consider the fundamental theorem of arithmetic for just a moment, you can see that it is one (of many) indication of how important prime numbers are.

The next interesting thing to note about prime numbers is that there are an infinite number of primes. This was first proven by Euclid; however, many other proofs of this have been constructed by mathematicians such as Leonard Euler and Christian Goldbach. Though we know there are infinite prime numbers, how many are found in a given range of numbers is a more difficult question. For example, if you consider the number of primes between 1 and 10 there are four (2, 3, 5, and 7). Then, between 11 and 20 there are four more (11, 13, 17, and 19). However, there are only 2 between 30 and 40 (31 and 37). In fact, as we go further and further through the positive integers, prime numbers become more sparse. It has been determined that the number of prime numbers less than x is approximately $x/\ln(x)$. Consider the number 10. The natural logarithm of 10 is approximately 2.302, $10/2.302 = 4.34$ and we have already seen there are 4 primes between 1 and 10. Euclid's famous work "Elements" written around 300 BCE contained significant discussion of prime numbers including theorems regarding there being an infinite number of primes.

Finding Prime Numbers

Now you understand what prime numbers are, you know they are infinite, and you know the general distribution of prime numbers. You will discover later in this book; prime numbers are a key part of cryptographic algorithms such as RSA. So how does one generate a prime number? The obvious first thought, which may have occurred to you, is to simply generate a prime number and then check to see if it is prime. You can check by simply trying to divide that random number n by every number up to the square root of n. So, if I ask you if 67 is prime you first try to divide it by 2, that does not work; then, next by 3, that does not work, then by 4, that does not work, and so forth. In just a few moments, you will find that yes indeed 67 is a prime number. That wasn't so hard was it? And for a computer it would be much faster. Unfortunately, in cryptographic algorithms, like RSA, we need much larger prime numbers. Modern RSA keys are a minimum of 2048 bits long and need two prime numbers to generate. Those prime numbers need to be very large. Therefore, simply generating a random number and using brute force to check if it is prime simply is not possible by hand and could take quite some time by computer. For this reason, there have been numerous attempts to formulate some algorithm that would consistently and efficiently generate prime numbers. Let us look at just a few.

Mersenne Prime

Marin Mersenne was a French theologian who lived from September 8, 1588 to September 1, 1648. He made significant contributions to acoustics and music theory as well as mathematics. Mersenne believed that the formula $2^p - 1$ would produce prime numbers. Investigation has shown that this formula will produce prime numbers for $p = 2, 3, 5, 7, 13, 17, 19, 31, 67, 127, 257$, but not for many other numbers such as $p = 4, 6, 8, 9, 10, 11$, etc. His conjecture resulted in a prime of the form $2p - 1$ being named a Mersenne prime. By 1996, there were 34 known Mersenne primes, with the last eight discoveries made on supercomputers. Clearly, the formula can generate primes, but frequently does not. Any number of the format $2^p - 1$ is considered a Mersenne number. It so happens that if p is a composite number, then so is $2^p - 1$. If p is a prime number, then $2^p - 1$ might be prime (Riesel 2012). Mersenne originally thought all numbers of that form would be prime numbers.

Fermat Prime

Pierre de Fermat, a mathematician of the seventeenth century, made many significant contributions to number theory. Fermat also proposed a formula which he thought could be used to generate prime numbers. He believed that numbers of the format shown in the following equation would be prime.

$$2^{2^n} + 1$$

Several such numbers, particularly the first few powers of 2, are indeed prime. As you can see here:

$$2^{2^0} + 1 = 3 \qquad \text{prime}$$
$$2^{2^1} + 1 = 5 \qquad \text{prime}$$
$$2^{2^2} + 1 = 17 \qquad \text{prime}$$
$$2^{2^3} + 1 = 257 \qquad \text{prime}$$
$$2^{2^4} + 1 = 65,537 \qquad \text{prime}$$

Unfortunately, just like Mersenne primes, not all Fermat numbers are prime. This was an attempt to produce a reliable algorithm for creating prime numbers that unfortunately failed.

Eratosthenes' Sieve

In some cases, the goal is not to generate a prime, but to find out if a number is indeed a prime. We need some method(s) that are faster than simply trying to divide by every integer up to the square root of the integer in question. Eratosthenes' Sieve is one such method. Eratosthenes was the librarian at Alexandria, Egypt, in 200 B.C. He was also quite the mathematician and astronomer. He invented a method for finding prime numbers that is still used today. This method is called Eratosthenes' Sieve. Eratosthenes's Sieve essentially filters out composite integers in order to leave only the prime numbers (Helfgott 2020).

The method is actually rather simple:

To find all the prime numbers less than or equal to a given integer n by Eratosthenes' method:

1. Create a list of consecutive integers from 2 to n: $(2, 3, 4, \ldots, n)$.
2. Initially, let p equal 2, the first prime number.
3. Starting from p, count up in increments of p and mark each of these numbers

greater than p itself in the list. These will be multiples of p: $2p$, $3p$, $4p$, etc.; note that some of them may have already been marked.

4. Find the first number greater than p in the list that is not marked. If there was no such number, stop. Otherwise, let p now equal this number (which is the next prime), and repeat from step 3.

To illustrate, you can do this with a chart. Start with one such as the one shown in Fig. 4.2. For the purposes of demonstrating this concept we are only concerning ourselves with numbers through 100.

The first step is to cross out the number 2 because it is prime, then cross out all multiples of 2, since they are, by definition, composite numbers. Following this sequence will eliminate several integers that we are certain are not prime numbers. You can see this in Fig. 4.3.

Fig. 4.2 Numbers through 100

1	2	3	4	5	6	7	8	9	10
11	12	13	14	15	16	17	18	19	20
21	22	23	24	25	26	27	28	29	30
31	32	33	34	35	36	37	38	39	40
41	42	43	44	45	46	47	48	49	50
51	52	53	54	55	56	57	58	59	60
61	62	63	64	65	66	67	68	69	70
71	72	73	74	75	76	77	78	79	80
81	82	83	84	85	86	87	88	89	90
91	92	93	94	95	96	97	98	99	100

Fig. 4.3 Eliminating multiples of 2

1	2̶	3	4̶	5	6̶	7	8̶	9	1̶0̶
11	1̶2̶	13	1̶4̶	15	1̶6̶	17	1̶8̶	19	2̶0̶
21	2̶2̶	23	2̶4̶	25	2̶6̶	27	2̶8̶	2̶9̶	3̶0̶
31	3̶2̶	33	3̶4̶	35	3̶6̶	37	3̶8̶	39	4̶0̶
41	4̶2̶	43	4̶4̶	45	4̶6̶	47	4̶8̶	49	5̶0̶
51	5̶2̶	53	5̶4̶	55	5̶6̶	57	5̶8̶	59	6̶0̶
61	6̶2̶	63	6̶4̶	65	6̶6̶	67	6̶8̶	69	7̶0̶
71	7̶2̶	73	7̶4̶	75	7̶6̶	77	7̶8̶	79	8̶0̶
81	8̶2̶	83	8̶4̶	85	8̶6̶	87	8̶8̶	89	9̶0̶
91	9̶2̶	93	9̶4̶	95	9̶6̶	97	9̶8̶	99	1̶0̶0̶

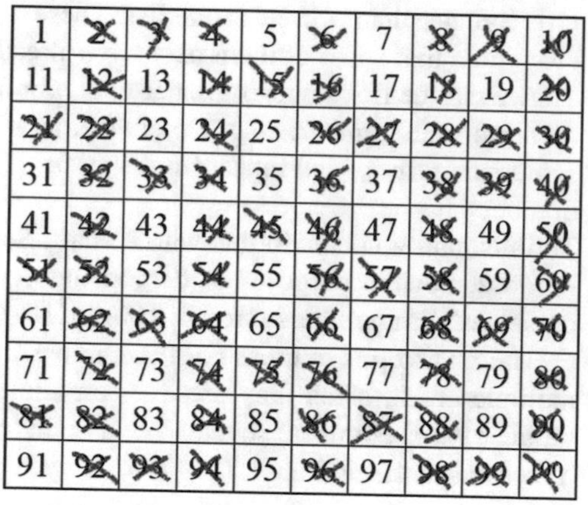

Fig. 4.4 Eliminating multiples of 3

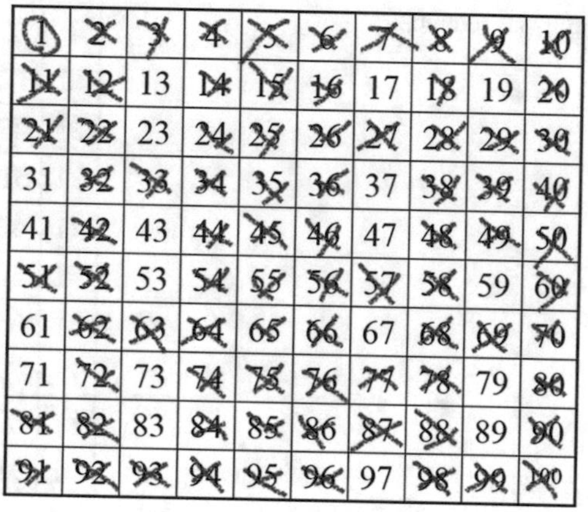

1	2	3	4	5	6	7	8	9	10
11	12	13	14	15	16	17	18	19	20
21	22	23	24	25	26	27	28	29	30
31	32	33	34	35	36	37	38	39	40
41	42	43	44	45	46	47	48	49	50
51	52	53	54	55	56	57	58	59	60
61	62	63	64	65	66	67	68	69	70
71	72	73	74	75	76	77	78	79	80
81	82	83	84	85	86	87	88	89	90
91	92	93	94	95	96	97	98	99	100

Fig. 4.5 The prime numbers that are left

1	2	3	4	5	6	7	8	9	10
11	12	13	14	15	16	17	18	19	20
21	22	23	24	25	26	27	28	29	30
31	32	33	34	35	36	37	38	39	40
41	42	43	44	45	46	47	48	49	50
51	52	53	54	55	56	57	58	59	60
61	62	63	64	65	66	67	68	69	70
71	72	73	74	75	76	77	78	79	80
81	82	83	84	85	86	87	88	89	90
91	92	93	94	95	96	97	98	99	100

Now, we do the same thing with 3 (since it is also a prime number) and for all multiples of 3. You can see this in Fig. 4.4.

You will notice that many of those are already crossed out, since they were also multiples of 2. You can skip 4, since it is a multiple of 2, and go straight to 5 and multiples of 5. Some of those (like 15) are also multiples of three or multiples of 2 (like 10, 20, etc.) and are already crossed out. You can skip 6, 8, and 9 since they are multiples of numbers that have already been used, which leaves 7 and multiples of 7. Given that we are stopping at 100, you will find if you use 11 and multiples of 11, that will finish your application of Eratosthenes' Sieve. Once you have crossed out multiples of 5, 7, and 11, what is left are prime numbers. Remember 1 is not prime. You can see this in Fig. 4.5.

This can be a tedious process and is often implemented via a computer program. If you are going past 100, you just keep using the next prime number and multiples of it (13, 17, etc.).

Other Sieves

There have been a number of more modern sieve systems developed. The Sieve of Atkin is one such sieve. It was developed in 2003 by Arthur Atkin and Daniel J. Bernstein. The sieve of Sundaram is another such algorithm for finding prime numbers. It was developed in the 1930s by the Indian mathematician S.P. Sundaram. The sieve of Sundaram is relatively simple, so we will go into more detail on it here.

The first step is to start with a list of integers from 1 to some number n. Then, remove all numbers of the form $i + j + 2ij$. The numbers i and j are both integers with $1 \leq i < j$ and $i + J + 2ij \leq n$. After you have removed those numbers, the remaining numbers are doubled and incremented by one, giving a list of the odd prime numbers below $2n + 2$. Like the sieve of Eratosthenes, the sieve of Sundaram operates by removing composite numbers, leaving behind only the prime numbers.

Lucas–Lehmer Test

This is a test designed to determine if a given number n is a prime number. It is a deterministic test of primality that works on Mersenne numbers. That means it is only effective for finding out if a given Mersenne number is prime. Remember a Mersenne number is one of the form $2^p - 1$. In the 1800s, French mathematician Edouard Lucas developed a primality test. In the early twentieth century, American mathematician Derrick Lehmer provided a proof of this test. The test is now called the Lucas–Lehmer test (Riesel 2012).

The term deterministic might not be entirely clear to some readers. Throughout the study of cryptography, you will encounter the terms deterministic and probabilistic. You will see this in reference to primality tests, pseudo random number generators, and other algorithms. The meanings of these two terms is fairly simple. The basic concept is that a probabilistic algorithm has a probability of giving the desired result, whereas a deterministic algorithm will give the desired result. In reference to primality testing, a probabilistic test can only tell you that a given number is probably prime, a deterministic test will tell you that it is definitely prime (or not).

A number, called the Lucas–Lehmer number, is part of a sequence of numbers where each number is the previous number squared minus 2. Therefore, if you start with 4, you will get the sequence shown here.

$$S_1 = 4$$
$$S_2 = 4^2 - 2 = 14$$
$$S_3 = 14^2 - 2 = 194$$
$$S_4 = 194^2 - 2 = 37,634, \text{etc.}$$

The test is as follows:

For some $p \geq 3$, which is itself a prime number, $2^p - 1$ is prime if and only if S_{p-1} is divisible by $2^p - 1$.

Example: $2^5 - 1 = 31$ so since $S_4 = 37,634$ is divisible by 31, 31 must be a prime number.

This may be one of those areas where you are tempted to question the veracity of a claim I have made; in this instance, the reliability of the Lucas–Lehmer test. As I stated at the beginning of this chapter, we will not be demonstrating mathematical proofs. You can spend some time with this test, using various values of S to see for yourself that it does indeed work. That is certainly not the same as working out a proof but could give you some level of comfort that the test does, indeed, work.

Relatively Prime

The concept of prime numbers, and how to generate them, is critical to your study of cryptography. There is another topic you should be familiar with, that is the concept of numbers that are relatively prime, also often called co-prime. The concept is actually pretty simple. Two numbers x and y are relatively prime (co-prime) if they have no common divisors other than 1. So, for example the numbers 8 and 15 are co-prime. The factors of 8 are 1, 2, and 4. The factors of 15 are 1,3, and 5. Since these two numbers (8 and 15) have no common factors other than 1, they are relatively prime. There are lots of interesting facts associated with co-prime numbers. Many of these facts were discovered by Leonard Euhler.

Leonard Euler asked a few simple questions about co-prime numbers. The first question he asked was, given an integer n, how many numbers are co-prime to n? That number is called Euler's Totient or simply the totient. The symbol for the totient of a number is shown in Fig. 4.6.

The term for the integers that are smaller than n and have no common factors with n (other than 1) is *totative*. For example, 8 is a totative of 15. Another term for Euler's totient is the Euler phi function.

The story of Euler's totient is emblematical of much of mathematics. Euler introduced this function in 1793. For many years after his death, there was no practical application for this concept. In fact, there was no practical use for it until modern asymmetric cryptography in the 1970s, almost 200 years later. This is one example of pure mathematics that later is found to have very important practical applications.

$$\varphi(n)$$

Fig. 4.6 Euler's Totient

The next question Euler asked was, what if a given integer, n, is a prime number? How many totatives will it have (i.e., what is the totient of n)? Let us take a prime number as an example, for example, 7. Given that it is prime, we know that none of the numbers smaller than it have any common factors with it. Thus, 2, 3, 4, 5, and 6 are all totatives of 7. Since 1 is a special case, it is also a totative of 7, so we find there are 6 numbers that are totatives of 7. It turns out that for any n that is prime, the Euler's totient of n is $n-1$. Therefore, if $n = 13$, then the totient of n is 12. And then Euler asked yet another question about co-prime numbers. Let us assume you have two prime numbers m and n. We already know that the totient of each is $m-1$ and $n-1$, respectively. But what happens if we multiply the two numbers together? If we multiply $m * n$ and get k. What is the totient of k? Well, you can simply test every number less than k and find out. That might work for small values of k but would be impractical for larger values. Euler discovered a nice shortcut. He found that if you have two prime numbers (m and n) and multiplied them together to get k, the totient of that k was simply the totient of $m *$ the totient of n. Put another way, the totient of k is $(m-1)(n-1)$.

Let us look at an example. Assume we have to prime numbers $m = 3$ and $m = 5$.

1. If we multiply them together, we have $k = 15$.
2. totient of $n = n-1$ or 2.
3. The totient of m is $m-1$ or 4.
4. 2 * 4 gives us 8
5. The totient of k (15) is 8.

15 is a small enough number, so we can test it. What are its factors (other than 1)? 3 and 5. Which means it has common factors with any multiple of 3 or 5 smaller than 15: 3, 6, 9, 12; 5, 10. So what numbers does it not have common factors with? 1, 2, 4, 7, 8, 11, 13, and 14. And if you take a moment to count those, you will see there are 8. Of course, Euler relied on rigorous mathematical proofs, which we are foregoing in this chapter, but hopefully this example provides you with a comfort level regarding calculating totients.

Now, all of this may sound rather obscure, even unimportant to some readers. However, these facts are a very big part of the RSA algorithm, which we will be examining in some depth in Chap. 10. In fact, when you first see the RSA algorithm, you will see several elements we have discussed in this section.

Important Operations

Some mathematical operations are critical to even a basic understanding of modern cryptographic algorithms. Readers with a solid foundation of mathematics should already be familiar with these operations, but in case you are not, we will briefly introduce you to them in this section.

Divisibility Theorems

There are some basic facts about divisibility that you should be aware of. These define some elementary facts regarding the divisibility of integers. For integers a, b, and c it is true that

if $a \mid b$ and $a \mid c$, then $a \mid (b + c)$. Example: $3 \mid 6$ and $3 \mid 9$, so $3 \mid 15$.
if $a \mid b$, then $a \mid bc$ for all integers c. Example: $5 \mid 10$, so $5 \mid 20$, $5 \mid 30$, $5 \mid 40$, ...
if $a \mid b$ and $b \mid c$, then $a \mid c$. Example: $4 \mid 8$ and $8 \mid 24$, so $4 \mid 24$.

These are basic divisibility theorems. You will see them applied in later chapters in specific cryptographic algorithms. It is recommended that you commit them to memory before proceeding.

Summation

You have probably seen the summation symbol frequently: $\sum$. If not, you will see it later in this book. It is basically a way of adding a sequence of numbers in order to get their sum. Any intermediate result, at any given point in the sequences, is a partial sum (sometimes called a running total or prefix sum). The numbers that are to be added up are often called addends or sometimes summands (Neukirch 2013). They can be any numbers, but in cryptography we will most often be interested in summing integers.

A basic example is shown in Fig. 4.7.

What this figure is stating is that we are dealing with some integer i. Begin with the value of i set to 1. Add up the i values from 1 until you reach 100. It is more convenient to use the summation symbol rather than expressly write out the summation of a long series of values. The notation you see used in Fig. 4.7 is often called the capital Sigma notation as the $\sum$ is the sigma symbol.

The letter to the right of the summation symbol is the variable we are dealing with. The number under the summation is the initial value of that variable. Finally, the number over the summation symbol (often denoted with an n) is the stopping point. You can manually add these numbers up $1 + 2 + 3 + 4 \ldots 100$ if you wish. However, there is a formula that will compute this summation:

$$(n\,(n + 1))/2$$

In our example that is

$$(100\,(101))/2 = 5050$$

Fig. 4.7 Summation

$$\sum_{i=1}^{100} i.$$

A similar operation can be done with multiplication using the pi symbol $\prod$. This symbol works exactly like the capital Sigma notation; however, rather than adding up the series to arrive at a sum, you multiply the series in order to arrive at a product.

Logarithms

Logarithms play a very important role in asymmetric cryptography. There is an entire class of asymmetric algorithms that are based on what is called the discrete logarithm problem. Most readers encountered logarithms during secondary school. The logarithm of a number is simply the exponent to which some fixed value (called the base) must be raised to produce that number. Let us assume base 10 and ask what is the logarithm of 10,000? Well, what number must 10 be raised to in order to produce the number 10,000? The answer is 4, $10^4 = 10,000$.

In our example, we used base 10. Most people are quite comfortable with base 10. It is known as the common logarithm in mathematics. However, you can use any base you might wish. For example, if you wanted to ask what the logarithm is of 16 in base 2, you would write that as $\log_2(16)$. Remember that you are asking what power must 2 be raised to, to get 16. The answer is 4, $2^4 = 16$.

Natural Logarithm

The natural logarithm, denoted as ln, is very widely used in math and science. Natural logarithms use the irrational number e (approximately 2.718) as the base. The natural logarithm of some number x is usually written as ln x but can be written as $\log_e x$. As with any logarithm, the natural log is what power e must be raised to in order to give some particular number.

The basic concept of the natural logarithm was developed by Gregoire de Saint Vincent and Alphonse de Sarasa in the early seventeenth century. They were working on problems involving hyperbolic sectors. For those readers who are new to logarithms and/or natural logs, this brief treatment might not leave you fully confident in your understanding. Fortunately, there are two online tutorials you may wish to consult to get more on these topics. The first is Purple Math http://www.purplemath.com/modules/logs3.htm. The second is Better Explained http://betterexplained.com/articles/demystifying-the-natural-logarithm-ln/

Discrete Logarithm

Now that you have a basic grasp of logarithms and even natural logs, we can turn our attention to discussing discrete logarithms. These play an important role in algorithms we will explore in Chaps. 10 and 11. To understand discrete logarithms, keep in mind the definition of a logarithm. It is the number to which some base must be raised to get another number. Discrete logarithms ask this same question but do so in regard to a finite group. In Chap. 5, we will take up the concept of groups in much more detail.

Put more formally, a discrete logarithm is some integer k that solves the equation $x^k = y$, where both x and y are elements of a finite group (Vinogradov 2016). Computing a discrete logarithm is a very difficult problem, which is why discrete logarithms form the basis for some modern cryptographic algorithms. We will take up this topic again in Chap. 10 when we discuss specific algorithms and after we have discussed groups in Chap. 5. For now, you should note that a discrete logarithm is, essentially, a logarithm within some finite group.

Modulus Operations

Modulus operations are important in cryptography, particularly in RSA. Let us begin with the simplest explanation, then later delve into more details. To use the modulus operator, simply divide A by N and return the remainder.

Thus 5 mod 2 = 1.
And 12 mod 5 = 2.

This explains how to use the modulus operator, and in many cases, this is as far as many programming textbooks go. But this still leaves the question of why we are doing this. Exactly what is a modulus? One way to think about modulus arithmetic is to imagine doing any integer math you might normally do but bound your answers by some number. A classic example is the clock. It has numbers 1 through 12, so any arithmetic operation you do has to have an answer that is 12 or less. If it is currently 4 o'clock and I ask you to meet me in 10 h, simple math would say I am asking you to meet me at 14 o'clock. But that is not possible, our clock is bounded by the number 12! The answer is simple, take your answer and use the mod operator with a modulus of 12 and look at the remainder.

14 mod 12 = 2.

I am actually asking you to meet me at 2 o'clock (whether that is a.m. or p.m. depends on which the original 4 o'clock was, but that is really irrelevant to understanding the concepts of modular arithmetic.) This is an example of how you use modular arithmetic every day. The clock example is widely used; in fact most textbooks rely on that example. It is intuitive and most readers will readily grasp the essential point of modular arithmetic. You are performing arithmetic bound by some number, which is called the modulus. In the clock example, the modulus is 12, but one can perform modular arithmetic using any integer as the modulus.

While the basic concepts of modular arithmetic dates back to Euclid who wrote about modular arithmetic in his book Elements, the modern approach to modular arithmetic was published by Carl Gauss in 1801.

Congruence

Congruence in modulus operations is a very important topic and you will see it applied frequently in modern cryptographic algorithms. We saw this earlier in Chap. 1, and stated then that we would explain it more detail later. Now is the time to more fully explore this concept.

Two numbers a and b are said to be "congruent modulo n" if

$$(a \bmod n) = (b \bmod n) \rightarrow a \equiv b(\bmod n)$$

The symbol $\equiv$ is used to denote congruence in mathematics. In many programming languages the % symbol is used to perform the modulo operation (i.e., to divide the first number by the second, and only return the remainder).

It should be noted that, if two numbers are congruent modulo n, then the difference between a and b will be some multiple of n. To make this clear, let us return to the clock analogy used earlier in this chapter. 14 and 2 are congruent modulo. And it turns out that the difference between 14 and 2 is 12, a multiple of n (1×12). We also know that 14, or 1400 h on the 24-h clock is 2 o'clock. So, when we say that 14 and 2 are congruent modulo 12, we are stating that, at least in reference to modulo 12, 14 and 2 are the same. Look at another example; 17 and 5 are congruent modulo 12. And the difference between them is 12. And again, using the 24-h clock to test our conclusions, we find that 1700 h is the same as 5 pm.

There are some basic properties of congruences that you should be aware of:

$$a \equiv b \;(\bmod\; n) \text{ if } n \mid (a - b)$$
$$a \equiv b \;(\bmod\; n) \text{ implies } b \equiv a(\bmod\; n)$$
$$a \equiv b \;(\bmod\; n) \text{ and } b \equiv c(\bmod\; n) \text{ imply } a \equiv c(\bmod\; n)$$

Congruence Classes

A congruence class is a group of numbers that are all congruent for some given modulus. Another name for congruence classes is residue class. So, let us consider an example where the modulus is 5. What numbers are congruent modulo 5? Let's begin with 7. 7 mod 5 = 2. So now we ask what other numbers mod 5 = 2. We will arrive at an infinite series of numbers 12, 17, 22, 27, 32, etc. (Weil 2013). You should also note that this works the other way (i.e., with integers smaller than the modulus, including negatives). We know that 7 mod 5 = 2, but 2 mod 5 = 2 (the nearest multiple of 5 would be 0; 0 *5, thus 2 mod 5 = 2). Make sure you fully understand why 2 mod 5 = 2, then we can proceed to examine negative numbers. Our next multiple of 5 after 0, going backwards is −5 (5 * −1). So, −3 mod 5 = 2. We can now expand the elements of our congruence class to include −3, 0, 2, 7, 12, 17, 22, 27, 32, etc.

You should also note that for any modulus n, there are n congruence classes. If you reflect on this for just a moment you should see that the modulus itself creates a boundary on the arithmetic done with that modulus, so it is impossible to have more congruence classes than the integer value of the modulus. Modulus operations are very important to several asymmetric cryptography algorithms that we will be examining in Chap. 10. Therefore, make certain you are comfortable with these concepts before proceeding.

Famous Number Theorists and Their Contributions

When learning any new topic, it is often helpful to review a bit of its history. In some cases, this is best accomplished by examining some of the most important figures in that history. Number theory is replete with prominent names. In this section, I have restricted myself to just a brief biography of those mathematicians whose contributions to number theory are most relevant to cryptography.

Fibonacci

Leonardo Bonacci (also known as Fibonacci) lived from 1170 until 1250 and was an Italian mathematician. He made significant contributions to mathematics. He is most famous for the Fibonacci numbers. Fibonacci numbers represent an interesting sequence of numbers.

1, 1, 2, 3, 5, 8, 13, 21, etc.

The sequence is derived from adding the two previous numbers. To put that in a more mathematical format.

$$Fn = Fn - 1 + Fn - 2$$

Therefore, in our example we see that 2 is 1 + 1; 3 is 2 + 1; 5 is 3 + 2; 8 is 5 + 3; etc.

The sequence continues infinitely. Fibonacci numbers were actually known in India, but unknown in Europe until Fibonacci discovered them. He published this concept in his book Liber Abaci. This number series has many interesting applications; for our purposes, the most important is its use in pseudo random number generators, which we will be exploring in Chap. 12.

Fibonacci was the son of a wealthy merchant and was able to travel extensively. He spent time traveling the Mediterranean area studying under various Arab mathematicians. In his aforementioned book Liber Abaci, he also introduced Arabic numerals, which are still widely used today.

Fermat

Pierre de Fermat (pronounced 'fermā') lived in the first half of the seventeenth century. He was trained as a lawyer and was an amateur mathematician, though he made significant contributions to mathematics. He did work in analytical geometry and probability but is probably best known for several theorems in number theory, which we will take a brief look at.

Fermat's little theorem states that if p is a prime number, then for any integer a, the number $a^p - a$ is an integer multiple of p. What does this mean? It may be best for us to examine an example. Let us take $a = 2$ and $p = 5$. This gives us 2^5 or 32. 32–2 is 30, which is indeed a multiple of p (5). This actually turns into a primality test. If you think P is a prime number, then test it. If it is prime, then a^{p-a} should be an integer multiple of p. If it is not, then p is not prime.

Let us test this on some numbers we know are not prime. Use $p = 0$. $2^8 - 2 = 256–2 = 250$. 250 is not an integer multiple of 9; therefore, 9 is not prime. This is an example of how various aspects of number theory that might not at first appear to have practical applications can indeed be applied to practical problems. This theorem is usually expressed in a modulus format, which we discussed previously in this chapter.

Fermat first articulated this theorem in 1640, wording his theorem as follows:

p divides $a^{p-1} - 1$ whenever p is prime and a is co-prime to p.

Fermat is perhaps best known for Fermat's Last Theorem, sometimes called Fermat's conjecture. This theorem states that no three positive integers (x, y, z) can satisfy the equation $x^n + y^n = c^n$ for any n that is greater than 2. Put another way if $n > 2$, then there is no integer to the n power that is the sum of two other integers to the n power. Fermat first proposed this in 1637, actually in handwritten notes in the margin of his copy of Arithmetica. He claimed to have a proof that would not fit in the margin, but to date no one has ever found Fermat's proof. However, in subsequent years, mathematicians have provided proofs of this theorem. Andrew Wiles in 1994 provided such a proof.

Euler

Leonhard Euler is an important figure in the history of mathematics and number theory. Let us begin by properly pronouncing his name. To an American, the spelling might seem to indicate it should be pronounced like 'you ler' but in fact it is pronounced 'oil er'. Euler is considered one of the best mathematicians of all time, and certainly the pinnacle of eighteenth-century mathematics.

Euler was born on April 15, 1707 in Switzerland. His father was a pastor of a church. As a child, his family was friends with the Bernoulli family. One of the Bernoulli's, Johann Bernoulli, was a prominent mathematician of the time. Euler's contributions to mathematics are very broad. He worked in algebra, number theory, physics, and geometry. The number e you probably encountered in elementary calculus courses (approximately 2.7182), which is the base for the natural logarithm, is named after Euler, and the Euler–Mascheroni constant is also named after Euler.

Leonard Euler was responsible for the development of infinitesimal calculus and provided proofs for Fermat's little theorem as well as inventing the totient function you encountered earlier in this chapter. He also made significant contributions to graph theory. We will be discussing graph theory later in this chapter in reference to discrete mathematics. Incidentally, Euler also provided a proof of Fermat's little theorem. In addition to these achievements, Euler is widely considered the father of modern graph theory.

Goldbach

Christian Goldbach lived from March 18, 1690, until November 20, 1765. He was trained as a lawyer but is remembered primarily for his contributions to number theory. It is reported that he had met with a number of famous mathematicians, including Euler. While he made other contributions to number theory, he is best remembered for Goldbach's conjecture, which, at least as of this writing, is an unsolved mathematical problem. The conjecture can be worded quite simply:

Every integer greater than 2 can be expressed as the sum of two prime numbers.

For small numbers, this is easy to verify (though the number 3 immediately poses a problem as mathematicians don't generally consider 1 to be prime).

$$4 = 2 + 2 \quad \text{both primes}$$
$$5 = 3 + 2 \quad \text{both primes}$$
$$6 = 3 + 4 \quad \text{both primes}$$

While this has been shown for very large numbers, and no exception has been found, it has not been mathematically proven.

Discrete Mathematics

As you saw in our discussion of number theory, some numbers are continuous. For example, real numbers are continuous. What I mean by that statement is that there is no clear demarcation point. As one example, 2.1 and 2.2 can be further divided into 2.11, 2.12, 2.13...2.2. And even 2.11 and 2.11 can be further divided into 2.111, 2.112, 2.113...2.12. This process can continue infinitely. The calculus deals with continuous mathematics. One of the most basic operations in calculus, the limit, is a good example of this. Discrete mathematics, however, is concerned with those mathematical constructs that have clearly defined (i.e., discrete) boundaries. You can probably already surmise that integers are a major part of discrete mathematics, but so are graphs and sets.

Set Theory

Set theory is an important part of discrete mathematics. A set is a collection of objects which are called the members or elements of that set. If we have a set, we say that some objects belong (or do not belong) to this set, are (or are not) in the set. We say also that sets consist of their elements.

As with much of mathematics, terminology and notation is critical. So, let us begin our study in a similar fashion, building from simple concepts to more complex. The most simple one I can think of is defining an element of a set. We say that x is a member of set A (Balakrishnan, 2012). This can be denoted as.

$$x \in A$$

Sets are often listed in brackets. For example, the set of all odd integers <10 would be shown as follows:

$$A = \{1, 3, 5, 7, 9\}$$

And a member of that set would be denoted as follows:

$$3 \in A$$

Negation can be symbolized by a line through a symbol. For example:

$$2 \notin A$$

2 is not an element of set A.

If a set is not ending, you can denote that with ellipses. For example, the set of all odd numbers (not just those less than 10) can be denoted

$$A = \{1, 3, 5, 7, 9, 11, \ldots\}$$

You can also denote a set using an equation or formula that defines membership in that set.

Sets can be related to each other; the most common relationships are briefly described here:

Union: If you have two sets A and B, elements that are a member of A, B, or both represent the union of A and B, symbolized as: $A \bigcup B$.

Intersection: If you have two sets A and B, elements that are in both A and B are the intersection of sets A and B, symbolized as $A \cap B$. If the intersection of set A and B is empty (i.e., the two sets have no elements in common) then the two sets are said to be disjoint.

Difference: If you have two sets A and B, elements that are in one set but not in both are the difference between A and B. This is denoted as $A\backslash B$.

Compliment: Set B is the compliment of set A if B has no elements that are also in A. This is symbolized as $B = A^c$.

Double Compliment: the compliment of a set compliment is that set. In other words, the compliment of A^c is A. That may seem odd at first read but reflect on the definition of the compliment of a set for just a moment. The compliment of a set has no elements in that set. So, it stands to reason that to be the compliment of the compliment of a set, you would have to have all elements within the set.

These are basic set relationships. Now a few facts about sets:

Order is irrelevant: $\{1, 2, 3\}$ is the same as $\{3, 2, 1\}$ or $\{3, 1, 2\}$ or $\{2, 1, 3\}$.

Subsets: Set A could be a subset of Set B. For example, if set A is the set of all odd numbers <10 and set B is the set of all odd numbers <100, then set A is a subset of set B. This is symbolized as $A \subseteq B$.

Power set: As you have seen, sets may have subsets. Let us consider set A as all integers less than 10. Set B is a subset; it is all prime numbers <10. Set C is a subset of A; it is all odd numbers <10. Set D is a subset of A; it is all even numbers less than 10. We could continue this exercise making arbitrary subsets such as $E = \{4, 7\}$, $F \{1, 2, 3\}$, etc. The set of all subsets for a given set is called the power set for that set.

Sets also have properties that govern the interaction between sets. The most important of these properties are listed here:

Commutative Law: The intersection of set A with set B is equal to the intersection of set B with Set. The same is true for unions. Put another way, when considering intersections and unions of sets, the order the sets are presented in is irrelevant. That is symbolized as seen here:

$$(a) \ A \cap B = B \cap A$$
$$(b) \ A \cup B = B \cup A$$

Associative Law: Basically, if you have three sets and the relationships between all three are all unions or all intersections, then the order does not matter. This is symbolized as shown here:

$$(a) \ (A \cap B) \cap C = A \cap (B \cap C)$$
$$(b) \ (A \cup B) \cup C = A \cup (B \cup C)$$

Distributive Law: The distributive law is a bit different than the associative, and order does not matter. The union of set A with the intersection of B and C is the same as taking the union of A and B intersected with the union of A and C. This is symbolized as you see here:

$$(a) \ A \cup (B \cap C) = (A \cup B) \cap (A \cup C)$$
$$(b) \ A \cap (B \cup C) = (A \cap B) \cup (A \cap C)$$

De Morgan's Laws: These govern issues with unions and intersections and the compliments thereof. These are more complex than the previously discussed properties. Essentially, the compliment of the intersection of set A and set B is the union of the compliment of A and the compliment of B. The symbolism of De Morgan's Laws are shown here:

$$(a) \ (A \cap B)^c = A^c \cup B^c$$
$$(b) \ (A \cup B)^c = A^c \cap B^c$$

These are the basic elements of set theory. You should be familiar with them before proceeding.

Logic

Logic is an important topic in mathematics as well as in science and philosophy. In discrete mathematics, logic is a clearly defined set of rules for establishing (or refuting) the truth of a given statement. While we won't be delving into mathematical proofs in this book, should you choose to continue your study of cryptography, at some point you will need to examine proofs. The formal rules of logic are the first steps in understanding mathematical proofs.

Logic is a formal language for deducing knowledge from a small number of explicitly stated premises (or hypotheses, axioms, facts). Some of the rules of logic we will examine in this section might seem odd or counterintuitive. Remember that logic is about the structure of the argument, not so much the content. Before we begin, I will need to define a few basic terms for you:

Axiom: An axiom is something assumed to be true. Essentially one has to start from somewhere. Now you could, like the philosopher Descartes begin with "I think therefore I am" (e cogito ergo sum) but that would not be practical for most mathematical purposes. So, we begin with some axiom that is true by definition. For example, any integer that is only evenly divisible by itself and 1 is considered to be prime. That is an axiom. Another word for axiom is premise.

Statement: This is a simple statement that something is true. For example, "3 is a prime number" is a statement, "16 is 3 cubed" is a statement.

Argument: An argument is just a sequence of statements. Some of these statements, the premises, are assumed to be true and serve as a basis for accepting another statement of the argument, called the conclusion.

Proposition: This is the most basic of statements. For example, if I tell you that I am either male or female, that is a proposition. If I then tell you I am not female, that is yet another proposition. If you follow these two propositions, you arrive at the conclusion that I am male. A proposition is a statement that is either true or not. Some statements have no truth value and therefore cannot be propositions. For example, "stop doing that!" does not have a truth value and therefore is not a proposition.

The exciting thing about formal logic is that if one begins with a premise that is true, and faithfully adheres to the rules of logic, then one will arrive at a conclusion that is true. Preceding from an axiom to a conclusion is deductive reasoning. This is the sort of reasoning often used in mathematical proofs. Inductive reasoning is generalizing from a premise. Inductive reasoning can suffer from overgeneralizing. We will be focusing on deductive logic.

Propositional Logic

Previously, you saw the definition for a proposition. You may also recall that I told you that mathematical logic is concerned with the structure of an argument, regardless of the content. This leads us to a common method of presenting logic in textbooks, that is to ignore actual statements and to put letters in their place. Often the letters p and q can be used. P and q might represent any propositions you might wish to make. For example, p could denote "The author of this book has a gray beard" and q could denote "The author of this book has no facial hair". Considering these two propositions it is not possible for both to be true.

P or q is true.
It is impossible that p and q are true.
It is certainly possible that not p is true.
And it is possible that not q is true.

There are symbols for each of these statements (as well as other statements you could make). It is important to become familiar with basic logic symbolism. Table 4.1 gives you an explanation of the most common symbols used in deductive logic.

Table 4.1 is not an exhaustive list of all symbols in logic, but these are the most commonly used symbols. Make certain you are familiar with these symbols before proceeding with the rest of this section. Several of the symbols given above are what are termed connectives. That just means they are used to connect two propositions. There are some basic rules for connectives, shown in Table 4.2.

A few basic facts about connectives are given here:

Expressions either side of a conjunction are called conjuncts ($p \land q$).
Expressions either side of a disjunction are called disjuncts ($p \lor q$).
In the implication $p \Rightarrow q$, p is called the antecedent and q is the consequence.

Table 4.1 Symbols used in deductive logic

Symbol	Meaning
$\land$	This is the and symbol. Sometimes represented with a. or &. For example, $p \land q$ means that both p and q are true. This is often called the conjunction of p and q.
$\lor$	This is the or symbol. Sometimes represented with a + or ‖. For example, $p \lor q$ means that either p is true, or q is true, or both are true. This is often called the disjunction of p and q.
$\vee$	This is the exclusive or symbol. For example, $p \vee q$ means that either p is true, or q is true, or both are true.
$\exists$	This symbol means "there exits". For example: $\exists$ some p such that $p \land q$ Read that as "there exists some p such that p and q are both true". Examples: $\exists$ x 'there is an x' or 'there exists an x'
$\exists!$	This symbol means "there exists exactly one"
$\forall$	This symbol means "for every". Examples: $\forall$ x 'for every x', or 'for all x'
$\bot$	This symbol denotes a contradiction.
$\rightarrow$	This symbol means "implies" for example $p \rightarrow q$ means that "if p is true, that implies that q is true"
$\neg$	This symbol denotes negation or "not". Sometimes symbolized with !.

Table 4.2 Connectives

Connective	Interpretation
$\neg$	Negation $\neg p$ is true if and only if p is false
$\land$	A conjunction $p \land q$ is true if and only if both p and q are true
$\lor$	A disjunction $p \lor q$ is true if and only if p is true or q is true
$\Rightarrow$	An implication $p \Rightarrow q$ is false if and only if p is true and q is false

The order of precedence with connectives is: Brackets, Negation, Conjunction, Disjunction, and Implication.

Building Logical Constructs

Now that you are familiar with the basic concepts of deductive logic, and you know the essential symbols, lets practice converting some textual statements into logical structures. That is the first step to applying logic to determining the veracity of the statements. Consider this sample sentence:

Although both John and Paul are amateurs, Paul has a better chance of winning the golf game, despite John's significant experience.

Now let us restate that using the language of deductive logic. First, we have to break this long statement down into individual propositions and assign each to some variable.

p: John is an amateur.
q: Paul is an amateur.
r: Paul has a better chance of winning the golf game.
s: John has significant golf experience.

We can now phrase this in symbols.

$$(p \land q) \land r \land s$$

This sort of sentence does not lend itself easily to deducing truth, but it does give us an example of how to take a sentence and re-phrase it with logical symbols.

Truth Tables

Once you have grasped the essential concepts of propositions and connectives, the next step is to become familiar with truth tables. Truth tables provide a method for visualizing the truth of propositions.

Here is a simple one. Keep in mind that the actual statement is irrelevant, just the structure. But if it helps you to grasp the concept, we will have p be the proposition "you are currently reading this book".

Proposition	Counter
P	¬ P
T	F
F	T

What this table tells us is that if P is true, then not P cannot be true. It also tells us that if P is false then not P is true. This example is simple, in fact trivial. All this truth table is doing is formally and visually telling you that either you are reading this book, or you are not, and both cannot be true at the same time. With such a trivial example you might even wonder why anyone would bother using a truth table to illustrate such an obvious fact. In practice, one probably would not use a truth table for this example. However, due to the ease most readers will have with comprehending the simple fact that something is either true or not, this makes a good starting point to learn about truth tables. Some of the tables we will see in just a few moments will not be so trivial.

Next let us examine a somewhat more complex truth table, one that introduces the concept of a conjunction. Again, the specific propositions are not important.

p	q	p ∧ q
T	T	T
T	F	F
F	T	F
F	F	F

This truth table illustrates the fact that only if p is true and q is true, will p ∧ q be true.

What about a truth table for disjunction (the or operator)?

p	q	p ∨ q
T	T	T
T	F	T
F	T	T
F	F	F

As you can see that truth table shows us, quite clearly, what we stated earlier about the or (disjunction) connective.

The exclusive or operation is also interesting to see in a truth table. Recall that we stated either p is true, or q is true, but not both.

p	q	p ∨ q
T	T	F
T	F	T
F	T	T
F	F	F

Next, we shall examine implication statements. Consider p → q. This is read as p implies q. The term implies, in logic, is much stronger than in vernacular speech. Here, it means that if you have p, you will have q.

p	q	p → q
T	T	T
T	F	F
F	T	F
F	F	T

The previous truth tables were so obvious as to not require any extensive explanation. However, this one may require a bit of discussion. The first row should make sense to you. If we are stating that p implies q, and we find that p is true and q is true, then yes, our statement p implies q is true. The second one is also obvious. If p is true, but q is not, then p implies q cannot be true. But what about the last two rows? We are stating that p implies q; however, in row three we see q is true, but p is not. Therefore, not p implied q, so our statement that p implies q is not true. In the fourth row if have not p and not q, then p implies q is still true. Nothing in p being false and q being false invalidates our statement that p implies q.

Combinatorics

Combinatorics is exactly what the name suggests, it is the mathematics behind how you can combine various sets. Combinatorics answers questions such as how many different ways can you choose from four colors to paint six rooms? How many different ways can you arrange four flowerpots? Before such questions can be answered. As you can probably surmise, Combinatorics, at least in crude form, is very old. Greek historian Plutarch mentions combinatorial problems, though he did not use the language of mathematics we use today. The Indian mathematician Mahavira, in the ninth century, actually provided specific formulas for working with combinations ad and permutations (Hunter 2017). It is not clear if he invented them, or simply documented what he had learned from other mathematicians.

There are general problems that combinatorics addresses. These include the existence problem, which asks if a specific arrangement of objects exits. The counting problem is concerned with finding the various possible arrangements of objects according to some requirements. The optimization problem is concerned with finding the most efficient arrangements.

Let us begin our discussion with a simple combinatorics problem. Let's assume that you are reading this textbook as part of a university course in cryptography. Let us further assume that the class has 30 students. Of those students, there are 15 computer science majors and 15 students majoring in something else, we will call these "other".

It should be obvious that there are 15 ways you can choose a computer science major from the class, and 15 ways you can choose an 'other'. This gives us 15×15 ways of choosing a student from the class. But what if we need to pick a student that is either a computer science major or an 'other'. This gives us 15 + 15 ways of choosing. You may be wondering why one method gave us multiplication and the other gave us addition. The first two rules you will learn about in combinatorics are the multiplication rule and the addition rule.

The multiplication rule: Assume there is a sequence of r events that we will call events E1, E2, etc. There are n ways in which a specific event can occur. The number of ways an event occurs does not depend on the previous event. This means we must multiple the possible events. In our example, the event was selecting one of the students, regardless of major. So, we might select one from computer science or we might select one from "other." The multiplication rule is sometimes called the rule of sequential counting.

The Addition Rule: Now suppose there are r events E1, E2, etc. such that there are n possible outcomes for a given event but further assume that no two events can occur simultaneously. This means we will use the addition rule. The addition rule is also sometimes called the rule of disjunctive counting. The reason for this name is that the addition rule (also called the sum rule) states that if set A and set B are disjoint sets, then addition is used.

Let us look at a few more examples of selecting from groups, illustrating both the addition rule (also sometimes called the sum rule) and the multiplication rule (sometimes called the product rule). I find that when a topic is new to a person, it is often helpful to view multiple examples in order to fully come to terms with the subject matter.

What happens when we need to pick from various groups? Assume you are about to take a trip and you wish to bring some reading material along with you. Assume you have five history books, three novels, and three biographies. You need to select one book from each category. There are five different ways to pick a history book, three different ways to select a novel, and three different ways to choose a biography. The total possible combinations is 5 + 3 + 3. Why did we use addition here? Because each choice can only be made once. You have five historical books to choose from, but once you have chosen one, you cannot select any more. Now you have three novels to choose from, but once you have chosen one, you cannot select any more.

Now let's look at an example where we use the multiplication or product rule. Let us assume you have three rooms to paint and four colors of paint. For each room, you can use any of the three colors regardless of whether it has been used before. So, we have 4 * 4 * 4 possible choices. Why did we use multiplication rather than addition? Because each choice can be repeated. Assume your three colors are green, blue, yellow, and red. You can select green for all three rooms if you wish, or green for the first room only. Or not use green at all.

The multiplication rule gets modified if selection removes a choice. Let's return the paint example. Only now, once you have selected a color you cannot use it anymore. In the first room, you have four colors to select from, but in the next room only three, and in the final room only two. So, the number of possible combinations is 4 * 3 * 2. Selection removed a choice each time. Selection with removal is actually pretty common. Assume you have five people you wish to line up for a picture. It should be obvious that you cannot 'reuse' a person. Once you have placed them, they are no longer available for selection. Thus, the possible combinations are $5 * 4 * 3 * 2 * 1$.

Permutations

Permutations are ordered arrangements of things. Because order is taken into consideration, permutations that contain the same elements but in different orders are considered to be distinct. When a permutation contains only some of the elements in a given set, it is called an r-permutation. What this means for combinatorics is that the order of the elements in the set matters. Consider the set {a, b, c}. How many ways can you combine these elements if order matters?

a b c;   a c b;   b a c;   b c a;   c a b;   c b a

There are six ways to combine this set, if order matters. If you add one more letter (d) then there are 24 combinations:

abcd	acbd	bcad	bacd	cabd	cbad
abdc	acdb	bcda	badc	cadb	cbda
adbc	adcb	bdca	bdac	cdab	cdba
dabc	dacb	dbca	dbac	dcab	dcba

Fortunately, you don't have to actually write out the possibilities each time in order to determine how many combinations of a given permutation there are. The answer is n! For those not familiar with this symbol, it is the factorial symbol. It means $n * n-1 * n-2 * n-3$; etc. For example, 4! is $4 * 3 *2 * 1$.

Consider ordering a subset of a collection of objects. If there is a collection of n objects to choose from, and we are selecting r of the objects, where $0 < r < n$, then we call each possible selection a r-permutation from the collection. Consider you have a set of three letters {a, b, c} and you wish to select two letters. In other words, you wish to select an r-permutation from the collection of colored balls. How many r-permutations are possible?

ab   ba   ac   ca   bc   cb

What if our original collection was four letters {a, b, c, d} and you wished to select a two-letter r-permutation?

ab	ac	ad	ba	bc	bd
ca	cb	cd	da	db	dc

These examples should be illustrating to you that the number of r-permutations of a set of n objects is $n!/(n-r)!$.

To verify this, go back to our first example of an r-permutation. We had a set of three objects, and we wished to select two. 3! is $3 * 2 * 1 = 6$. The divisor $(n-r)!$ is $(3-2)!$ Or 1. So, we have six r-permutations. In the second example, $n = 4$. $4! = 4 * 3 * 2 * 1 = 24$. The divisor $(n-r)!$ is $(4-2)!$ Or 2. So, we have 12 r-permutations.

Clearly, we could continue exploring combinatorics and delving further into various ways to combine subsets and sets. But at this point, you should grasp the basic concepts of combinatorics. This field is relevant to cryptography in a variety of ways, particularly in optimizing algorithms. And on a personal note, I have found combinatorics to be a fascinating branch of discrete mathematics with numerous practical applications.

Conclusions

This chapter covered the basic mathematical tools you will need to understand later cryptographic algorithms. In this chapter, you were introduced to number systems. Most importantly, you were given a thorough introduction to prime numbers, including how to generate prime numbers. The concept of relatively prime or co-prime was also covered. That is an important topic and one you should thoroughly understand as it is applied in the RSA algorithm.

You also saw same basic math operations such as summation and logarithms and were introduced, briefly, to discrete logarithms. The discussion of modulus operations is critical for your understanding of asymmetric algorithms. The chapter concluded with an introduction to discrete mathematics including set theory, logic, combinatorics, and graph theory. Each topic was only briefly covered in this chapter; however, if you thoroughly and completely understand what was discussed in this chapter, it will be sufficient for you to continue your study of cryptography. The key is to ensure that you do thoroughly and completely understand this chapter before proceeding. If needed, re-read the chapter and consult the various study aids referenced in this chapter.

Test Your Knowledge

1. The statement ': Any number $n \geq 2$ is expressible as a unique product of 1 or more prime numbers' describes what?
2. $\log_3 27 = $ ____,
3. Are 15 and 28 relatively prime? ____.

References

Balakrishnan, V. K. (2012). Introductory discrete mathematics. Courier Corporation.

Bolker, E. D. (2012). Elementary number theory: an algebraic approach. Courier Corporation.

Fel'dman, N. I., & Nesterenko, Y. V. (2013). Number theory IV: transcendental numbers (Vol. 44). Springer Science & Business Media.

Helfgott, H. (2020). An improved sieve of Eratosthenes. Mathematics of Computation, 89(321), 333–350.

Hua, L. K. (2012). Introduction to number theory. Hunter, D. J. (2017). Essentials of discrete mathematics. Jones & Bartlett Publishers. Springer Science & Business Media.

Neukirch, J. (2013). Algebraic number theory (Vol. 322). Springer Science & Business Media.

Riesel, H. (2012). Prime numbers and computer methods for factorization (Vol. 126). Springer Science & Business Media.

Weil, André. Basic number theory. Vol. 144. Springer Science & Business Media, 2013.

Vinogradov, I. M. (2016). Elements of number theory. Courier Dover Publications.

Chapter 5
Essential Algebra

Abstract Algebra is much more than what is normally taught in secondary school. In this chapter, we explore the basic facts of abstract algebra and linear algebra. You will find elements of abstract algebra used in many cryptographic algorithms including symmetric algorithms such as AES and asymmetric algorithms such as RSA. We also explore linear algebra which is widely used in some cryptographic algorithms such as NTRU. Finally, we discuss algorithm analysis in a basic manner. The concepts in this chapter provide a foundation for understanding these algorithms. As with other chapters on mathematics, the goal of this current chapter is to provide you just enough information for you to understand algorithms presented later in this book.

Introduction

Particular aspects of algebra are essential to cryptographic algorithms you will be exploring later in this book. As with the previous chapter, this one poses a few problems. First, as those readers with a strong education in mathematics might be wondering, how does one cover such a broad topic in a single chapter? Secondly, those readers without a robust background in mathematics might be wondering if they can learn enough, grasp enough of the concepts, to later apply them to cryptography.

Let me address these questions before you start reading. As with Chap. 4, in this chapter, I am going to focus only on those items you absolutely need in order to later understand the cryptographic algorithms we are going to cover. Certainly, each of these topics could be given much more coverage. Entire books have been written on abstract algebra, linear algebra, and algorithms. In fact, some rather sizeable tomes have been written just on the topic of algorithm analysis. And linear algebra can be a course unto itself at both the undergraduate and graduate levels. However, it is not required that you have that level of knowledge in order to understand the cryptography you will encounter later in this book. However, I would certainly recommend exploring these topics in more depth at some point, particularly if you have an interest in going further in cryptography. I am also going to continue with the

© The Author(s), under exclusive license to Springer Nature Switzerland AG 2021
W. Easttom, *Modern Cryptography*, https://doi.org/10.1007/978-3-030-63115-4_5

practice of not delving into mathematical proofs. This means that you, as the reader, must accept a few things simply on faith. Or you can refer to numerous mathematical texts that do indeed provide proofs for these concepts.

To those readers with a less rigorous mathematical background, keep in mind that you only need to have a firm grasp of the concepts presented in this chapter. You do not need to become a mathematician, though I would be absolutely delighted if this book motivated you to study mathematics more deeply. I definitely think one needs to be a mathematician (at some level) in order to create cryptographic algorithms, but one can understand existing algorithms with a bit less mathematical depth. And that is the ultimate goal of this book, to help you understand existing cryptography. And as with the preceding chapter, I will be providing you with links to resources where you can get additional information and explanations if you feel you need it.

In This Chapter We Will Cover

Groups, Rings, and Fields.
Diophantine Equations.
Basic Linear Algebra.
Algorithms.
The History of Algebra.

Groups, Rings, and Fields

The initial algebraic concept we will explore in this chapter is one that might not seem like algebra at all to some readers. Readers with a less rigorous mathematical background might think of algebra as solving linear and quadratic equations as you probably did in secondary school. However, that is only an application of algebra. A far more interesting topic in algebra, and one far more pertinent to cryptography, is the study of sets and operations on those sets (Vinogradov 2016).

One of the major concepts studied in abstract algebra is that of special sets of numbers and operations that can be done on those numbers. Mathematics students frequently struggle with these concepts, I will endeavor to make them as simple as I can without leaving out any important details. The concepts of groups, rings, and fields are just sets of numbers with associated operations.

First think about a set of numbers. Let us begin with thinking about the set of real numbers. This is an infinite set, as I am sure you are aware of. Now ask what operations can you do on numbers that are a member of this set wherein the result will still be in the set? You can certainly add two real numbers, and the answer will always be a real number. You can multiply two real numbers and the answer will always be a real number. What about the inverse of those two operations? You can subtract two real numbers and the answer will always be a real number. You can divide two real numbers and the answer will always be a real number (of course division by zero is undefined). Now at this point, you may think all of this is absurdly obvious, you might even think it odd I would spend a paragraph discussing it. However, let us turn our attention to sets of numbers wherein all of these facts are not true.

Think about the set of all integers. That is certainly an infinite set, just like the set of real numbers. You can certainly add any two integers and the sum will be another integer. You can multiply any two integers, and the product will still be an integer. So far, this sounds just like the set of real numbers. Now, let us turn our attention to the inverse operations. You can certainly subtract any integer from another integer and the answer is still an integer. But what about division? There are infinitely many scenarios where you cannot divide one integer by another and still have the answer be an integer. Certainly dividing 6 by 2 gives you an integer, as would dividing 10 by 5 and 21 by 3 and infinitely more examples. But what if I divide 5 by 2? The answer is not an integer, it is instead a rational number. Or what if I divide 20 by 3, the answer is not an integer. And there are infinitely many other examples wherein I cannot divide. Therefore, if I wish to limit myself only to integers, I cannot use division as an operation.

Imagine for a moment that you wish to limit your mathematics to an artificial world in which only integers exist. Set aside, for now, any considerations of why you might do this, and just focus on this thought experiment for just a moment. As we have already demonstrated, in this artificial world you have created, the addition operation exists and functions as it always has. So does the inverse of addition, subtraction. The multiplication operation behaves in the same fashion you have always seen it. However, in this imaginary world, the division operation simply does not exist, since it has the very real possibility of producing non-integer answers. And such answers do not exist in our imaginary world of "only integers."

Before we continue on with more specific examples from the world of abstract algebra, let us look at one more hypothetical situation that should help clarify these basic points. What if we have limited ourselves to only natural numbers, counting numbers. Certainly, I can add any two counting numbers and the answer will always be a natural number. I can also multiply any two natural numbers and I can rest assured that the product will indeed by another natural number. But what of the inverse of these operations? I can certainly subtract some natural numbers and have an answer that is still a natural number. But there are infinitely many cases where this is not true. For example, if I attempt to subtract 7 from 5, the answer is a negative number, which is not a natural number. In fact, any time I attempt to subtract a larger natural number from a smaller natural number, the result will not be a natural number. Furthermore, division is just as tricky with natural numbers as it is with integers. There are infinitely many cases where the answer will not be a natural number. So, in this imaginary world of only natural numbers, addition and multiplication work exactly as you would expect them too. However, their inverse operations, subtraction won't always work, and division will often not work.

Abstract algebra concerns itself with structures just like this. These structures (groups, rings, fields, etc.) have a set of numbers, and certain operations that can be performed on those numbers. The only allowed operations in a given structure are those whose result would still be within the prescribed set of numbers. You have already been introduced to concepts like this in Chap. 4. Recall the modulo arithmetic you learned about in Chap. 4. The numbers used were integers bound by the modulo operation.

Don't be overly concerned with the term "abstract algebra." There are certainly practical applications of abstract algebra. In fact, some sources prefer to call this "modern algebra" but since it dates back a few centuries, even that may be a misnomer. You will see applications later in this book when we discuss asymmetric cryptography, particularly with Chap. 11, Elliptic Curve Cryptography. Even the symmetric algorithm, Advanced Encryption Standard (AES), which you will study in Chap. 7, applies some of these structures.

Groups

A group is an algebraic system consisting of a set which includes an identity element, one operation, and its inverse operation. Let us begin with explaining what an identity element is. An identity element is simply some number within a set that you can apply some operation to any other number in the set, and the other number will still be the same. Put more mathematically the identity element I can be expressed as follows:

$$a * I = a$$

where $*$ is any operation that we might specify, not necessarily multiplication. An example would be with respect to the addition operation, zero is the identity element. You can add zero to any member of any given group, and you will still have that same number. With respect to multiplication, 1 is the identity element. Any number multiplied by 1 is still the same number.

There are four properties any group must satisfy: closure, associativity, identity, and invertibility. Closure is the simplest of these properties. It simply means that an operation performed on a member of the set will result in a member of the set. This is what was discussed a bit earlier in this section. It is important that any operations allowed on a particular set will result in an answer that is also a member of the set. Associative property just means that you can rearrange the elements of a particular set of values in an operation without changing the outcome. For example $(2 + 2) + 3 = 7$. But even if I change the order and instead write $2 + (2 + 3)$ the answer is still 7. This is an example of the associative property. The identity element was already discussed. The invertibility property simply means that a given operation on an element in a set can be inverted. As we previously discussed, subtraction is the inversion of addition; division is the inversion of multiplication.

Thinking back to our example of the set of integers. Integers constitute a group. First there is an identity element, zero. There is also one operation (addition) and its inverse (subtraction). Furthermore, you have closure. Any element of the group (i.e., any integer) added to any other element of the group (i.e., any other integer) still produces a member of the group (i.e., the answer is still an integer).

Abelian Group

An abelian group or commutative group has an additional property. That property being the commutative property: $a + b = b + a$ if the operation is addition, $ab = ba$ if the operation is multiplication (Weil 2013).

This commutative requirement simply means that applying the group operation (whatever that operation might be) does not depend on the order of the group elements. In other words, whatever the group operation is, you can apply it to members of the group in any order you wish. To use a trivial example, consider the group of integers with the addition operation. Order does not matter:

$$4 + 2 = 2 + 4$$

Therefore, the set of integers with the operation of addition is an abelian group.

Cyclic Group

A cyclic group is a group that has elements that are all powers of one of its elements. For example, if you start with element x, then the members of a cyclic group would be

$$x^{-2}, x^{-1}, x^0, x^1, x^2, x^3, \ldots$$

Of course, the other requirements for a group, which we discussed previously in this section, would still apply to a cyclic group. The basic element x is considered to be the generator of the group, since all other members of the group are derived from it (Neukirch 2013). It is also referred to as a "primitive element" of the group. Integers could be considered a cyclic group with 1 being the primitive element (i.e., generator). All integers can be expressed as a power of 1, in this group. This may seem like a rather trivial example, but it is also one that is easy to understand.

Rings

A ring is an algebraic system consisting of a set, an identity element, two operations, and the inverse operation of the first operation (Bolker 2012). That is the formal definition of a ring, but it may seem a bit awkward to you at first read, so let us examine this concept a bit further, then look at an example. If you think back to our previous examples, it should occur to you that when we are limited to two operations, they are usually addition and multiplication. It may also occur to you that the easiest to implement an inverse of is addition, with subtraction being the inverse.

A ring is essentially just an abelian group that has a second operation. We have seen that the set of integers with the addition operation form a group, and furthermore they form an abelian group. If you add the multiplication operation, then the set of integers with both the addition and the multiplication operations form a ring.

Fields

A field is an algebraic system consisting of a set, an identity element for each operation, two operations, and their respective inverse operations. You can think of a group that has two operations rather than one, and it has an inverse for both of those operations. It is also the case that every field is a ring, but not every ring will necessarily be a field. For example, the set of integers are a ring, but not a field (the inverse of multiplication can produce answers that are not in the set of integers). Fields also allow for division (the inverse operation of multiplication), whereas groups do not.

A classic example of a field is the field of rational numbers. Each number can be written as a ratio (i.e., a fraction) such as x/y (x and y could be any integers you like) and the additive inverse is simply $-x/y$. The multiplicative inverse is just y/x.

Galois Fields

Galois Fields are also known as finite fields. You will learn more about Galois himself in the history section of this chapter. These types of fields are very important in cryptography and you will see them used in later chapters. These are called finite fields because they have a finite number of elements (Vinogradov 2016). If you think about some of the groups, rings, and fields we have previously discussed, all were infinite. The set of integers, rational numbers, and real numbers are all infinite. Galois fields are finite.

In a Galois field, there are some integers such that n repeated terms equal zero. Put another way there is some boundary on the field, thus making it finite. The smallest n that satisfies this is some prime number and that prime number is referred to as the characteristic of the field. You will often see a Galois field defined as follows:

$$GF(p^n)$$

In this case, the GF does not denote a function. Rather this statement is saying that there is a Galois field with p as the prime number (the characteristic we mentioned previously in this section) and the field has p^n elements. A Galois field is some finite set of numbers (from 0 to p^n-1) and some mathematical operations (usually addition and multiplication) along with the inverses of those operations.

Now you may immediately be thinking, how can a finite field even exist? If the field is indeed finite, would it not be the case that addition and multiplication operations could lead to answers that don't exist within the field? To understand how this works, think back to the modulus operation we discussed in Chap. 4. Essentially operations "wrap around" the modulus. If we consider the classic example of a clock, then any operation whose answer would be more than 12 simply wraps around. The same thing occurs in a Galois field. Any answer greater than p^n simply wraps around.

Let us examine an example, a rather trivial example but one that makes the point. Consider the Galois field GF (3^1). First, I hope you realize that 3^1 is the same as 3, and most texts would simply write this as GF(3). Thus, we have a Galois field defined by 3. In any case where addition or multiplication of the elements would cause us to exceed the number 3. We simply wrap around. This example is easy to work with because it only has three elements: 1, 2, 3. Considering operations with those elements, several addition operations pose no problem at all:

$$1 + 1 = 2 \qquad 1 + 2 = 3$$

However, others would be a problem.

$2 + 2 = 4$, except that we are working within a Galois field, which utilizes the modulus operations so $2 + 2 = 1$ (we wrapped around at three).
Similarly, $2 + 3 = 2$.

The same is true with multiplication.

$2 \times 2 = 1$ (we wrap around at 3).
$2 \times 3 = 0$ (again we wrap around at 3).

You can, I hope, see how this works. Now in cryptography we will deal with Galois fields that are larger than the trivial example we just examined, but the principles are exactly the same.

Diophantine Equations

A Diophantine equation is any equation for which you are interested only in the integer solutions to the equation. Thus, a linear Diophantine equation is a linear equation $ax + by = c$ with integer coefficients for which you are interested only in finding integer solutions. There are two types of Diophantine equations. Linear Diophantine equations have elements that are of degree 1 or zero. Exponential Diophantine equations have at least one term that has an exponent greater than 1.

The word Diophantine comes from Diophantus, a third-century C.E. mathematician who studied such equations. You have actually seen

Diophantine equations before, though you might not have realized it. The traditional Pythagorean theorem can be a Diophantine equation, if you are interested in only integer solutions as can be the case in many practical applications.

The simplest Diophantine equations are linear and are of the form

$$ax + by = c$$

where a, b, and c are all integers. Now there is an integer solution to this problem if and only if c is a multiple of the greatest common divisor of and b. For example: the Diophantine equation $3x + 6y = 18$ does have solutions (in integers) since gcd $(3,6) = 3$ which does indeed, evenly divide 18.

Linear Algebra

While linear algebra has applications in many fields, including quantum physics, it began as a way to solve systems of linear equations (thus the name). Linear equations are those for which all elements are of the first power. Thus, the following three equations are linear equations:

$$a + b = 10$$
$$3x + 5 = 44$$
$$4x + 2y - z = 35$$

However, the following are not linear equations:

$$2x^2 + 2 = 10$$
$$4y^2 + 4x + 3 = 17$$

The first three equations have all elements to the first power (often that with a number such as x^1 the 1 is simply assumed and not written). But in the second set of equations at least one element is raised to some higher power. Thus, they are not linear.

When linear algebra is used, numbers are presented in the form of a matrix. And in this section, we will examine how to perform a variety of operations using matrices. Before we can delve into matrix math, we will need to define what a matrix is. A matrix is a rectangular arrangement of numbers in rows and columns. Rows run horizontally and columns run vertically. The dimensions of a matrix are stated "$m \times n$" where "m" is the number of rows and "n" is the number of columns. Here is an example in Fig. 5.1.

A matrix is just an array that is arranged in columns and rows. You may have previously heard the term "vector." Vectors are simply matrices that have one

Fig. 5.1 A matrix

$$\begin{bmatrix} 1 & 2 \\ 2 & 0 \\ 3 & 1 \end{bmatrix}$$

Fig. 5.2 Matrix annotation

$$\begin{bmatrix} a_{ij} & a_{ij} \\ a_{ij} & a_{ij} \end{bmatrix}$$

Fig. 5.3 Matrix annotation continued

$$\begin{bmatrix} a_{11} & a_{12} \\ a_{21} & a_{22} \end{bmatrix}$$

column or one row. A vector can be considered a 1 X m matrix A vector that is vertical is called a column vector, one that is horizontal is called a row vector. Matrices are usually labeled based on column and row, this is shown in Fig. 5.2.

The letter i represents the row, and the letter j represents the column. A more concrete example is shown in Fig. 5.3.

This notation is commonly used for matrices including row and column vectors. There are different types of matrices, common ones are listed here:

Column matrix: a matrix with only one column.
Row matrix: a matrix with only one row.
Square matrix: a matrix that has the same number of rows and columns.
Equal matrices: two matrices are considered equal if they have the same number of rows and columns (the same dimensions) and all their corresponding elements are exactly the same.
Zero matrix: contains all zeros.

Each of these has a role in linear algebra. We will see that as we proceed.

Matrix Addition and Multiplication

If two matrices are of the same size, then they can be added to each other by simply adding each element together (Poole 2014). You start with the first row and first column in the first matrix and add that to the first row and the first column of the second matrix thus in the sum matrix. This is shown in Fig. 5.4.

Let us repeat this with a more concrete example, as shown in Fig. 5.5.

However, multiplication is a bit more complicated. You can only multiply two matrices if the number of columns in the first matrix is equal to the number of rows in the second matrix. First let us take a look at multiplying a matrix by a single number. This single number is generally referred to as a *scalar*. You simply multiply the scalar value by each element in the matrix. This is shown in Fig. 5.6.

Fig. 5.4 Matrix addition

$$\begin{bmatrix} a_{11} & a_{12} \\ a_{21} & a_{22} \end{bmatrix} + \begin{bmatrix} b_{11} & b_{12} \\ b_{21} & b_{22} \end{bmatrix} = \begin{bmatrix} A_{11} + b_{11} & a_{12} + b_{12} \\ A_{21} + b_{21} & a_{22} + b_{22} \end{bmatrix}$$

Fig. 5.5 Matrix addition example

$$\begin{bmatrix} 3 & 2 \\ 1 & 4 \end{bmatrix} + \begin{bmatrix} 2 & 3 \\ 2 & 1 \end{bmatrix} = \begin{bmatrix} 3+2 & 2+3 \\ 1+4 & 2+1 \end{bmatrix} = \begin{bmatrix} 5 & 5 \\ 5 & 3 \end{bmatrix}$$

Fig. 5.6 Scalar multiplication

$$c \begin{bmatrix} a_{ij} & a_{ij} \\ a_{ij} & a_{ij} \end{bmatrix} = \begin{bmatrix} ca_{ij} & ca_{ij} \\ ca_{ij} & ca_{ij} \end{bmatrix}$$

Fig. 5.7 Scalar multiplication example

$$2 \begin{bmatrix} 1 & 3 \\ 2 & 2 \end{bmatrix} = \begin{bmatrix} 2*1 & 2*3 \\ 2*2 & 2*2 \end{bmatrix} = \begin{bmatrix} 2 & 6 \\ 4 & 4 \end{bmatrix}$$

For a more concrete example, consider what is shown in Fig. 5.7.

Multiplication of two matrices is a bit more complex. The two matrices need not be of the same size. The requirement is that number of columns in the first matrix is equal to the number of rows in the second matrix. If that is the case, then you multiply each element in the first row of the first matrix, by each element in the second matrixes first column. Then you multiply each element of the second row of the first matrix by each element of the second matrixes second column. Let's first examine this using variables rather than actual numbers. We will also use square matrices to make the situation even simpler.

$$\begin{bmatrix} a & b \\ c & d \end{bmatrix} + \begin{bmatrix} e & f \\ g & h \end{bmatrix}$$

This is multiplied in the following manner

$$a*e + b*g \ (a_{11}*b_{11} + a_{12}*b_{21})$$
$$a*f + b*h \ (a_{11}*b_{12} + a_{12}*b_{22})$$
$$c*e + d*g \ (a_{11}*b_{11} + a_{12}*b_{21})$$
$$c*f + d*h \ (a_{11}*b_{11} + a_{12}*b_{21})$$

Thus, the product will be

$$(a*e + b*g) \quad (a*f + b*h)$$
$$(c*e + d*g) \quad (c*f + d*h)$$

It is worthwhile to memorize this process. Let us now see this implemented with a concrete example.

$$\begin{bmatrix} 1 & 2 \\ 3 & 1 \end{bmatrix}\begin{bmatrix} 2 & 2 \\ 1 & 3 \end{bmatrix}$$

We begin with

$$1 * 2 + 2 * 1 = 6$$
$$1 * 2 + 2 * 3 = 8$$
$$3 * 2 + 1 * 1 = 7$$
$$3 * 2 + 1 * 3 = 9$$

The final answer is

$$\begin{bmatrix} 6 & 8 \\ 7 & 9 \end{bmatrix}$$

This is why we previously stated that you can only multiply two matrices if the number of columns in the first matrix is equal to the number of rows in the second matrix.

It is important to remember that matrix multiplication, unlike traditional multiplication (with scalar values), is not commutative. Recall the commutative property states $a * b = b * a$. If a and b are scalar values then this is true, however, if they are matrices, this is not the case. For example, consider the matrix multiplication shown in Fig. 5.8.

Now if we simply reverse the order, you can see that an entirely different answer is produced. This is shown in Fig. 5.9.

This example illustrates the rather important fact that matrix multiplication is not commutative.

Fig. 5.8 Matrix multiplication is not commutative

$$\begin{bmatrix} 2 & 3 \\ 1 & 4 \end{bmatrix}\begin{bmatrix} 1 & 1 \\ 2 & 3 \end{bmatrix} = \begin{bmatrix} 8 & 11 \\ 9 & 13 \end{bmatrix}$$

Fig. 5.9 Matrix multiplication is not commutative Part B

$$\begin{bmatrix} 1 & 1 \\ 2 & 3 \end{bmatrix}\begin{bmatrix} 2 & 3 \\ 1 & 4 \end{bmatrix} = \begin{bmatrix} 3 & 7 \\ 7 & 15 \end{bmatrix}$$

Matrix Transposition

Matrix transposition simply reverses the order of rows and columns (Greub 2012). While we have focused on 2×2 matrices, the transposition operation is most easily seen with a matrix that has a different number of rows and columns. Consider the matrix shown in Fig. 5.10.

To transpose it, the rows and columns are switched creating a 2×3 matrix. The first row is now the first column. You can see this in Fig. 5.11.

If we label the first matrix A, then the transposition of that matrix is labeled A^T. Continuing with the original matrix being labeled A, there are a few properties of matrices that need to be described.

Property	Explanation
$(A^T)^T = A$	If you transpose, the transposition of A, you get back to A.
$(cA)^T = cA^T$	The transposition of a constant, c multiplied by an array A, is equal to multiplying the constant c by the transposition of A.
$(AB)^T = B^T A^T$	A multiplied by B then the product transposed is equal to B transposed multiplied by A transposed.
$(A + B)^T = A^T + B^T$	Adding the matrix A and the matrix B, then transposing the sum is equal to first transposing A and B, then adding those transpositions.
$A^T = A$	If a square matrix is equal to its transpose, it is called a symmetric matrix.

The preceding table is not exhaustive. Rather it is a list of some of the most common properties regarding matrices. These properties are not generally particularly difficult to understand. However, there is an issue with why one would do them. What do they mean? All too often introductory linear algebra texts focus so intensely on helping a student to learn how to do linear algebra that the meaning behind operations is lost. Let us take just a moment to explore what transpositions are. A transposition is rotating about the diagonal. Remember that matrices can be viewed graphically. Consider a simple row matrix.

Fig. 5.10 An exemplary matrix

$$\begin{bmatrix} 2 & 3 & 2 \\ 1 & 4 & 3 \end{bmatrix}$$

Fig. 5.11 Matrix transposition

$$\begin{bmatrix} 2 & 1 \\ 3 & 4 \\ 2 & 3 \end{bmatrix}$$

[1 2 4]

Transposing that row matrix creates a column matrix:

$$\begin{bmatrix} 1 \\ 2 \\ 4 \end{bmatrix}$$

Think of a vector as a line in some space. For now, we will limit ourselves to 2D and 3D space. You can then think of a matrix as a transformation on a line or set of lines.

Submatrix

A submatrix is any portion of a matrix that remains after deleting any number of rows or columns. Consider the 5×5 matrix shown in Fig. 5.12.

If you remove the second column and second row, as shown in Fig. 5.13, you get a submatrix.

You are left with the matrix shown in Fig. 5.14.

That matrix shown in Fig. 5.14 is a submatrix of the original matrix.

Fig. 5.12 A 5×5 matrix

$$\begin{bmatrix} 2 & 2 & 4 & 5 & 3 \\ 3 & 8 & 0 & 2 & 1 \\ 2 & 3 & 2 & 2 & 1 \\ 4 & 3 & 1 & 2 & 4 \\ 1 & 2 & 2 & 0 & 3 \end{bmatrix}$$

Fig. 5.13 Removing a row and column

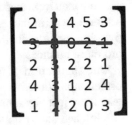

Fig. 5.14 The resulting
submatrix

$$\begin{bmatrix} 2 & 4 & 5 & 3 \\ 2 & 2 & 2 & 1 \\ 4 & 1 & 2 & 4 \\ 1 & 2 & 0 & 3 \end{bmatrix}$$

Identity Matrix

An identity matrix is actually rather simple. Think back to the identity property of groups. An identity matrix accomplishes the same goal, multiplying a matrix by its identity matrix, leaves it unchanged. To create an identity matrix just have all the elements along the main diagonal set to 1, and the rest to zero. Consider the matrix shown below.

$$\begin{bmatrix} 3 & 2 & 1 \\ 1 & 1 & 2 \\ 3 & 0 & 3 \end{bmatrix}$$

Now consider the identity matrix. It must have the same number of columns and rows, with its main diagonal set to all 1's and the rest of the elements all 0's. You can see the identity matrix shown here:

$$\begin{bmatrix} 1 & 0 & 0 \\ 0 & 1 & 0 \\ 0 & 0 & 1 \end{bmatrix}$$

If you multiple the original matrix by the identity matrix, the product will be the original matrix. You can see this in Fig. 5.15.

In our examples we have not only focused on square, 2×2 or 3×3 matrices, but also on matrices that only consist of integers. A matrix can consist of rational numbers, real numbers, even complex numbers. But since our goal was just to give you an introduction to the concepts of matrix algebra, we have not delved into those nuances.

One application of matrix algebra is with linear transformations. A linear transformation is sometimes called a linear mapping or even just a linear function. It is essentially just some mapping between two vector spaces that has the operations of addition and scalar multiplication.

Note: Wolfram Math defines a vector space as "A vector space V is a set that is closed under finite vector addition and scalar multiplication." Another way of putting this is a vector space is a collection of objects (in our case integers) called vectors, which may be added together and multiplied ("scaled") by numbers, called scalars.

FIG. 5.15 Multiply by the
identity matrix

$$\begin{bmatrix} 3 & 2 & 1 \\ 1 & 1 & 2 \\ 3 & 0 & 3 \end{bmatrix} \times \begin{bmatrix} 1 & 0 & 0 \\ 0 & 1 & 0 \\ 0 & 0 & 1 \end{bmatrix} = \begin{bmatrix} 3 & 2 & 1 \\ 1 & 1 & 2 \\ 3 & 0 & 3 \end{bmatrix}$$

Determinants

Next, we will turn our attention to another relatively easy computation, the determinant of a matrix (Wilkinson et al. 2013). The determinant of a matrix A is denoted by $|A|$. An example of a determinant in a generic form is shown below.

$$|A| \begin{bmatrix} a & b \\ C & d \end{bmatrix} = ad - bc$$

A more concrete example might help elucidate this concept.

$$|A| \begin{bmatrix} 2 & 3 \\ 1 & 2 \end{bmatrix} = (2)(2) - (3)(1) = 1$$

A determinant is a value which is computed from the individual elements of a square matrix. It provides a single number, also known as a scalar value. Only a square matrix can have a determinant. The calculation for a 2×2 matrix is simple enough, we will explore more complex matrices in just a moment. However, what does this single scalar value mean? There are many things one can do with a determinant, most of which we won't use in this text. It can be useful in solving linear equations, changing variables in integrals (yes linear algebra and calculus go hand in hand). But what is immediately useable for us is that if the determinant is non-zero then the matrix is invertible. This will be important later.

What about a 3×3 matrix, such as Fig. 5.16.

This calculation is substantially more complex. There are a few methods to do this. We will use one called "expansion by minors." This method depends on breaking the 3×3 matrix into 2×2 matrices. The 2×2 matrix formed by b_2, c_2, b_3, and c_3 is shown in Fig. 5.17.

This one was rather simple, as it fits neatly into a contiguous 2×2 matrix. But to find the next one, we have a bit different selection (Fig. 5.18).

The next step is to get the lower left corner square matrix (Fig. 5.19).

As with the first one, this one forms a very nice 2×2 matrix. Now what shall we do with these 2×2 matrices? The formula is actually quite simple and is shown in Fig. 5.20.

We take the first column, multiplying it by its cofactors, and with a bit of simple addition and subtraction, we arrive at the determinant for a 3×3 matrix. A more concrete example might be useful. Let us calculate the determinant for this matrix shown in Fig. 5.21.

Fig. 5.16 A 3×3 matrix

$$\begin{bmatrix} a_1 & b_1 & c_1 \\ a_2 & b_2 & c_2 \\ a_3 & b_3 & c_3 \end{bmatrix}$$

Fig. 5.17 3×3 matrix determinant part 1

$$\begin{bmatrix} a_1 & b_1 & c_1 \\ a_2 & b_2 & c_2 \\ a_3 & b_3 & c_3 \end{bmatrix}$$

Fig. 5.18 3×3 matrix determinant part 2

$$\begin{bmatrix} a_1 & b_1 & c_1 \\ a_2 & b_2 & c_2 \\ a_3 & b_3 & c_3 \end{bmatrix}$$

Fig. 5.19 3×3 matrix determinant part 3

$$\begin{bmatrix} a_1 & b_1 & c_1 \\ a_2 & b_2 & c_2 \\ a_3 & b_3 & c_3 \end{bmatrix}$$

$$\begin{bmatrix} a_1 & b_1 & c_1 \\ a_2 & b_2 & c_2 \\ a_3 & b_3 & c_3 \end{bmatrix} = a1 \det \begin{bmatrix} b_2 & c_2 \\ b_3 & c_3 \end{bmatrix} - a_2 \det \begin{bmatrix} a_2 & c_2 \\ a_3 & c_3 \end{bmatrix} + a_3 \det \begin{bmatrix} a_2 & b_2 \\ a_3 & b_3 \end{bmatrix}$$

Fig. 5.20 3×3 matrix determinant part 4

Fig. 5.21 3×3 matrix example

$$\begin{bmatrix} 3 & 2 & 1 \\ 1 & 1 & 2 \\ 3 & 0 & 3 \end{bmatrix}$$

This leads to

$$3 * \det \begin{bmatrix} 1 & 2 \\ 0 & 3 \end{bmatrix} = 3 * ((1*3) - (2*0))) = 3\,(3) = 9$$

$$2 * \det \begin{bmatrix} 1 & 2 \\ 3 & 3 \end{bmatrix} = 3 * ((1*3) - (2*3))) = 2\,(-3) = -6$$

$$1 * \det \begin{bmatrix} 1 & 1 \\ 3 & 0 \end{bmatrix} = 1 * ((1*0) - (1*3)) = 1\,(-3) = -3$$

And that leads us to $9 - (-6) + (-3) = 12$

Yes, that might seem a bit cumbersome, but the calculations are not overly difficult. We will end our exploration of determinants at 3×3 matrices. But yes, one can take the determinant of larger square matrices.

Eigenvalues and Eigenvectors

Let us move on to an interesting topic, that of eigenvalues and eigenvectors. Eigenvalues are a particular set of scalars associated with a linear system of equations (i.e., matrix equations) that are sometimes also known as characteristic roots, characteristic values, proper values, or latent roots (Poole 2014). Let us make this a bit clearer. Let us consider a column vector we will call v. Then also consider an $n \times n$ matrix we will call A. Then consider some scalar λ. If it is true that

$$Av = \lambda v$$

Then we say that v is an eigenvector of the matrix A and λ is an eigenvalue of the matrix A.

Let us look a bit closer at this. The prefix eigen is actually the German word which can be translated as specific, proper, particular, etc. Put in its most basic form an eigenvector of some linear transformation T is a vector that when T is applied to it does not change direction, it only changes scale. It changes scale by the scalar value λ, the eigenvalue (Lang 2012). Now we can revisit the former equation just a bit and expand our knowledge of linear algebra.

$$T(v) = \lambda v$$

This appears precisely like the former equation, but with one small difference. The matrix A is now replaced with the transformation T. Not only does this tell us about eigenvectors and eigenvalues, it tells us a bit more about matrices. A matrix, when applied to a vector, transforms that vectors. The matrix itself is an operation on the vector! Later in this book we will be seeing these transformations used quite frequently, particularly in reference to logic gates for quantum computers. So, make certain you are quite familiar with them before proceeding.

Let us add something to this. How do you find the eigenvalues and eigenvectors for a given matrix? Surely it is not just a matter of trial and error with random numbers. Fortunately, there is a very straight forward method, one that is actually quite easy at least for 2×2 matrices. Consider the following matrix:

$$\begin{bmatrix} 5 & 2 \\ 9 & 2 \end{bmatrix}$$

How do we find its eigenvalues?

Well the Cayley–Hamilton theorem provides insight on this issue. That theorem essentially states that a linear operator A is a zero of its characteristic polynomial (Wilkinson et al. 2013). For our purposes it means that

$$\det |A - \lambda I| = 0$$

We know what a determinant is, we also know that I is the identity matrix. The λ is the eigenvalue we are trying to find. The A is the matrix we are examining. Remember that in linear algebra one can apply a matrix to another matrix or vector, so a matrix is, at least potentially, an operator. Thus, we can fill in this equation:

$$\det \left| \begin{bmatrix} 5 & 2 \\ 9 & 2 \end{bmatrix} - \lambda \begin{bmatrix} 1 & 0 \\ 0 & 1 \end{bmatrix} \right| = 0$$

Now, we just have to do a bit of algebra, beginning with multiplying λ by our identity matrix which will give us:

$$\det \left| \begin{bmatrix} 5 & 2 \\ 9 & 9 \end{bmatrix} - \begin{bmatrix} \lambda & 0 \\ 0 & \lambda \end{bmatrix} \right| = 0$$

Which in turn leads to

$$\det \left| \begin{bmatrix} 5 - \lambda & 2 \\ 9 & 2 - \lambda \end{bmatrix} \right| = 0$$
$$= (5 - \lambda)(5 - \lambda) - 18$$
$$= 10 - 7\lambda + \lambda^2 - 18 = 0$$
$$\lambda^2 - 7\lambda - 8 = 0$$

This can be factored (note if the result here cannot be factored, things do get a bit more difficult, but that is beyond our scope here).

$$(\lambda - 8)(\lambda + 1) = 0$$

This means we have two eigenvalues

$$\lambda_1 = 8$$

$$\lambda_2 = -1$$

For a 2×2 matrix you will always get two eigenvalues. In fact, for any $n \times n$ matrix you will get n eigenvalues, but they may not be unique.

Now that you have the eigenvalues, how do you calculate the eigenvectors?

We know that

$$A = \begin{bmatrix} 5 & 2 \\ 9 & 2 \end{bmatrix}$$

$$\lambda_1 = 8$$

$$\lambda_2 = -1$$

We are seeking unknown vectors, so let us label the vector $\begin{bmatrix} X \\ Y \end{bmatrix}$.

Now recall the equation that gives us eigenvectors and eigenvalues

$$Av = \lambda v$$

Let us take one of our eigenvalues and plug it in

$$\begin{bmatrix} 5 & 2 \\ 9 & 2 \end{bmatrix} \begin{bmatrix} X \\ Y \end{bmatrix} = 8 \begin{bmatrix} X \\ Y \end{bmatrix}$$

$$\begin{bmatrix} 5x + & 2y \\ 9x + & 2y \end{bmatrix} = \begin{bmatrix} 8X \\ 8Y \end{bmatrix}$$

This gives us two equations:

$$5x + 2y = 8x$$

$$9x + 2y = 8y$$

Now we take the first equation and to a bit of algebra to isolate the y value. Subtract the $5x$ from each side to get:

$$2y = 3x$$

Then divide both sides by 2 to get

$$y = 3/2\,x$$

It should be easy to see that to solve this with integers (which is what we want) then $x = 2$ and $y = 3$ solve it. Thus, our first eigenvector is

$$\begin{bmatrix} 2 \\ 3 \end{bmatrix} \text{ with } \lambda_1 = 8$$

You can work out the other eigenvector for the second eigenvalue on your own using this method.

This section should provide you just enough linear algebra to move forward with your study of cryptography. Obviously, one can go much further. But if you fully understand this section, you will have a basic working knowledge of linear algebra.

Algorithms

All the cryptography you will study from this point on consist of various algorithms. Readers who have majored in computer science have probably taken a course in data structures and algorithms. This section will first introduce you to some concepts of algorithms, then discuss general algorithm analysis in the context of sorting algorithms. We will complete this section by studying a very important algorithm.

Basic Algorithms

Before we can begin a study of algorithms, we must first define what an algorithm is. An algorithm is simply a systematic way of solving a problem. A recipe for an apple pie is an algorithm. If you follow the procedure you get the desired results. Algorithms are a routine part of computer programming. Often the study of computer algorithms centers on sorting algorithms. The sorting of lists is a very common task and therefore a common topic in any study of algorithms.

It is also important that we have a clear method for analyzing the efficacy of a given algorithm. When considering any algorithm if the desired outcome is achieved, then clearly the algorithm worked. But the real question is how well did it work. If you are sorting a list of 10 items, the time it takes to sort the list is not of particular concern. However, if your list has 1 million items, then the time it takes to sort the list, and hence the algorithm you choose, is of critical importance. Similar issues exist in evaluating modern cryptographic algorithms. It is obviously imperative that they work (i.e., that they are secure), but they also have to work efficiently. E-commerce would not be so widespread if the cryptographic algorithms used provided for an unacceptable lag. Fortunately, there are well-defined methods for analyzing any algorithm.

When analyzing algorithms, we often consider the asymptotic upper and lower bounds. Asymptotic analysis is a process used to measure the performance of computer algorithms. This type of performance is based on a factor called computational complexity. Usually this is a measure of either the time it takes for an algorithm to work, or the resources (memory) needed. It should be noted that one usually can only optimize time or resources, but not both. The asymptotic upper bound is simply the worst-case scenario for the given algorithm. Whereas the asymptotic lower bound is a best case.

Some analysts prefer to simply use an average case, however knowing the best case and worst can be useful in some situations. In simple terms, both the asymptotic upper bound and lower bounds must be within the parameters of the problem you are attempting to solve. You must assume that the worst-case scenario will occur in some situations.

The reason for this disparity between the asymptotic upper and lower bounds has to do with the initial state of a set. If one is applying a sorting algorithm to a set that is at its maximum information entropy (state of disorder) then the time taken for the sorting algorithm to function will be the asymptotic upper bound. If, on the other hand, the set is very nearly sorted, then one may approach or achieve the asymptotic lower bound.

Perhaps the most common way to formally evaluate the efficacy of a given algorithm is Big O notation. This method is the measure of execution of an algorithm, usually the number of iterations required, given the problem size n. In sorting algorithms n is the number of items to be sorted. Stating some algorithm $f(n) = O(g(n))$ means it is less than some constant multiple of $g(n)$. The notation is read, "f of n is big oh of g of n". This means that saying an algorithm is 2 N, means it will have to execute 2 times the number of items on the list. Big O notation essentially measures the asymptotic upper bound of a function. Big O is also the most often used analysis (Cormen et al. 2009).

Big O notation was first introduced by the mathematician Paul Bachmann in his 1892 book *Analytische Zahlentheorie*. The notation was popularized in the work of another mathematician named Edmund Landau. Because Landau was responsible for popularizing this notation. It is sometimes referred to as a Landau symbol.

Omega notation is the opposite of Big O notation. It is the asymptotic lower bound of an algorithm and gives the best-case scenario for that algorithm. It gives you the minimum running time for an algorithm.

Theta notation combines Big O and Omega to give the average case (average being arithmetic mean in this situation) for the algorithm. In our analysis we will focus heavily on the Theta, also often referred to as the Big O running time. This average time gives a more realistic picture of how an algorithm executes.

Now that we have some idea of how to analyze the complexity and efficacy of a given algorithm, let's take a look at a few commonly studied sorting algorithms, and apply these analytical tools.

Sorting Algorithms

Sorting algorithms are often used to introduce someone to the study of algorithms. This is because they are relatively easy to understand, and they are so common. Since we have not yet explored modern cryptography algorithms, I will also use sorting algorithms to illustrate concepts of algorithm analysis.

Elementary implementations of the merge sort sometimes utilize three arrays. This would be one array for each half of the data set and one to store the final sorted list in. There are non-recursive versions of the merge sort, but they don't yield any significant performance enhancement over the recursive algorithm on most machines. It should also be noted that almost every implementation you will find of merge sort will be recursive.

You can see that this is relatively simple code. The name derives from the fact that the lists are divided, sorted, then merged, and this procedure is applied recursively. If one had an exceedingly large list and could separate the two sublists onto different processors, then the efficacy of the merge sort would be significantly improved.

Quick Sort

The quick sort is a very commonly used algorithm. Like the merge sort, it is recursive, meaning that it simply calls itself repeatedly until the list is sorted. Some books will even refer to the quick sort as a more effective version of the merge sort. The quick sort is the same as merge sort ($n \ln n$), however the difference is that this is also its best-case scenario. However, it has an $O(n^2)$, which indicates that its worst-case scenario is quite inefficient (Cormen et al. 2009). So, for very large lists, the worst-case scenario may not be acceptable.

This recursive algorithm consists of three steps (which bear a strong resemblance to the merge sort):

1. Pick an element in the array to serve as a pivot point.
2. Split the array into two parts. The split will occur at the pivot point. So, one array will have elements larger than the pivot and the other with elements smaller than the pivot. Clearly one or the other should also include the pivot point.
3. Recursively repeat the algorithm for both halves of the original array.

One very interesting aspect of this algorithm is that the efficiency of the algorithm is significantly impacted by which element is chosen as the pivot point. The worst-case scenario of the quick sort occurs when the list is sorted and the left-most element is chosen, this gives a complexity of, $O(n^2)$. Randomly choosing a pivot point rather than using the left-most element is recommended if the data to be sorted are not random. As long as the pivot point is chosen randomly, the quick sort has an algorithmic complexity of $O(n \log n)$.

Bubble Sort

The bubble sort is the oldest and simplest sort in use. By simple I mean that from a programmatic point of view it is very easy to implement. Unfortunately, it is also one of the slowest. It has an $O(n^2)$. This means that for very large lists, it is probably going to be too slow. As with most sort algorithms, its best case (lower asymptotic bound) is $O(n)$.

The bubble sort works by comparing each item in the list with the item next to it and swapping them if required. The algorithm repeats this process until it makes a pass all the way through the list without swapping any items (in other words, all items are in the correct order). This causes larger values to "bubble" to the end of the list while smaller values "sink" toward the beginning of the list, thus the name of the algorithm.

The bubble sort is generally considered to be the most inefficient sorting algorithm in common usage. Under best-case conditions (the list is already sorted), the bubble sort can approach a constant n level of complexity. General case is an $O(n^2)$.

Even though this is one of the slower algorithms available, it is seen more often simply because it is so easy to implement. Many programmers who lack a thorough enough understanding of algorithm efficiency and analysis will depend on the bubble sort.

Now that we have looked at three common sorting algorithms, you should have a basic understanding of both algorithms and algorithm analysis. Next, we will turn our attention to an important algorithm that is used in cryptography.

Euclidean Algorithm

The Euclidean algorithm is a method for finding the greatest common divisor of two integers. Now that may sound like a rather trivial task, but with larger numbers it is not. It also happens that this plays a role in several cryptographic algorithms you will see later in this book. The Euclidian algorithm proceeds in a series of steps such that the output of each step is used as an input for the next one. Let k be an integer that counts the steps of the algorithm, starting with zero. Thus, the initial step corresponds to $k = 0$, the next step corresponds to $k = 1$, $k = 2$, $k = 3$, etc.

Each step, after the first begins with two remainders r_{k-1} and r_{k-2} from the preceding step. You will notice that at each step the remainder is smaller than the remainder from the preceding step. So that r_{k-1} is less than its predecessor r_{k-2}. This is intentional, and central to the functioning of the algorithm. The goal of the kth step is to find a quotient qk and remainder r_k such that the equation is satisfied

$$r_{k-2} = qk\, r_{k-1} + r_k$$

where $r_k < r_{k-1}$. In other words, multiples of the smaller number $rk - 1$ are subtracted from the larger number r_{k-2} until the remainder is smaller than the r_{k-1}.

That explanation may not be entirely clear. So, let us look at an example.

$$\text{Let } a = 2322, b = 654.$$

$2322 = 654 \cdot 3 + 360$ (i.e., the 360 is the remainder). This now tells us that the greatest common denominator of the two initial numbers, gcd(2322, 654) is equal to the gcd(654, 360). These are still a bit unwieldy, so let us proceed with the algorithm.

$654 = 360 \cdot 1 + 294$ (the remainder is 294). This tells us that the gcd(654, 360) is equal to the gcd(360, 294). The following steps we continue this process until there simply is no further to go:

$$360 = 294 \cdot 1 + 66 \quad \gcd(360,\ 294) = \gcd(294,\ 66)$$
$$294 = 66 \cdot 4 + 30 \quad \gcd(294,\ 66) = \gcd(66,\ 30)$$
$$66 = 30 \cdot 2 + 6 \quad \gcd(66,\ 30) = \gcd(30,\ 6)$$
$$30 = 6 \cdot 5 \quad \gcd(30,\ 6) = 6$$

Therefore, gcd(2322, 654) = 6.

This process is quite handy and can be used to find the greatest common divisor of any two numbers. And, as I previously stated, this will play a role in cryptography you will study later in this book.

If you still need a bit more on the Euclidian algorithm there are some very helpful resources on the internet. Rutgers University has a nice simple page in this issue http://www.math.rutgers.edu/~greenfie/gs2004/euclid.html as does the University of Colorado at Denver http://www-math.ucdenver.edu/~wcherowi/courses/m5410/exeucalg.html

Designing Algorithms

Designing an algorithm is a formal process. Algorithms are developed to provide systematic ways of solving certain problems. Therefore, it should be no surprise that there are systematic ways of designing algorithms. In this section, we will examine a few of these methods.

The divide-and-conquer approach to algorithm design is a commonly used approach. In fact, it could be argued that it is the most commonly used approach. It works by recursively breaking down a problem into sub-problems of the same type, until a point is reached where these sub-problems become simple enough to be solved directly. Once the sub-problems are solved, the solutions to the sub-problems are then combined to provide a solution to the original problem. In short you keep subdividing the problem until you find manageable portions that you can solve. Then after solving those smaller, more manageable sub-problems, you combine those solutions in order to solve the original problem.

When approaching difficult problems, such as the classic Tower of Hanoi puzzle, the divide-and-conquer method provides an efficient way to solve the problem. For many such problems the paradigm offers the only workable way to find a solution. The divide-and-conquer method often leads to algorithms that are not only effective, but that are also efficient.

The efficiency of the divide-and-conquer method can be examined by considering the number of sub-problems being generated. If the work of splitting the problem and combining the partial solutions is proportional to the problem's size n, then there are a finite number p of sub-problems of size $\sim n/p$ at each stage. Furthermore, if the base cases require $O(1)$ (i.e., constant-bounded) time, then the divide-and-conquer algorithm will have $O(n \log n)$ complexity. This is often used in sorting problems in order to reduce the complexity from $O(n^2)$. Recall from the first paper on algorithm analysis that an O of n log n is fairly good for most sorting algorithms.

In addition to allowing one to devise efficient algorithms for solving complex problems, the divide-and-conquer approach is well suited for execution in multi-processor machines. The reason for this is that the sub-problems can be assigned to different processors in order to allow each processor to work on a different sub-problem. This leads to sub-problems being solved simultaneously thus increasing the overall efficacy of the process.

The second method is the "greedy approach." The textbook *An Introduction to Algorithms by Cormen and Leiserson* defines greedy algorithms as those that select what appears to be the most optimal solution in a given situation. In other words, a solution is selected that is ideal for a specific situation, but not be the most effective solution for the broader class of problems. The greedy approach is used with optimization problems. In order to give a precise description of the greedy paradigm, we must first consider a more detailed description of the environment in which most optimization problems occur. In most optimization problems, one will have the following:

- A collection of candidates. That collection might be a set, list, array, or other data structure. How the collection is stored in memory is irrelevant.
- A set of candidates which have previously been used.
- A predicate solution that is used to test whether or not a given set of candidates provide an efficient solution. This does not check to see if those candidates provide an optimal solution, just whether or not they provide a working solution.
- Another predicate solution (feasible) to test if a set of candidates can be extended to a solution.
- A selection function, which chooses some candidate which has not yet been used.
- A function which assigns a value to a solution.

Essentially, an optimization problem involves finding a subset S from a collection of candidates C where that subset satisfies some specified criteria. For example, the criteria may be that it is a solution such that the function is optimized by S. Optimized may denote any number of factors, such as minimized or maximized. Greedy methods are distinguished by the fact that the selection function assigns a numerical value to each candidate C and chooses that candidate for which:

SELECT(C) is largest
or SELECT(C) is smallest

All greedy algorithms have this same general form. A greedy algorithm for a particular problem is specified by describing the predicates, and the selection function.

Consequently, greedy algorithms are often very easy to design for optimization problems.

The general form of a greedy algorithm is as follows:

functionselect (C: candidate_set) returncandidate;
functionsolution (S: candidate_set) returnboolean;
functionfeasible (S: candidate_set) returnboolean;

The aforementioned book *An Introduction to Algorithms* by Cormen and Leiserson is an excellent book on algorithms. It is often used as a textbook for university courses. It might be a bit beyond that mathematical abilities of some readers, but if you are looking for a thorough discussion of algorithms, I recommend that book.

Conclusions

Several topics have been explored in this chapter. Groups, rings, and fields were the first topics. These algebraic structures, particularly Galois fields, are very important to cryptography. Make certain you are quite familiar with them before proceeding to Chap. 6. Next you saw Diophantine equations. These are so common in algebra that at least a basic familiarity of them is quite important.

The next thing we explored in this chapter was matrix algebra. Matrices are not overly complex, and you saw the fundamentals explained in this chapter. You should be comfortable with matrices and the basic arithmetic operations on matrices. You will see this mathematics used in some cryptographic algorithms.

This chapter also introduced you to algorithms. An understanding of the sorting algorithms is not as critical as having a conceptual understanding of algorithms and algorithm analysis. Finally, you were introduced to some basics of the history of algebra. As with Chaps. 3 and 4, the topics in this chapter are foundational, and necessary for your understanding of subsequent chapters.

It is also important to reiterate something stated in the introduction to this chapter. Each of the topics presented here is the subject of entire books. You have only seen a basic introduction in this chapter, just enough information to allow you to understand the cryptography discussed later in this book. If you wish to delve even deeper into cryptography, you will benefit from further study in these various topics.

Test Your Knowledge

1. A _____ is any equation for which you are interested only in the integer solutions to the equation
2. A matrix which has all 1 s in its main diagonal and the rest of the elements zero is a what?

   (a) Inverse matrix.
   (b) Diagonal matrix.
   (c) Identity matrix.
   (d) Revers matrix.

3. Omega notation is ____?

   (a) The asymptotic upper bound of an algorithm.
   (b) The asymptotic lower bound of an algorithm.
   (c) The average performance of an algorithm.
   (d) Notation for the row and column of a matrix.

4. A _____ is an algebraic system consisting of a set, an identity element, two operations, and the inverse operation of the first operation.

   (a) Ring.
   (b) Field.
   (c) Group.
   (d) Galois Field.

5. A _____ is an algebraic system consisting of a set, an identity element for each operation, two operations, and their respective inverse operations.

   (a) Ring.
   (b) Field.
   (c) Group.
   (d) Galois Field.

6. Is the set of integers is a group with reference to multiplication? _____.
7. Is the set of natural numbers a subgroup of the set of integers with reference to addition? _____.
8. If $a \mid b$ and $a \mid c$, then _____.
9. A group that also has the commutative property is an abelian group.
10. _____ is a process used to measure the performance of computer algorithm.

References

Bolker, E. D. (2012). Elementary number theory: an algebraic approach. Courier Corporation.
Cormen, T. H., Leiserson, C. E., Rivest, R. L., & Stein, C. (2009). Introduction to algorithms. MIT Press.
Greub, W. H. (2012). Linear algebra (Vol. 23). Springer
Lang, S. (2012). Introduction to linear algebra. Springer
Neukirch, J. (2013). Algebraic number theory (Vol. 322). Springer
Poole, D. (2014). Linear algebra: A modern introduction. Cengage Learning.
Vinogradov, I. M. (2016). Elements of number theory. Courier Dover Publications.
Weil, André. Basic number theory. Vol. 144. Springer 2013.
Wilkinson, J. H., Bauer, F. L., & Reinsch, C. (2013). Linear algebra (Vol. 2). Springer.

Chapter 6
Feistel Networks

Abstract Feistel networks have existed since the 1970s. The term Feistel network defines a structure for creating symmetric ciphers. As you will see in this chapter, there are numerous algorithms that utilize this structure. This makes it an important structure to understand. We will first examine the general structure of the Feistel network, then explore numerous specific examples. Some algorithms will be covered in more detail than others, based on how widely used those algorithms are.

Introduction

This chapter is the first of two chapters regarding symmetric ciphers. The focus of this chapter will be on Feistel networks, but before we get to that rather specific topic, it is important to establish what a symmetric cipher is and the different types of symmetric ciphers. The first issue is to determine what is a symmetric cipher? Fortunately defining a symmetric cipher is very easy. It is an algorithm that uses the same key to decrypt the message as was used to encrypt the message. You can see a demonstration of this in Fig. 6.1.

The process is rather straight forward and functions much like a key in a physical door. The same key is used to unlock the door that was used to lock it. This is probably the modality of cryptography that is easiest for cryptography novices to learn. There are some significant advantages that symmetric cryptography has over asymmetric cryptography. For example, symmetric cryptography is always faster than asymmetric (which we will examine in Chaps. 10 and 11) and it is just as secure with a smaller key.

Now that you understand what symmetric cryptography is, at least in principle, we need to consider the various types of symmetric ciphers. While the classic cryptography discussed in Chaps. 1 and 2 could be considered to be symmetric ciphers, for the purposes of our current discussion we are only contemplating modern ciphers. With that in mind, there are two major classifications of symmetric algorithms. They are block ciphers and stream ciphers. A block cipher literally encrypts the data in blocks. 64-bit blocks are quite common, although some

A key has to be exchanged before communications begin

Bob   Alice

Bob encrypts a
message with   Eve uses that
the key and   key to decrypt
sends it to Alice   the message

Fig. 6.1 Symmetric cryptography

algorithms (like AES) use larger blocks. For example, AES uses a 128-bit block. Stream ciphers encrypt the data as a stream, one bit at a time.

Modern block ciphers use a combination of substitution and transposition to achieve both confusion and diffusion (recall those concepts from Chap. 3). Remember that substitution is changing some part of the plain text for some matching part of ciphertext. The Caesar and Atbash ciphers are simple substitution ciphers. Transposition is the swapping of blocks of ciphertext. For example, if you have the text "I like icecream" You could transpose or swap every three-letter sequence (or block) with the next and get:

"ikeI l creiceam"

Of course, modern block ciphers accomplish substitution and transposition at the level of bits, or rather blocks of bits, and our example used characters. However, the example above illustrates the concept. Within the topic of block ciphers are two major subtypes: Feistel networks (e.g., DES) and substitution—permutation networks (e.g., AES). In this chapter, we will examine several Feistel Ciphers. Some, like DES, we will examine in rather thorough detail. Others, we will simply describe in general terms. In Chap. 7, we will discuss substitution—permutation networks as well as stream ciphers.

It is not critical that you memorize every algorithm in this chapter, even those that are described in detail. It is recommended that you pay particular attention to DES and Blowfish. Then study the other algorithms enough to see how that differ from DES and Blowfish. The goal of this chapter is to provide you with a solid understanding of Feistel ciphers.

As with previous chapters, I will also provide you with places to go get more information. In some cases, these will be books, in others they will be websites. I am aware that websites may not always be up and available, but they are also easy and free for you to access. If a given URL is no longer available, then I suggest you simply use your favorite search engine to search the topic.

Cryptographic Keys

Before we start looking at Feistel Ciphers, you need to have a basic understanding of cryptographic keys and how they are used. With all block ciphers, there are really two types of keys. The first is the cipher key and the second is the round key. The cipher key is a random string of bits that is generated and exchanged between parties that wish to send encrypted messages to each other. It should be noted that it is not actually a random number but rather a pseudorandom number, but we will discuss that in more detail in Chap. 12. For now, you just need to understand that the cipher key is a number generated by some algorithm, such that it is at least somewhat random. The more random it is, the better the key. When someone says a particular algorithm has a certain key size, they are referring to this cipher key. For example, DES uses a 56-bit cipher key. How that key is exchanged between the two parties is not relevant at this point. Later, when we discuss asymmetric cryptography, and again when we cover SSL/TLS, we will cover key exchange.

In addition to that cipher key, all block ciphers also have a second algorithm called a key schedule. This key schedule takes the cipher key and derives from it a unique key, called a round key, for each round of the cipher. For example, DES has a 56-bit cipher key, but derives 48-bit round keys for each of its 16 rounds. The key schedule is usually rather simple, mostly consisting of shifting bits.

The purpose of using a key schedule is so that each round uses a key that is slightly different than the previous round. Since both the sender (who encrypts the message) and the receiver (who must decrypt the message) are using the same cipher key as a starting point, and are using the same key schedule, they will generate the same round keys. But this leaves the question of why do it this way? If you wanted a different key for each round, why not just generate that many cipher keys? For example, why not just create 16, DES keys? There are two reasons. The first being the time needed. Using pseudorandom number generators to generate keys is computationally more intensive than most key scheduling algorithms, and thus much slower. Secondly, there is the issue of key exchange. If you generate multiple cipher keys, then you have to exchange all of those keys. It is much easier to generate a single cipher key, exchange that key, then derive round keys from the cipher key.

Feistel Function

At the heart of many block, ciphers is a Feistel function. Since this function forms the basis for so many symmetric ciphers, it makes it one of the most influential developments in symmetric block ciphers. It is also known as a Feistel network or

Fig. 6.2 Feistel –
simple view

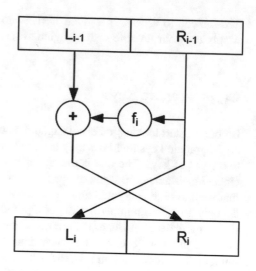

a Feistel cipher. This function is named after its inventor, the German-born physicist and cryptographer Horst Feistel.

While most known for his contributions to DES, Horst Feistel made other important contributions to cryptography. He invented the Lucifer cipher, which was the pre-cursor to DES. He also worked for the US Airforce on IFF (Identify Friend or Foe) devices. Feistel had a bachelor's degree from MIT and a masters from Harvard, both degrees were in physics.

A Feistel cipher is a particular class of block ciphers. A simple view of the Feistel function is shown in Fig. 6.2.

The process is actually not that difficult to understand if you take it one step at a time. Trying to grasp the entire process at once might prove difficult if this is new to you, so take it one small step at a time.

This function starts by splitting the block of plain text data into two parts (traditionally termed L_0 and R_0). The specific size of the plain text block varies from one algorithm to another. Sizes of 64 bit and 128 bit are very commonly used in many block ciphers. If you are using a 64-bit block, this leaves you with 32 bits in L_0 and 32 bits in R_0. The 0 simply indicates that this is the initial round. The next round these will be L_1 and R_1 then the next round they will be L_2 and R_2, and so on for as many rounds as that particular cipher uses (Nyberg 1996).

After the initial block of plain text is split into two halves, the round function F is applied to 1 of the two halves (Courtois et al. 2018). The term "round function" simply means a function performed with each iteration, or round, of the Feistel cipher. The details of the round function F can vary with different implementations. Usually these are relatively simple functions, to allow for increased speed of the algorithm. For now, you can simply treat the round function as a black box. We will get to the details of each round function in the specific algorithms we will examine (DES, Blowfish, etc.). In every case, however, at some point in the round function,

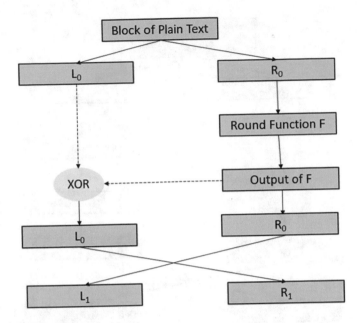

Fig. 6.3 Feistel function

the round key is XOR'd with the block of text input into the round function. This is, in fact, usually the first step of the round function.

The output of the round function F is then XOR'd with the other half (the half that was not subjected to the round function). What this means is that, for example, you take L_0, pass it through the round function F, then take the result and XOR it with R_0. Then the halves are transposed. So L_0 gets moved to the right and R_0 gets moved to the left (Courtois et al. 2018).

This process is repeated a given number of times. The main difference between various Feistel functions is the exact nature of the round function F, and the number of iterations. You can see this process in Fig. 6.3.

This basic structure can provide a very robust basis for cryptographic algorithms. The swapping of the two halves guarantees some transposition occurs, regardless of what happens in the round function. The Feistel function is the basis for many algorithms including DES, CAST-128, BlowFish, TwoFish, RC5, FEAL, MARS, TEA, and others. But it was first used in IBM's Lucifer algorithm (the precursor to DES).

You might wonder how many rounds do you need to make a Feistel Secure? Michael Luby and Charles Rackoff analyzed the Feistel cipher construction and proved that if the round function is a cryptographically secure pseudorandom function, then 3 rounds is sufficient to make the block cipher a pseudorandom permutation, while 4 rounds is sufficient to make it a "strong" pseudorandom permutation.

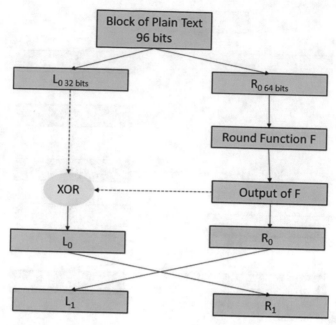

Fig. 6.4   Unbalanced Feistel function

Unbalanced Feistel

There is a variation of the Feistel network called an Unbalanced Feistel cipher. Unbalanced Feistel ciphers use a modified structure where L_0 and R_0 are not of equal lengths. This means that L_0 might be 32 bits and R_0 could be 64 bits (making a 96-bit block of text). This variation is actually used in the Skipjack algorithm. You can see this process in Fig. 6.4.

Pseudo-Hadamard Transform

This is a technique that is applied in several symmetric ciphers, so it is important to be familiar with it. Fortunately, it is rather simple to understand. A pseudo-Hadamard transform (often simply called a PHT) is a transformation of a bit string that is designed to produce diffusion in the bit string. The bit string must be of even length because it is split into two equal halves. So, for example, a 64-bit string is divided into two 32-bit halves. To compute the transform of a you add $a + b$ (mod 2^n). To compute the transform of b you add $a + 2b$ (mod 2^n). The n is the number of bits of each half, in our example 32. Put in more formal mathematical notation.

$$a' = a + b \,(\mathrm{mod}\ 2^n)$$

$$b' = a + 2b \,(\mathrm{mod}\ 2^n)$$

The key is that this transform is reversible, as you can see here:

$$a = 2a' - b' \,(\mathrm{mod}\ 2^n)$$

$$b = a' - b' \,(\mathrm{mod}\ 2^n)$$

That is it, PHT is a rather simple transform, and because of its simplicity, it is computationally fast, making it attractive for cryptographic applications.

MDS Matrix

An MDS matrix (maximum distance separable) is a matrix that represents a function. It is used in cryptography to accomplish diffusion. It is used in several cryptographic algorithms, including Twofish. A basic definition of the MDS matrix is: An $m \times n$ matrix over a finite field is an MDS matrix if it is the transformation matrix of a linear transformation. This is a greatly simplified definition. This definition is, however, adequate for you to understand how algorithms like Twofish are designed. For those readers seeking a more mathematically rigorous definition of MDS matrices, consider this quote:

> MDS code over a field is a linear mapping from a field elements to b field elements, producing a composite vector of a + b elements with the property that the minimum number of non-zero elements in any non-zero vector is at least b + 1. Put another way, the 'distance' (i.e., the number of elements that differ) between any two distinct vectors produced by the MDS mapping is at least b + 1. It can easily be shown that no mapping can have a larger minimum distance between two distinct vectors, hence the term maximum distance separable. MDS mappings can be represented by an MDS matrix consisting of a × b elements.

Lucifer

The Lucifer cipher was the first incarnation of a Feistel cipher. There were variations, perhaps the most widely known used a 128-bit cipher key and operated on 128-bit blocks. The original description by Horst Feistel had a 48-bit cipher key applied to a 48-bit block. Yet another variant used a 64-bit cipher key on a 32-bit block. Regardless of the key size or block size, Lucifer operated exactly as the previous description of the Feistel cipher. The 128-bit block of plain text is separated into two 64-bit blocks (L and R). The R block is subjected to the round function. Unlike DES, Lucifer did not have initial or final permutations (we will discuss those in the section on DES). The round function has only a few, relatively simple steps.

The Key schedule algorithm produces a 64-bit round key (often called a subkey) for each round as well as 8 bits that are called ICBs (interchange control bits). The 64-bit block to be encrypted is treated as a series of 8 bytes. The 8 bits of the ICB is matched to each of the bytes of the 64-bit block. What this means is that you look at the first bit of the ICB, if it is a 0 then the left nibble and right nibble of the first byte of the block are swapped. If that bit is a 1, then the two nibbles are left unchanged. Then you look at the second bit of the ICB, if it is a 0 then the nibbles of the second byte of the block are swapped. This continues through all 8 bits of the ICB and all 8 bytes of the block. Since the ICB is generated each round with the round key, it changes each round. This gives a transposition in the round function that is based on the key.

It should be noted that there are a number of block ciphers that have key-dependent functionality. In some cases, like Lucifer, it is simply deciding whether or not to execute a basic transposition. In other algorithms, the round key may be used to generate the s-boxes. Key dependency does make an algorithm more resistant to some types of cryptanalysis. Once this step is done (swapping based on the ICB) the output bits are XOR'd again with the round key. The output of that is then input into the s-boxes.

The key-scheduling algorithm is simple. Each round the 128-bit cipher key is shifted in a shift register. The left 64 bits are taken to form the round key and the right 8 bits form the ICB. After each round the 128 bits are shifted 56 bits to the left.

As you can see Lucifer is relatively simple, however consider how much change is actually taking place. First individual bytes of plain text have their nibbles (4-byte halves) swapped based on the ICB. Then the result is XOR'd with the round key. Then that output is subjected to an s-box. In these three steps, the text is subjected to transposition as well as multiple substitutions. Then consider that the most widely known variant of Lucifer (the 128-bit key and 128-bit block) uses 16 rounds. So, the aforementioned transposition and substitutions will be executed on a block of plain text 16 times.

To get a good idea of just how much this alters the plain text, let us consider a simplified example. In this case, we won't have any s-boxes. Just transposition and XOR operation. Let us begin with a single word of plain text:

Attack

If we convert that to ASCII codes, then convert those to binary we get

01000001 01110100 01110100 01100001 01100011 01101011

Now let's assume we have a cipher key that is just 8 random bits. We are not even going to alter this each round, as most real algorithms do. Our key for this example is:

11010011 00110110

Furthermore, to make this similar to Lucifer but simpler, we will simply swap nibbles on every other byte. And again, to make this simpler we will do the entire text at one time, instead of one block at a time. So round one is like this

Plaintext   01000001 01110100 01110100 01100001 01100011 01101011
Swapped   00010100 01110100 01000111 01100001 00110110 01101011
(only every other block is swapped)

XOR with key produces.

11000111 01000010 10010100 01010111 11100101 01011101

Converted to ASCII is

199 66 148 87 229 93

You can find ASCII tables all over the internet, using one to convert this to actual text yields:

199 = a symbol, not even a letter or number.
66 = B.
148 = a symbol, not even a letter or number.
87 = W.
229 = a symbol, not even a letter or number.
93 =].

Consider that this is the output after one round, and it was an overly simplified round at that. This should give you some indication of how effective such encryption can be.

DES

The Data Encryption Standard is a classic in the annals of cryptography. It was selected by the National Bureau of Standards as an official Federal Information Processing Standard (FIPS) for the United States in 1976. While it is now considered outdated and is not recommended for use, it was the premier block cipher for many years and bears study. The primary reason it is no longer considered secure is its small key size. The algorithm itself is quite sound. Many cryptography textbooks and university courses use this as the basic template for studying all block ciphers. We will do the same and give this algorithm more attention than most of the others in this chapter.

DES uses a 56-bit cipher key applied to a 64-bit block. There is actually a 64-bit key, but one bit of every bite is actually used for error correction, leaving just 56 bits for actual key operations.

DES is a Feistel cipher with 16 rounds and a 48-bit round key for each round. Recall from the earlier discussion on keys that a round function is a subkey that is derived from the cipher key each round, according to a key schedule algorithm. DES's general functionality follows the Feistel method of dividing the 64-bit block into two halves (32 bits each, this is not an unbalanced Feistel cipher), applying the round function to one half, then XOR'ing that output with the other half.

142

The first issue to address is the key schedule. How does DES generate a new subkey each round? The idea is to take the original 56-bit key and to slightly permute it each round, so that each round is applying a slightly different key, but one that is based on the original cipher key. To generate the round keys, the 56-bit key is split into two 28-bit halves and those halves are circularly shifted after each round by one or two bits. This will provide a different subkey each round. During the round key generation portion of the algorithm (recall that this is referred to as the *key schedule*) each round, the two halves of the original cipher key (the 56 bits of key the two endpoints of encryption must exchange) are shifted a specific amount. The amount is shown in this table:

Round	How far to shift to the left
1	1
2	1
3	2
4	2
5	2
6	2
7	2
8	2
9	1
10	2
11	2
12	2
13	2
14	2
15	2
16	1

Once the round key has been generated for the current round, the next step is to address the ½ of the original block that is going to be input into the round function. Recall that the two halves are each 32 bit. The round key is 48 bits. That means that the round key does not match the size of the half block it is going to be applied to. You cannot really XOR a 48-bit round key with a 32-bit ½ block, unless you simply ignore 16 bits of the round key. If you did so, you would basically be making the round key effectively shorter and thus less secure, so this is not a good option. The 32-bit half needs to be expanded to 48 bits before it is XOR'd with the round key. This is accomplished by replicating some bits so that the 32 bit half becomes 48 bits.

This expansion process is actually quite simple. The 32 bits that are to be expanded are broken into 4-bit sections. The bits on each end are duplicated. If you divide 32 by 4 the answer is 8. So, there are 8 of these 4-bit groupings. If you duplicate the end bits of each grouping, that will add 16 bits to the original 32, thus providing a total of 48 bits.

It is also important to keep in mind that it was the bits on each end that were duplicated, this will be a key item later in the round function. Perhaps this example will help you to understand what is occurring at this point. Let us assume 32 bits as shown here:

11110011010101111111000101011001

Now divide that into 8 sections each of 4 bits, as shown here:

1111 0011 0101 1111 1111 0001 0101 1001

Now each of these has its end bits duplicated, as you see here:

1111 becomes 111111
0011 becomes 000111
0101 becomes 001011
1111 becomes 111111
1111 becomes 111111
0001 becomes 000011
0101 becomes 001011
1001 becomes 110011

The resultant 48-bit string is now XOR'd with the 48-bit round key. That is the extent of the round key being used in each round. It is now dispensed with, and on the next round another 48-bit round key will be derived from the two 28-bit halves of the 56-bit cipher key.

Now we have the 48-bit output of the XOR operation. That is now split into 8 sections of 6 bits each. For the rest of this explanation we will focus on just one of those 6-bit sections, but keep in mind that the same process is done to all 8 sections.

The 6-bit section is used as the input to an s-box. An s-box is a table that takes input and produces an output based on that input. In other words, it is a substitution box that substitutes new values for the input. The s-boxes used in DES are published, the first of which is shown in Fig. 6.5.

Notice this is simply a lookup table. The two bits on either end are shown in the left-hand column and the four bits in the middle are shown in the top row. They are matched and the resulting value is the output of the s-box. For example, with the previous demonstration numbers we were using, our first block would be 111111. Therefore, you find 1xxxx1 on the left and x1111x on the top. The resulting value is 13 in decimal or 1101 in binary.

	x0000x	x0001x	x0010x	x0011x	x0100x	x0101x	x0110x	x0111x	x1000x	x1001x	x1010x	x1011x	x1100x	x1101x	x1110x	x1111x
0yyyy0	14	4	13	1	2	15	11	8	3	10	6	12	5	9	0	7
0yyyy1	0	15	7	4	14	2	13	1	10	6	12	11	9	5	3	8
1yyyy0	4	1	14	8	13	6	2	11	15	12	9	7	3	10	5	0
1yyyy1	15	12	8	2	4	9	1	7	5	11	3	14	10	0	6	13

Fig. 6.5 The first DES s-box

You may be wondering, why take in six bits and only output four? Is that not losing information? Recall during the expansion phase we simply duplicated the outermost bits, so when we come to the s-box phase and drop the outermost bits, no data are lost. As we will see in Chap. 8 this is a compression s-box. Those are difficult to design properly.

At the end of this you have produced 32-bits that are the output of the round function. Then in keeping with the Feistel structure, they get XOR'd with the 32 bits that were not input into the round function, and the two halves are swapped. DES is a 16-round Feistel cipher, meaning this process is repeated 16 times.

There are only two parts still left to discuss regarding DES. The first is the initial permutation called the IP, and the final permutation, which is an inverse of the IP. This is shown in Fig. 6.6, which is an excerpt from FIPS Publication 46-3.

The initial permutation that was mentioned earlier is simply a transposition of the bits in the 64-bit plain text block. This is done before the rounds of DES are executed, then the reverse transposition is done after the rounds of DES have completed. Basically, the first 58th bit is moved to the first bit spot, the 50th bit to the second bit spot, the 42nd bit to the third bit spot, etc.

Earlier in this chapter, we looked at a very simplified version of the Lucifer cipher and saw that it was quite effective at altering the plain text. It should be obvious to you that DES is at least as effective as Lucifer. The small key size is really the only reason that DES is no longer considered secure. Furthermore, DES is probably the most widely studied symmetric cipher, certainly the most often cited Feistel cipher. It is important that you fully grasp this cipher before moving on to the rest of this chapter. You may need to re-read this section a few times to make sure you fully grasp the details of DES.

For a thorough discussion of the technical details of DES refer to the actual government standard documentation: U.S. DEPARTMENT OF COMMERCE/ National Institute of Standards and Technology FIPS Publication 46-3 http://csrc. nist.gov/publications/fips/fips46-3/fips46-3.pdf. For a less technical but interesting animation of DES you may try http://kathrynneugent.com/animation.html

3DES

Eventually, it became obvious that DES would no longer be secure. The U.S. Federal government began a contest seeking a replacement cryptography algorithm. However, in the meantime, 3DES was created as an interim solution. Essentially it does DES three times, with three different keys.

Triple DES uses a "key bundle" which comprises three DES keys, K1, K2, and K3. Each key is a standard 56-bit DES key. There were some variations that would use the same key for K1 and K3, but three separate keys are considered the most secure.

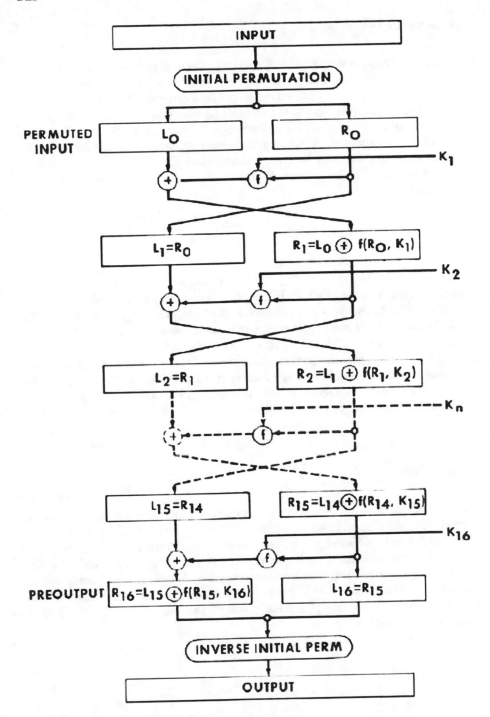

Fig. 6.6 DES overview from FIPS 46-3

S-Box and P-Box

An s-box is just a substitution box. It defines that each of the input bits is substituted with a new bit (Easttom, 2018). You have seen an s-box in the description of DES. It simply is a table that substitutes some output for a given input.

A p-box, or permutation box is a variation on the s-box. Instead of each input bit being mapped to a bit found in a lookup table, the bits that are input are also transposed or permuted. Some may be transposed, and others left in place. For example, a 6-bit p-box may swap the first and fourth bits, swap the second and third bit, but leave the fifth bit in place. S-boxes and p-boxes will be discussed in detail in Chap. 8.

DEAL

This algorithm designed by Lars Knudsen is derived from DES, to be an improvement on DES. The name is an acronym for Date Encryption Algorithm With Larger Blocks. Like DES it is a Feistel cipher and uses the same round function as DES. Unlike DES it has a 128-bit block size with key sizes of 128, 192, or 256 bits. The number of rounds is contingent on the key size: The two smaller key sizes using 6 rounds and the 256-bit key using 8 rounds. It was submitted as a candidate for the AES Contest, but was rejected due to slow speed. That slow speed may not be an issue today with more advanced computers being widely available.

McGuffin

This algorithm is not widely used, but it is noteworthy because it is an example of a generalized unbalanced Feistel network. McGuffin is based on DES, but the plain text block is not split into equal halves. Instead, either 48 bits of the 64-bit block go into the round function, thus the unbalanced Feistel cipher (Blaze and Schneier 1994). The algorithm still uses the expansion, then follows it with 8 compression s-boxes, just like DES. The key schedule is actually a modified version of the algorithm itself. Variations of McGuffin using 32 rounds with a 128-bit key have also been proposed.

GOST

GOST is a DES-like algorithm developed by the Soviets in the 1970s. It was classified but released to the public in 1994. It uses a 64-bit block and a key of 256 bits. It is a 32-round Feistel cipher (Courtois and Misztal 2011). GOST is an acronym for gosudarstvennyy standart, which translates into English as "state standard."

GOST was developed in the Soviet Union in the 1970s as a classified algorithm. It was declassified and made public in 1994. The official designation is GOST 28147-89. It was meant as an alternative to the U.S. DES algorithm and has some similarities to DES. GOST uses a 64-bit block and a 256-bit key with 32 rounds. Like DES, GOST uses substitution boxes or s-boxes; however, unlike DES, the original s-boxes were classified. The round key for GOST is relatively simple: add a 32-bit subkey to the block, modulo 2^{32}, then take the output and submit it to the s-boxes. The output of the s-boxes is routed to the left by 11 bits.

The key schedule is as follows: Break the 256-bit cipher key into 8 separate 32-bit keys. Each of these keys will be used 4 times.

The s-boxes are 8 boxes each 4×4. The s-boxes are implementation-dependent. That means that they are generated, for example, using a pseudorandom number generator. Since the s-boxes are not standardized, each party using GOST needs to ensure that the other party is using the same s-box implementation. The s-boxes each take in 4 bits and produce 4 bits. This is different from DES where the s-boxes take in 6 bits and produce 4 bits of output.

The round function is very simple, so to compensate for this, the inventors of GOST use 32 rounds (twice that of DES) and the s-boxes are secret and implementation-specific. While the diversity of s-boxes can make certain types of cryptanalysis more difficult on GOST, it also means the individuals implementing GOST must be very careful in their creation of the s-boxes. We will be examining s-boxes in detail in Chap. 8.

Blowfish

Blowfish is a symmetric block cipher. This algorithm was published in 1993 by Bruce Schneier. Schneier has stated that, "Blowfish is unpatented, and will remain so in all countries. The algorithm is hereby placed in the public domain and can be freely used by anyone." This fact has made Blowfish quite popular in open source utilities.

This cryptography algorithm was intended as a replacement for DES. Like DES it is a 16-round Feistel cipher working on 64-bit blocks. However, unlike DES it can have varying key sizes ranging from 32 bits to 448 bits (Schneier 1993). Early in the cipher, the cipher key is expanded. Key expansion converts a key of at most 448 bits into several subkey arrays totaling 4168 bytes.

The cipher itself is a Feistel cipher, so the first step is to divide the block into two halves. For example, a 64-bit block is divided into L_0 and R_0. Now there will be 16 rounds.

The s-boxes are key-dependent (as we will see later in this section). That means rather than be standardized as they are in DES, they are computed based on the key. There are four s-boxes each is 32 bits in size. Each s-box has 256 entries. So that is

s1/0, s1/1, s1/2. . .s1/255
s2/0, s2/1, s2/2. . .s2/255
s3/0, s3/1, s3/2. . .s3/255
s4/0, s4/1, s4/2. . .s4/255

Blowfish uses multiple subkeys (also called round keys). One of the first steps in the cipher is to compute those subkeys from the original cipher key. This begins with an array called the P array. Each element of the P array is initialized with the hexadecimal digits of Pi, after the three. In other words, only the values after the decimal point. Each element of the P array is 32 bits in size. Here are the first few digits of Pi in decimal:

3.14159265358979323846264

Here is the same number in hexadecimal:

2B992DDFA23249D6

There are 18 elements in the P array, each is 32 bits in size. They are labeled simply P1, P2,. . .P18. The four s-boxes, each 32 bits in size are also initialized with the digits of Pi.

Then the first 32 bits of the cipher key are XOR'd with P1 then the next 32 bits are XOR'd with P2, etc. This continues until the entire P array has been XOR'd with bits from the cipher key. When you're done the P array elements are each hexadecimal digits of Pi XOR'd with segments of the cipher key.

The next step may be confusing for some readers, so pay close attention. You then take an all zero string and encrypt it with the Blowfish algorithm using the subkeys you just generated (i.e., the P array). Now replace P1 and P2 with the output of that encryption you just did. Then take that output and encrypt it with Blowfish, this time using the new modified keys (the P1 and P2 were modified by the last step). And now replace P3 and P4 with the output of that step. This process continues until all the entries of the P array and all four s-boxes have been replaced with the output of this continually evolving Blowfish algorithm.

Despite the complex key schedule, the round function itself is rather simple:

The 32-bit half is divided into four bytes, designated a, b, c, and d. Then this formula is applied.

The F function is: $F(xL) = ((S1,a + S2,b \bmod 2^{32}) \text{ XOR } S3,c) + S4,d \bmod 2^{32}$. You can see this process in Fig. 6.7.

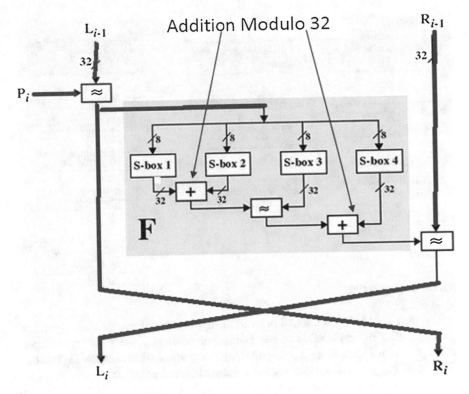

Fig. 6.7 Blowfish round function

Twofish

Twofish was one of the five finalists of the AES contest (which we will explore in more detail in Chap. 7). It is related to the block cipher Blowfish, and Bruce Schneier also was part of the team that worked on this algorithm. Twofish is a Feistel cipher that uses a 128-bit block size and key sizes of 128, 192, and 256 bits. It also has 16 rounds, like DES. Like Blowfish, Twofish is not patented and is in the public domain and can be used without restrictions by anyone who wishes to use it.

Like Blowfish, Twofish uses key-dependent S-boxes and has a fairly complex key schedule. There are four s-boxes, each 8 bit by 8 bit. The cipher key is actually split in half. Half is used as a key; the other half is used to modify the key-dependent s-boxes.

Twofish also uses a process called whitening. Key whitening is simply generating a second key and XOR'ing it with the block. This can be done before or after the round function. In the case of Twofish it is done both before and after. You can see this in Fig. 6.8.

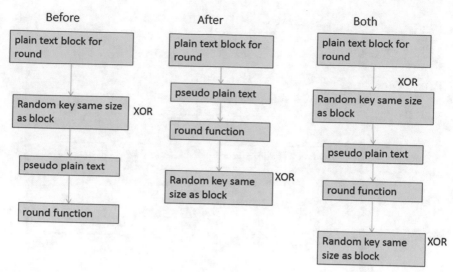

Fig. 6.8 Key whitening

The plain text is key whitened prior to the application of the Twofish algorithm, and again after the application of the Twofish algorithm. In the Twofish documentation, this is referred to as input whitening and output whitening. With Twofish, 128 bits of key material is used for the input and output whitening.

Twofish has some aspects that are not found in most other Feistel ciphers. The first of which is the concept of the cycle. Every two rounds are considered a cycle. The design of Twofish is such that in one complete cycle (two rounds) every bit of plaintext has been modified once.

The plaintext is split into four 32-bit words. The first step is the input whitening, then there will be 16 rounds of Twofish. Each round the two 32-bit words on the left side are put into the round function (remember all Feistel ciphers divide the plain text, submitting only part of it to the round function). Those two 32-bit words are submitted to the key-dependent s-boxes, and then mixed using an MDS matrix. Then the results from this process are combined using the pseudo-Hadamard transform (discussed earlier in this chapter). After this is complete the results are XOR'd with the right two 32-bit words (that were previously left untouched) and swapped for the next round.

There is one nuance to the XOR of the left and right halves. The two 32-bit words on the right are given a slight rotation. One 32-bit word is rotated 1 bit to the left, then after the XOR the other 32-bit word is rotated 1 bit to the right.

An overview of Twofish can be seen in Fig. 6.9. Note that this image is the diagram originally presented in Bruce Schneier's paper on Twofish.

If you wish to delve into more detail with Twofish, I recommend the paper that Schneier wrote on Twofish (found in the endnotes to this chapter). You may also wish to consult "The Twofish Encryption Algorithm: A 128-Bit Block Cipher" by Kelsey, Whiting, Wagner, & Ferguson. ISBN-10: 0471353817.

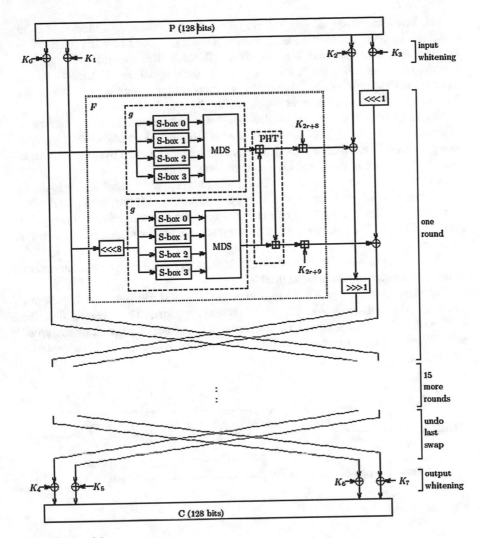

Fig. 6.9 Twofish

Skipjack

This algorithm was developed by the NSA and was designed for the clipper chip. While the algorithm was originally classified, it was eventually declassified and released to the public. The clipper chip was a chip with built-in encryption, however a copy of the decryption key would be kept in a key escrow in case law enforcement need to decrypt data without the device owner's cooperation. This feature made the process highly controversial. The skipjack algorithm was designed for voice encryption, and the clipper chip was meant for secure phones.

Skipjack uses an 80-bit key to encrypt or decrypt 64-bit data blocks. It is an unbalanced Feistel network with 32 rounds. Skipjack's round function operates on 16-bit words (Knudsen and Wagner 2001). Skipjack uses two different types of rounds. These are called the A and B rounds. In the original specifications A and B were termed "stepping rules." First step A is applied for 8 rounds, then step B for 8 rounds. Then the process is repeated, totaling 32 rounds.

Skipjack makes use of a substitution box that is rather interesting. The documentation refers to it as the G function or G permutation. It takes a 16-bit word as input along with a 4-byte subkey. This G permutation includes a fixed byte S-table that is termed the F-table in the Skipjack documentation.

A is accomplished by taking the 16-bit word termed W1 and submits it to the G box. The new W1 is produced by XOR'ing the output of the G box, with the counter, then with w4. The W2 and W3 words shift one register to the right, becoming W3 and W4. This is shown in Fig. 6.10, which is taken directly from the specification for Skipjack.

The B round is very similar, with slight differences shown in Fig. 6.11 (also taken from the specification documentation).

The Skipjack algorithm itself is considered to be robust. However, the key escrow issue made it unacceptable to many civilian vendors. They feared that the U.S. government would, at will, be able to eavesdrop on communications encrypted with Skipjack. For this reason, the Skipjack algorithm did not gain wide acceptance and usage.

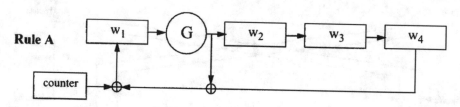

Fig. 6.10 Skipjack A round

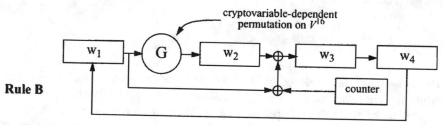

Fig. 6.11 Skipjack B round

CAST

There are two well-known versions of CAST, CAST-128 and CAST-256. CAST-256 was a candidate in the AES Contest, and is based on the earlier CAST-128. CAST-128 is used in some versions of PGP (Pretty Good Privacy). The algorithm was created by Carlisle Adams and Stafford Tavares. CAST-128 can use either 12 or 16 rounds, working on a 64-bit block. The key sizes are in 8-bit increments, ranging from 40 bits to 128 bits, but only in 8-bit increments. The 12-round version of CAST-128 is used with key sizes less than 80 bits, and the 16-round version is used with key sizes of 80 bits are longer. CAST-128 has 8 s-boxes each 32 bits in size.

FEAL

FEAL is an acronym for Fast data Encipherment ALgorithm. It was designed by Akihiro Shimizu and Shoji Miyaguchi and published in 1987. There are variations of the FEAL cipher, but all use a 64-bit block and essentially the same round function. FEAL-4 uses 4 rounds, FEAL-8 uses 8 rounds, FEAL-N uses N rounds, chosen by the implementer. This algorithm has not been done well under cryptanalysis. Several weaknesses have been found in the algorithm and it is not considered secure.

MARS

MARS was IBM's submission to the AES contest and became one of the final five finalists, though it was ultimately not selected. Mars was designed by a team of cryptographers that included Don Coppersmith. Coppersmith is a mathematician who was involved in the creation of DES as well as working on various other cryptological topics such as cryptanalysis of RSA. The algorithm uses a 128-bit block with a key size that varies between 128 and 448 bits, in increments of 32 bits.

MARS divides the input into the round function into four 32-bit words labeled A, B, C, and D. The algorithm has three phases. There is the 16-round phase that is the core of the algorithm. That phase is preceded and followed by an 8-round phase that does forward and backward mixing.

The general overview of MARS is shown in Fig. 6.12.

As previously mentioned, MARS can use a cipher key of 128 to 448 bits, this is divided into 4 to 14 32-bit words. MARS uses a key schedule that expands the user-supplied cipher key into a key array of 40, 32-bit words.

Fig. 6.12 General
overview of MARS

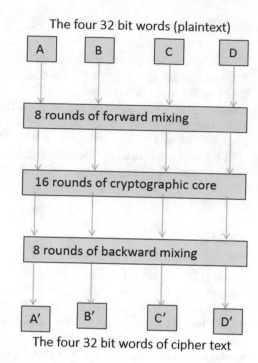

The four 32 bit words of cipher text

MARS uses a single, but very large s-box. The s-box for mars is a table of 512, 32-bit words. The operations are applied to the 32-bit words. The original specification for MARS actually described the algorithm entirely in an image, which is shown in Fig. 6.13:

You can see that the A, B, C, and D are the 32-bit words that are input. The R and M are intermediate steps in the algorithm. Most of the algorithm consists of bit shifting (denoted by the >>> and <<< symbols) and XOR'ing. The S represents the s-box (or a portion thereof). This is a fairly secure algorithm, even though it was not chosen in the AES contest, it, like any of the five finalists, is a secure algorithm that you should feel confident using.

TEA

TEA or Tiny Encryption Algorithm was created by David Wheeler and Roger Needham and first publicly presented in 1994. It is a simple algorithm, easy to implement in code. It is a Feistel cipher that uses 64 rounds (note this is a suggestion, it can be implemented with fewer or more rounds). The rounds should be even, since they are implemented in pairs called cycles.

TEA uses a 128-bit key operating on a 64-bit block. It also uses a constant that is defined as 2^{32}/the golden ratio. This constant is referred to as delta and each round a

// **Forward Mixing**
(A,B,C,D) = (A,B,C,D) + (K[0],K[1],K[2],K[3])
For i = 0 to 7 do {
  B = (B ⊕ S$_0$[A]) + S$_1$[A>>>8]
  C = C + S$_0$[A>>>16]
  D = D ⊕ S$_1$[A>>>24]
  A = (A>>>24) + B(if i=1,5) + D(if i=0,4)
  (A,B,C,D) = (B,C,D,A)
}
// **Cryptographic Core**
For i = 0 to 15 do {
  R = ((A<<<13) × K[2i+5]) <<< 10
  M = (A + K[2i+4]) <<< (low 5 bits of (R>>>5))
  L = (S[M] ⊕ (R>>>5) ⊕ R) <<< (low 5 bits of R)
  B = B +L(if i<8) ⊕ R(if i≥8)
  C = C + M
  D = D ⊕ R(if i<8) + L(if i≥8)
  (A,B,C,D) = (B,C,D,A<<<13)
}
// **Backwards Mixing**
For i = 0 to 7 do {
  A = A - B(if i=3,7) - D(if i=2,6)
  B = B ⊕ S$_1$[A]
  C = C - S$_0$[A<<<8]
  D = (D - S$_1$[A<<<16]) ⊕ S$_0$[A<<<24]
  (A,B,C,D) = (B,C,D,A<<<24)
}
(A,B,C,D) = (A,B,C,D) - (K[36],K[37],K[38],K[39])

NOTE: S$_0$[X] and S$_1$[X] use low 8 bits of X. S[X] uses low 9 bits of X.
S is the concatenation of S$_0$ and S$_1$.

Fig. 6.13 MARS in detail

multiple of delta is used. The 128-bit key is split into 4 different 32-bit subkeys labeled K[0], K[1], K[2], and K[3]. Rather than use the XOR operation, TEA uses addition and subtraction, but done mod 2^{32}. The block is divided into two halves R and L. R is put through the round function.

The round function takes the R half and performs a left shift of 4. Then the result of that operation is added to K[0] (keep in mind that all addition is being done mod 2^{32}). The result of that operation is added to Delta (recall that Delta is the current multiple of the 2^{32}/the golden ratio). The result of that operation is then shifted right 5 and added to K[1]. That is the round function. As with all Feistel ciphers, the result of the round function is XOR'd with L and then L and R are swapped for the next round.

If you are not familiar with the Golden Ratio, it is a very interesting number. Two quantities are said to be in the golden ratio if their ratio is the same as the ratio of their sum to the larger of the two quantities. This can be expressed as $(a + b)/a = a/b$. The ratio is 1.6180339887.... It is an irrational number. This number appears in many places, including in the paintings of Salvador Dali. Mathematicians throughout history including Pythagoras and Euclid have been fascinated by the Golden Ratio.

The Tiny Encryption Algorithm is a rather simple Feistel cipher and is often included in programming courses because writing it in code is relatively simple. If you are so inclined, you can find numerous examples of TEA in various programming languages (Java, C, C++, etc.) on the internet.

XTEA

As you may surmise, XTEA is eXtended TEA. It is a block cipher designed by Roger Needham and David Wheeler with the intent to correct some weaknesses in TEA. XTEA uses a 64-bit block with a 128-bit key and 64 rounds, just like TEA. However, it does rearrange the TEA algorithm somewhat, and uses a more complex key schedule algorithm.

LOKI97

LOKI97 was a candidate in the AES contest, but was not chosen. The algorithm was developed by Lawrie Brown. There have been other incarnations of the LOKI cipher, namely LOKI89 and LOKI91. The earlier versions used a 64-bit block.

LOKI97 uses 16 rounds (like DES) and a 128-bit block. But it operates with key sizes of 128, 192, or 256 bit (as does AES). It has a rather complex round function, and the key schedule is accomplished via an unbalanced Feistel cipher. The algorithm is available without royalty.

The key schedule is treated as four 64-bit words $K4_0$, $K3_0$, $K2_0$, and $K1_0$. Depending on the key size being used the four 64-bit words may need to be generated from the key provided. In other words, if there is a cipher key less than 256 bits, then there is not enough material for four 64-bit words, but the missing material for 192- and 128-bit cipher keys is generated. The key schedule itself uses 48 rounds to generate the subkeys.

The round function is probably one of the most complex for a Feistel cipher, certainly more complex than the algorithms we have examined in this chapter. More detail can be found in the paper "Introducing the new LOKI97 Block Cipher," by Lawrie Brown and Josef Pieprzyk. This paper was published in June 1998 and is available on the internet.

Camellia

Camellia is a Japanese cipher that uses a block size of 128 bits and key sizes of 128, 192, or 256 bit (like AES). With a 128-bit key, Camellia uses 18 rounds, but with the 192- or 256-bit key it uses 24 rounds (Aoki et al., 2000). Note both of these (18 and 24) are multiples of 6. The reason is that every six rounds there is a transformation layer applied that is called the FL function. Camellia uses four s-boxes and uses key whitening on the input and output (key whitening is described later in this chapter). Some browsers such as Chrome prefer to use Camellia for encryption. They will attempt to negotiate with the server, and if the server can support Camellia, the browser will select it over AES.

ICE

ICE, or Information Concealment Engine, was developed in 1997. It is similar to DES, however, uses a bit permutation during its round function that is key-dependent. ICE works with 64-bit blocks using a 64-bit key and 16 rounds. The algorithm is in the public domain and not patented. There is a faster variant of ICE, called Thin-ICE, it is faster because it uses only 8 rounds.

Simon

Simon is actually a group of block ciphers. What makes it interesting is that it was released by the NSA in 2013. It has many variations, for example, there is a variation that uses 32-bit blocks with a 64-bit key and 32 rounds, however, if you use a 64-bit block you can choose between 96- and 128-bit keys and either 42 or 44 rounds. The largest block size, 128 bit, can use key sizes of 128, 192, or 256 bit with either 68, 69, or 72 rounds.

IDEA

The International Data Encryption Algorithm was designed by James Massey and Xuejia Lai. It is a block cipher and was first published in 1991. IDEA uses 64-bit blocks and a 128-bit cipher key. The process has 8 identical rounds, then an output round (Meier 1993). This algorithm is interesting because it uses a process we have not yet examined. That process is the Lai-Massey scheme.

Each round the block is broken into two equal pieces, much like the Feistel cipher. There are some permutations done to the left and right sides. Thus, it is not truly a Feistel cipher, but more of a pseudo-Feistel. IDEA has done reasonably well under cryptanalysis.

MISTY1

This algorithm was developed by Mitsuru Matsui in 1995. It was selected for the European NESSIE project and has been used in some implementations. However, in 2015, it was broken with integral cryptanalysis. The name MISTY is an acronym with two meanings. The first is the names of the developers: Matsui Mitsuru, Ichikawa Tetsuya, Sorimachi Toru, Tokita Toshio, and Yamagishi Atsuhiro. The second meaning can be "Mitsubishi Improved Security Technology." This algorithm is patented, but freely available for academic use.

KASUMI

The KASUMI algorithm was designed specifically for use in UMTS (Universal Mobile Telecommunications System) and GSM (Global System for Mobile Communications) cellular communications. It is a block cipher. With UMTS, KASUMI is used in the confidentiality and integrity algorithms, labeled f8 and f9 in the UMTS standard. In GSM, KASUMI is used in the A5/3 key stream generator.

The KASUMI algorithm uses a 128-bit key. During the key schedule algorithm that cipher key is divided into 8 subkeys of 16 bits each (Wallén 2000). Round keys are derived from the subkeys, often using bitwise rotation to the subkeys. KASUMI divides a 64-bit block of plain text into two 32-bit halves. It is not purely a Feistel cipher, but very similar.

MAGENTA

This algorithm is a block cipher and it is also a Feistel cipher. It was developed for Deutsche Telekom by Michael Jacobson and Klaus Huber. The name is an acronym for Multifunctional Algorithm for General-purpose Encryption and Network Telecommunication Applications. As you can probably surmise from both the name and the company it was developed for, this algorithm is intended for use in telecommunications (Jacobson et al. 1998). It uses a block size of 128 bits with key sizes of 128, 192, and 256 bits which is reminiscent of AES (which you will see in the next chapter). However, it is Feistel cipher that can use either 6 or 8 rounds.

Speck

This is actually a group of related block ciphers, each a slight variation of the others. It was released in 2013 by the U.S. NSA with the intention of Speck, along with a related algorithm Simon, be used for Internet of Things devices. Due to the restricted processing power of IoT devices, Speck and Simon are as concerned about performance as they are about security.

The block size for Speck is always two words, but the words themselves could be 16, 24, 32, 48, or 64 bits. The key is either 2, 3, or 4 words. Due to performance issues, the round function is rather simple. It has two rotations that consist of adding the right word to the left, then XOR'ing the key into the left word, then XOR'ing the left word into the right word. The block and key size will determine the number of rounds, shown in the following table (all sizes are in bits).

Word size	Block size	Key size	Rounds
16	$2 \times 16 = 32$	$4 \times 16 = 64$	22
24	$2 \times 24 = 48$	$3 \times 24 = 72$	22
24		$4 \times 24 = 96$	23
32	$2 \times 32 = 64$	$3 \times 32 = 96$	26
32		$4 \times 32 = 128$	27
48	$2 \times 48 = 96$	$2 \times 48 = 96$	28
48		$3 \times 48 = 144$	29
64	$2 \times 64 = 128$	$2 \times 64 = 128$	32
64		$3 \times 64 = 192$	33
64		$4 \times 64 = 256$	34

This cipher is constructed to work on systems with limited resources, yet still provide adequate security.

Symmetric Methods

There are methods that can be used to alter the way a symmetric cipher works. Some of these are meant to increase the security of the cipher. Others to change a block cipher into a stream cipher. We will examine a few of the more common methods here in this section.

ECB

The most basic encryption mode is the electronic codebook (ECB) mode. The message is divided into blocks and each block is encrypted separately. The problem is that if you submit the same plain text more than once, you always get the same ciphertext. This gives attackers a place to begin analyzing the cipher to attempt to derive the key. Put another way, ECB is simply using the cipher exactly as it is described without any attempts to improve its security.

CBC

When using cipher block chaining (CBC) mode, each block of plaintext is XOR'd with the previous ciphertext block before being encrypted. This means there is significantly more randomness in the final ciphertext. This is much more secure than electronic codebook mode and is the most common mode. This process is shown in Fig. 6.14.

There really is no good reason to use ECB over CBC, if both ends of communication can support CBC. Cipher block chaining is a strong deterrent to known plain text attacks, a cryptanalysis method we will examine in Chap. 17.

Fig. 6.14 Cipher block chaining

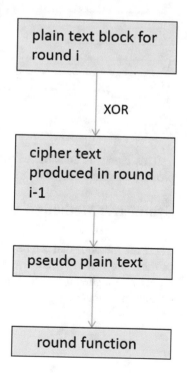

The only issue with CBC is the first block. There is no preceding block of ciphertext to XOR the first plaintext block with. It is common to add an initialization vector to the first block so that it has something to be XOR'd with. The initialization vector is basically a pseudorandom number, much like the cipher key. Usually, an IV is only used once and is thus called a nonce (Number Only used Once). The CBC mode is actually fairly old. It was introduced by IBM in 1976.

PCBC

The propagating cipher block chaining mode was designed to cause small changes in the ciphertext to propagate indefinitely when decrypting, as well as when encrypting. This method is sometimes called plaintext cipher block chaining. The PCBC mode is a variation on the CBC mode of operation. It is important to keep in mind that the PCBC mode of encryption has not been formally published as a federal standard.

CFB

In Cipher Feedback mode, the previous ciphertext block is encrypted then the ciphertext produced is XOR'd back with the plaintext to produce the current ciphertext block. Essentially it loops back on itself, increasing the randomness of the resultant ciphertext. While CFB is very similar to cipher block chaining, its purpose is a bit different. The goal is to take a block cipher and turn it into a stream cipher. Output Feedback mode is another method used to transform a block cipher into a synchronous stream cipher. We will examine both of these in more detail when we discuss stream ciphers in Chap. 7.

Galois/Counter Mode

The Galois Counter Mode (GCM) uses counter mode with Galois authentication. Counter mode was introduced by Whitfield Diffie and Martin Hellman in 1979 and is used to turn a block cipher into a stream cipher. Counter mode is combined with Galois field multiplication used for authentication.

Conclusions

In this chapter, you have seen the different types of symmetric ciphers and studied Feistel ciphers in detail. It is critical that you fully understand the general Feistel structure. It is also important that you have a strong understanding of DES and Blowfish before proceeding to the next chapter. Other algorithms were presented in this chapter. While you need not memorize every single algorithm, the more algorithms you are familiar with, the better.

You have also learned some new mathematics in this chapter, such as the pseudo-Hadamard transform. Just as importantly you have seen improvements to symmetric ciphers with techniques such as cipher block chaining.

Test Your Knowledge

1. _____ has an 80-bit key and is an unbalanced cipher
2. _____ is a Feistel cipher using variable length key sizes from 32 bits to 448 bits.
3. The following formulas describe what?

$$a' = a + b \ (\text{mod} \ 2^n)$$
$$b' = a + 2b \ (\text{mod} \ 2^n)$$

_____

4. Which of the following is a Russian cipher much like DES?

   (a) Blowfish.
   (b) CAST.
   (c) FEAL.
   (d) GOST.

5. What is the proper term for the algorithm used to derive subkeys (round keys) from the cipher key?

   (a) Key algorithm.
   (b) Subkey generator.
   (c) Key schedule.
   (d) Round key generator.

6. Which algorithm described in this chapter was an unbalanced Feistel cipher used with the clipper chip?

   _____.

7. _____ used input whitening and output whitening as well as a pseudo-Hadamard transform.

8. _____ divides the input into the round function into four 32-bit words labeled A, B, C, and D, then uses three phases: The 16-round phase that is the core of the algorithm; pre and post phases of forward and backward mixing.

9. With a 128-bit key, Camellia uses ___ rounds, but with the 192- or 256-bit key it uses ___ rounds.

10. With _____ block of plaintext is XOR'd with the previous ciphertext block before being encrypted.

References

Aoki, K., Ichikawa, T., Kanda, M., Matsui, M., Moriai, S., Nakajima, J., & Tokita, T. (2000, August). Camellia: A 128-bit block cipher suitable for multiple platforms—design and analysis. In International Workshop on Selected Areas in Cryptography (pp. 39–56). Springer, Berlin, Heidelberg.

Blaze, M., & Schneier, B. (1994, December). The MacGuffin block cipher algorithm. In International Workshop on Fast Software Encryption (pp. 97–110). Springer, Berlin, Heidelberg.

Courtois, N., Drobick, J., & Schmeh, K. (2018). Feistel ciphers in East Germany in the communist era. Cryptologia, 42(5), 427–444.

Courtois, N., & Misztal, M. (2011). Differential Cryptanalysis of GOST. IACR Cryptol. ePrint Arch., 2011, 312.

Easttom, C. (2018). A Generalized Methodology for Designing Non-Linear Elements in Symmetric Cryptographic Primitives. In Computing and Communication Workshop and Conference (CCWC), 2018 IEEE 8th Annual. IEEE.

Jacobson Jr, M. J., Huber, K., & AG, D. T. (1998). The MAGENTA Block Cipher Algorithm. NIST AES Proposal, 94.

Knudsen, L., & Wagner, D. (2001). On the structure of Skipjack. Discrete Applied Mathematics, 111(1–2), 103–116.

Meier, W. (1993, May). On the security of the IDEA block cipher. In Workshop on the Theory and Application of Cryptographic Techniques (pp. 371–385). Springer, Berlin, Heidelberg.

Nyberg, K. (1996, November). Generalized Feistel networks. In International conference on the theory and application of cryptology and information security (pp. 91–104). Springer, Berlin, Heidelberg.

Schneier, B. (1993, December). Description of a new variable-length key, 64-bit block cipher (Blowfish). In International Workshop on Fast Software Encryption (pp. 191–204). Springer, Berlin, Heidelberg.

Wallén, J. (2000). Design principles of the Kasumi block cipher. In Proceedings of the Helsinki University of Technology Seminar on Network Security.

Chapter 7
Substitution–Permutation Networks

Abstract This chapter continues our exploration of symmetric ciphers. Symmetric ciphers are such a substantial part of modern cryptographic implementations, that having two chapters devoted to the topic is necessary. In this chapter, we focus extensively on AES because it is such a prominent algorithm. However, we explore a variety of other algorithms that have a similar structure. Coupled with Chap. 6, this should provide the reader with a solid working knowledge of symmetric ciphers. The algorithms explored in this chapter are an important part of symmetric cryptography. AES may be the most widely used symmetric algorithm, and thus warrants significant attention.

Introduction

A substitution–permutation network (sometimes simply called an SPM) is a series of operations linked together. The first thing that is noticeably different from a Feistel cipher is that the plaintext block is not split in half, but instead is operated on as a whole. The block is subjected to alternating rounds or layers of operations including substitution or permutation boxes, in order to generate the ciphertext. There are several prominent algorithms that are structured in this manner, so we will explore those. We also explore stream ciphers in this chapter.

In This Chapter We Will Cover

AES.
Serpent.
Stream Ciphers.
RC4.
FISH.
PIKE.

Replacing DES

DES (Data Encryption Standard) is actually a quite well-designed algorithm. Now, you may be aware that it is no longer recommended and is considered insecure. The issue is not the algorithm, but rather the key is simply too small. Unquestionably, in the late 1970s, when DES was first published, and even in the early 1980s that key size was actually adequate. However, as computing power increased, it became more practical for a computer to brute force the DES key. It became apparent that a replacement was needed for DES. The United States National Institute of Standards began a process in 1997 to 2000 in order to find a replacement for DES.

On January 2, 1997, when the NIST announced the contest the search was on to find a replacement for DES. However, it was not until September 12, 1997 that the NIST announced the general criteria for selection as the new standard. The requirements were that contestant algorithms had to be block ciphers that supported a block size of 128 bits as well as key sizes of 128, 192, and 256 bits. During the next several months, 15 different algorithms were submitted from a variety of countries. The algorithms that were submitted were: CAST-256, CRYPTON, DEAL, DFC, E2, FROG, HPC, LOKI97, MAGENTA, MARS, RC6, Rijndael, SAFER+, Serpent, and Twofish. Several of these submitted algorithms, while not ultimately selected, were solid algorithms. In Chap. 6 we examined MARS and CAST-256 and in this chapter you will see a few of the others, although there will be a substantial focus on AES itself.

Over the period of the contest, the algorithms were subjected to a variety of tests, including common cryptanalysis attacks. Two conferences were held, the first AES1 in August 1998, and the second AES2 in March 1999. In August of 1999, it was announced that the process had narrowed down the candidate list to just five algorithms: MARS, RC6, Rijndael, Serpent, and Twofish. These are often referred to in the literature as the AES finalists. It should be noted that all five of these are robust algorithms and are widely used today. Then in October of 2000, the NIST announced that the Rijndael cipher had been chosen. In computer security literature you will see the same algorithm referred to as Rijndael or just AES, however cryptography literature usually refers to Rijndael.

AES

Advanced Encryption Standard was ultimately chosen as a replacement for DES. AES is also known as Rijndael block cipher. It was officially designated as a replacement for DES in 2001 after a 5-year process involving 15 competing algorithms. AES is designated as FIPS 197. FIPS is an acronym for Federal Information Processing Standard. Other algorithms that did not win that competition include such well-known algorithms as Twofish. The importance of AES cannot be overstated. It

11011001	01110010	10110000	11101010
01011111	00011001	11011001	10011001
10011100	11011101	00011001	11111101
11011001	10001001	11011001	10001001

Fig. 7.1 The Rijndael matrix

is widely used around the world and is perhaps the most widely used symmetric cipher. Of all the algorithms in this chapter, AES is the one you should give the most attention too.

AES can have three different key sizes, they are: 128, 192, or 256 bits. The three different implementations of AES are referred to as AES 128, AES 192, and AES 256. The block size, however, is always 128-bit. It should be noted that the original Rijndael cipher allowed for variable block and key sizes in 32-bit increments (Daemen and Rijmen 1998). It should be noted that the algorithm, the Rijndael algorithm, supports other key and block sizes. Rijndael supports key and block sizes of 128, 160, 192, 224, and 256 bit. However, the AES standard specifies a block size of only 128 bits and key sizes of 128, 192, and 256 bit.

This algorithm was developed by two Belgian cryptographers, Joan Daemen and Vincent Rijmen. John Daeman is a Belgian cryptographer who has worked extensively on the cryptanalysis of block ciphers, stream ciphers, and cryptographic hash functions. Vincent Rijmen is also a Belgian cryptographer who has helped design the WHIRLPOOL cryptographic hash (which we will study in Chap. 9) as well as working on ciphers such as KHAZAD, Square, and SHARK.

Rijndael uses a substitution–permutation matrix rather than a Feistel network. The Rijndael cipher works by first putting the 128-bit block of plain text into a 4-byte X 4-byte matrix (Daemen and Rijmen 1998). This matrix is termed the state and will change as the algorithm proceeds through its steps. Thus, the first step is to convert the plain text block into binary, then put it into a matrix as shown in Fig. 7.1.

Rijndael Steps

The algorithm consists of a few relatively simple steps that are used during various rounds. The steps are described here:

AddRoundKey—each byte of the state is combined with the round key using bitwise XOR. This is where Rijndael applies the round key generated from the key schedule.

SubBytes—a non-linear substitution step where each byte is replaced with another according to a lookup table. This is where the contents of the matrix are put through the s-boxes. Each of the s-boxes is 8 bits.

ShiftRows—a transposition step where each row of the state is shifted cyclically a certain number of steps. In this step the first row is left unchanged. Every byte in

Fig. 7.2 Shift rows

Initial State					After Shift Rows			
1a	1b	1c	1d		1a	1b	1c	1d
2a	2b	2c	2d		2b	2c	2a	2a
3a	3b	3c	3d		3c	3d	3a	3b
4a	4b	4c	4d		4d	4a	4b	4c

Fig. 7.3 Mix columns

$$\begin{bmatrix} 2 & 3 & 1 & 1 \\ 1 & 2 & 3 & 1 \\ 1 & 1 & 2 & 3 \\ 3 & 1 & 1 & 2 \end{bmatrix} \begin{bmatrix} a_0 \\ a_1 \\ a_2 \\ a_3 \end{bmatrix} = \begin{bmatrix} b_0 \\ b_1 \\ b_2 \\ b_3 \end{bmatrix}$$

the second row is shifted one byte to the left (with the far left wrapping around). Every byte of the third row is shifted two to the left, and every byte of the fourth row is shifted three to the left (again with wrapping around. This is shown in Fig. 7.2.

Notice that in Fig. 7.2 the bytes are simply labeled by their row then a letter, for example, 1a, 1b, 1c, 1d.

MixColumns—a mixing operation which operates on the columns of the state, combining the four bytes in each column. In the MixColumns step, each column of the state is multiplied with a fixed polynomial. Each column in the state (remember the matrix we are working with) is treated as a polynomial within the Galois Field (2^8). The result is multiplied with a fixed polynomial c $(x) = 3x^3 + x^2 + x + 2$ modulo $x^4 + 1$.

The MixColumns step can also be viewed as a multiplication by the particular matrix in the finite field $GF(2^8)$ (Daemen and Rijmen 1999). This is often shown as matrix multiplication, as you see in Fig. 7.3.

Essentially, you take the four bytes and multiply them by the matrix, yielding a new set of four bytes.

Rijndael Outline

With the aforementioned steps in mind, this is how those steps are executed in the Rijndael cipher. For 128-bit keys, there are 10 rounds. For 192-bit keys there are 12 rounds. For 256-bit keys there are 14 rounds.

Key Expansion—The first step is that the round keys are derived from the cipher key using Rijndael's key schedule. The key schedule is described in more detail later in this chapter.

Initial Round

This initial round will only execute the AddRoundKey step. This is simply XOR'ing with the round key. This initial round is executed once, then the subsequent rounds will be executed.

Rounds

This phase of the algorithm executes several steps, in the following order:

SubBytes.
ShiftRows.
MixColumns.
AddRoundKey.

Final Round

This round has everything the rounds phase has, except no mix columns.

SubBytes.
ShiftRows.
AddRoundKey.

In the AddRoundKey step, the subkey is xord with the state. For each round, a subkey is derived from the main key using Rijndael's key schedule; each subkey is the same size as the state.

Rijndael S-Box

The s-box of Rijndael is fascinating to study. We will look more deeply into them in Chap. 8, S-box design. However, a brief description is needed here. The s-box is generated by determining the multiplicative inverse for a given number in GF $(2^8) = GF(2)[x]/(x^8 + x^4 + x^3 + x + 1)$, Rijndael's finite field (zero, which has no inverse, is set to zero). In other words, the s-boxes are based on a mathematical formula. In fact, there are variations of the standard Rijndael s-box. It will still operate as any other s-box, taking in bits as input and substituting it for some other bits. You can see the standard Rijndael s-box in Fig. 7.4.

```
  | 0  1   2   3   4   5   6   7   8   9   a   b   c   d   e   f
---|--|--|--|--|--|--|--|--|--|--|--|--|--|--|--|--|
00 |63 7c  77  7b  f2  6b  6f  c5  30  01  67  2b  fe  d7  ab  76
10 |ca 82  c9  7d  fa  59  47  f0  ad  d4  a2  af  9c  a4  72  c0
20 |b7 fd  93  26  36  3f  f7  cc  34  a5  e5  f1  71  d8  31  15
30 |04 c7  23  c3  18  96  05  9a  07  12  80  e2  eb  27  b2  75
40 |09 83  2c  1a  1b  6e  5a  a0  52  3b  d6  b3  29  e3  2f  84
50 |53 d1  00  ed  20  fc  b1  5b  6a  cb  be  39  4a  4c  58  cf
60 |d0 ef  aa  fb  43  4d  33  85  45  f9  02  7f  50  3c  9f  a8
70 |51 a3  40  8f  92  9d  38  f5  bc  b6  da  21  10  ff  f3  d2
80 |cd 0c  13  ec  5f  97  44  17  c4  a7  7e  3d  64  5d  19  73
90 |60 81  4f  dc  22  2a  90  88  46  ee  b8  14  de  5e  0b  db
a0 |e0 32  3a  0a  49  06  24  5c  c2  d3  ac  62  91  95  e4  79
b0 |e7 c8  37  6d  8d  d5  4e  a9  6c  56  f4  ea  65  7a  ae  08
c0 |ba 78  25  2e  1c  a6  b4  c6  e8  dd  74  1f  4b  bd  8b  8a
d0 |70 3e  b5  66  48  03  f6  0e  61  35  57  b9  86  c1  1d  9e
e0 |e1 f8  98  11  69  d9  8e  94  9b  1e  87  e9  ce  55  28  df
f0 |8c a1  89  0d  bf  e6  42  68  41  99  2d  0f  b0  54  bb  16
```

Fig. 7.4 Rijndael s-box

Rijndael Key Schedule

As with other ciphers we have seen Rijndael has a key schedule in order to generate round keys from the original cipher key. The key schedule has three operations that will be combined to create the key schedule.

The Operations

Rotate: The first operation is simply to take a 32-bit word (in hexadecimal) and to rotate it 8 bits (1 byte) to the left.

Rcon: This is the term that the Rijndael documentation uses for the exponentiation of 2 to a user-specified value. However, this operation is not performed with regular integers, but in Rijndael's finite field. In polynomial form, 2 is $2 = 00000010 = 0$ $x^7 + 0\ x^6 + 0\ x^5 + 0\ x^4 + 0\ x^3 + 0\ x^2 + 1\ x + 0$.

Inner Loop: The key schedule has an inner loop that consists of the following steps:

1. The input is a 32-bit word and at an iteration number i. The output is a 32-bit word.
2. Copy the input over to the output.
3. Use the above-described rotate operation to rotate the output eight bits to the left.
4. Apply Rijndael's s-box on all four individual bytes in the output word.

5. On just the first (leftmost) byte of the output word, exclusive OR the byte with 2 to the power of $(i-1)$. In other words, perform the rcon operation with i as the input, and exclusive or the rcon output with the first byte of the output word.

The Actual Key Schedule

The actual key schedule for Rijndael is one of the more complex key schedules found in symmetric ciphers.

Since the key schedule for 128-bit, 192-bit, and 256-bit encryption is very similar, with only some constants changed, the following key size constants are defined here:

- n has a value of 16 for 128-bit keys, 24 for 192-bit keys, and 32 for 256-bit keys,
- b has a value of 176 for 128-bit keys, 208 for 192-bit keys, and 240 for 256-bit keys (with 128-bit blocks as in AES, it is correspondingly larger for variants of Rijndael with larger block sizes).

Once you have the appropriate constants, then the steps of the Rijndael key schedule can be executed:

1. The first n bytes of the expanded key are simply the encryption key.
2. The rcon iteration value i is set to 1.
3. Until we have the desired b bytes of the expanded key, we do the following to generate n more bytes of the expanded key:
4. We do the following to create 4 bytes of the expanded key:

   (a) We create a 4-byte temporary variable, t.
   (b) We assign the value of the previous four bytes in the expanded key to t.
   (c) We perform the key schedule core (see above) on t, with i as the rcon iteration value.
   (d) We increment i by 1.
   (e) We XOR t with the four-byte block n bytes before the new expanded key. This becomes the next 4 bytes in the expanded key.

5. We then do the following three times to create the next 12 bytes of the expanded key:

   (a) We assign the value of the previous 4 bytes in the expanded key to t.
   (b) We XOR t with the four-byte block n bytes before the new expanded key. This becomes the next 4 bytes in the expanded key.

6. If we are processing a 256-bit key, we do the following to generate the next 4 bytes of the expanded key:

   (a) We assign the value of the previous 4 bytes in the expanded key to t.
   (b) We run each of the 4 bytes in t through Rijndael's s-box.

(c) We XOR t with the 4-byte block n bytes before the new expanded key. This becomes the next 4 bytes in the expanded key.

7. If we are processing a 128-bit key, we do not perform the following steps. If we are processing a 192-bit key, we run the following steps twice. If we are processing a 256-bit key, we run the following steps three times:

(a) We assign the value of the previous 4 bytes in the expanded key to t.
(b) We XOR t with the four-byte block n bytes before the new expanded key. This becomes the next 4 bytes in the expanded key.

This should give you a solid working knowledge of AES. Given the widespread use of this algorithm, it is important that you take the time to thoroughly understand this algorithm before proceeding.

Serpent

This algorithm was invented by Ross Anderson, Eli Biham, and Lars Knudsen. It was submitted to the AES competition but was not selected, in large part due to the fact that its performance is slower than AES. However, in the ensuing years since the AES competition, computational power has increased dramatically. This has led some experts to reconsider the use of Serpent on modern systems.

Ross Anderson worked with Eli Biham on the Tiger cryptographic hash and other algorithms. He also designed the PIKE stream cipher. Eli Biham is an Israeli cryptographer and well known for his extensive contributions to cryptography including inventing the topic of differential cryptanalysis, along with Adi Shamir (we will be studying differential cryptanalysis in Chap. 17).

Serpent has a block size of 128 bits and can have a key size of 128, 192, or 256 bits, much like AES (Biham et al. 1998). The algorithm is also a substitution–permutation network like AES. It uses 32 rounds working with a block of four 32-bit words. Each round applies one of eight 4-bit to 4-bit s-boxes 32 times in parallel. Serpent was designed so that all operations can be executed in parallel.

Serpent S-Boxes

The inventors of DES stated in their original proposal that they had initially considered simply using the s-boxes from DES due to the fact that those s-boxes had been thoroughly studied. However, they abandoned that idea in favor of using s-boxes optimized for more modern processors.

Serpent Key Schedule

There are 33 sub-keys or round keys, each 128 bits in size. The key length could have been variable, but for the purposes of the AES competition, the key sizes were fixed at 128, 192, and 256 bits.

The Serpent Algorithm

The cipher begins with an initial permutation (IP), much as DES does. It then has 32 rounds each round consists of a key mixing operation (note that the key mixing is simply XORing the round key with the text), s-boxes, and a linear transformation (except for the last round). In the final round, that linear transformation is instead replaced with a key mixing step. Then there is a final permutation (FP). The IP and FP do not have any cryptographic significance, instead they simply optimize the cipher.

The cipher uses one s-box per round. Then during R_0 the S_0 s-box is used, then during R_1, the S_1 s-box is used. Since there are only 8 s-boxes, each one is used four times. So that R_{16} uses S_0 again.

It should be obvious that Serpent and Rijndael have some similarities. The following table shows a comparison of the two algorithms.

	Serpent	Rijndael
Rounds	32	10, 12, or 14
Key size	128, 192, 256	128, 192, 256
Round function operations	XOR, s-boxes, and a linear transformation	s-boxes, row shifting, column mixing, and XOR
S-box	8 s-boxes each 4 X 4	A single 8×8 s-box
Speed	Slower	Faster

Clearly Serpent has more rounds the Rijndael, but the round functions of the two algorithms are different enough that simply having more rounds does not automatically mean Serpent is more secure. Serpent is slower due to more rounds, and it is particularly slower on computers that do not support multi-processing. That was the norm for personal computers at the time of the AES competition. Now, however, multiprocessing computers are ubiquitous. This indicates that today, Serpent can be a secure and efficient algorithm.

Square

Square was actually the predecessor algorithm to the Rijndael cipher. It was developed by Joan Daemen, Vincent Rijmen, and Lars Knudsen. It uses a 128 block with a 128-bit key working in 8 rounds (Daemen et al. 1997). This algorithm was first published in 1997. Given the success of AES, the Square cipher has largely become a footnote in the history of cryptography.

You have already become acquainted with Rijmen and Daemen earlier in this book. Lars Knudsen is a well-known Danish cryptographer. He has done extensive work with analyzing block ciphers, cryptographic hash functions, and message authentication codes. He received his doctorate from Aarhus University in 1994.

SHARK

SHARK was invented by a team of cryptographers including Vincent Rijmen, Joan Daemen, Bart Preneel, Antooon Bosslaers, and Erik De Win. SHARK uses a 64-bit block with a 128-bit key and operates in 6 rounds (the original SHARK used 6 rounds). It has some similarities to the Rijndael cipher including the use of s-boxes that are based on $GF(2^8)$. Remember that GF is a Galois Field defined by a particular prime number raised to some power. Like Rijndael (and unlike DES) the s-boxes take a fixed number of bits and put out the same number of bits (recall that DES s-boxes took in 6 bits and produced 4 bits).

The original paper for SHARK described two different ways to create a key schedule algorithm (recall the key schedule creates round keys from the cipher key). The first method took n bits of the round input were XOR'd with n bits of the cipher key. The result was the round key. The second method used an Affine Transformation.

The term Affine Transformation might be new to some readers. An Affine Transformation is a function between affine spaces which preserves points, straight lines and planes. This is often applied to geometry but works well with matrix mathematics. For more details on Affine Transformations, this textbook chapter from Clemson University is a good resource http://people.cs.clemson.edu/~dhouse/courses/401/notes/affines-matrices.pdf

The specific Affine Transformation used in SHARK is as follows. Let K_i be a key-dependent invertible $(n \times n)$ matrix over $GF(2^m)$. Then the operation on that matrix is shown in Fig. 7.5.

The general flow of the SHARK algorithm is shown in Fig. 7.6, which is a figure from the original SHARK paper.

$$Y = \kappa_i \cdot X \oplus K_i$$

Fig. 7.5 The SHARK Affine Transformation

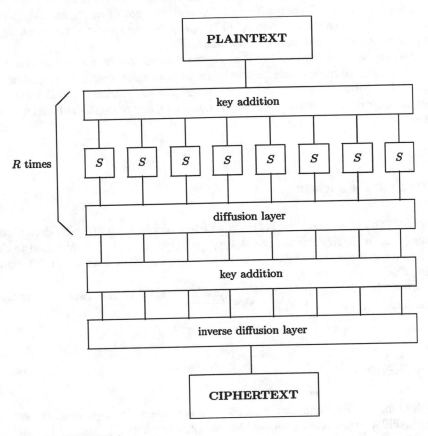

Fig. 7.6 The SHARK Cipher

SAFER

SAFER is an acronym for Secure And Fast Encryption Routine. It is actually a family of ciphers invented by a team that included James Massey (one of the creators of IDEA a rather well-known block cipher). The older versions include SAFER K and SAFER SK, the newer versions are SAFER+ and SAFER++. The first SAFER cipher was published in 1993 and used a key size of 64 bits. It was eponymously named SAFER K-64. There was also a 128-bit version named SAFER K-128.

Cryptanalysis uncovered weaknesses in the original design, specifically in the key schedule. Thus, an improved key schedule was designed, and the variants were named SAFER SK-64 and SAFER SK-128. The SK is an acronym that means Safer Key schedule. SAFER+ was an improvement that was published in 1998 and submitted to the AES competition. It did not become a finalist, however some Bluetooth implementations used SAFER+ for generating keys and as a message authentication code, but not for encrypting traffic.

The key size used with SAFER obviously depends on the particular variant of SAFER being discussed. The number of rounds can range from 6 to 10 with more rounds being used with larger key sizes. However, all the various SAFER variations use the same round function. They differ in key size, key schedule, and total number of rounds.

The Round Function

The 64-bit block of plain text is converted into 8 blocks, each of 8 bits. Each round consists of an XOR with the round key, the output of which is submitted to the s-boxes, and the output of that is subjected to another XOR with a different round key. This is shown in Fig. 7.7.

After this XOR'ing and s-boxes, the text is subjected to three pseudo-Hadamard transforms. This is shown in Fig. 7.8.

Key Schedule

Each of the SAFER variations has a different key scheduling algorithm. For example, SAFER+ expands a 16, 24, or 32-byte cipher key into 272, 400, or 528 subkey bytes (again bytes not bits). An overview of the SAFER structure is shown in Fig. 7.9. This figure is taken from the original SAFER paper.

KHAZAD

The KHAZAD algorithm was designed by Vincent Rijmen (one of the creators of the Rijndael cipher) and Paulo Barreto. The name is not an acronym but rather is derived from a fictional dwarven kingdom from J.R.R. Tolkien's books: Khazad-dum. KHAZAD uses 8 rounds on a 64-bit block, applying a 128-bit key (Barreto and Rijmen 2000). This algorithm is not patented and is free to anyone who wishes to use it.

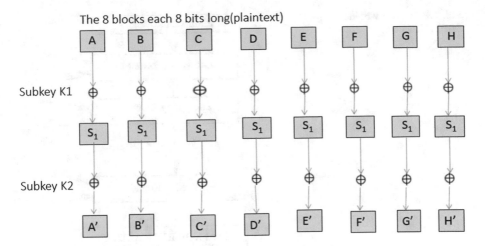

Fig. 7.7 The SAFER round function

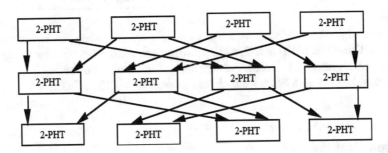

Fig. 7.8 The SAFER PHT's

NESSIE

In Europe, there was a project similar to the AES competition. This project was named NESSIE or New European Schemes for Signatures, Integrity, and Encryption (Barreto and Rijmen 2000). The research project lasted from 2000 to 2003 with the goal of not just finding a new standard symmetric cipher, but rather secure cryptographic primitives. You will see cryptographic primitives discussed frequently in this book. The term is easy to understand. Cryptographic primitives are basic cryptographic algorithms that are used to build cryptographic protocols. Cryptographic hashes, symmetric algorithms, and asymmetric algorithms all qualify as cryptographic primitives. Essentially all the algorithms you have seen so far are cryptographic primitives. These primitives are combined to create cryptographic protocols such as SSL and TLS which we will study in Chap. 13.

Fig. 7.9 Overview of the
SAFER function

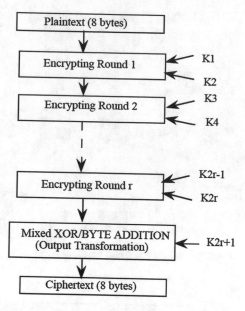

Several algorithms of various types were selected including the Block ciphers: MISTY1, Camellia, SHACAL-2, and AES/Rijndael. Also, RSA was one of the three asymmetric algorithms selected. For cryptographic hashes and message authentication codes, HMAC and SHA were chosen as well as WHIRLPOOL.

Stream Ciphers

A stream cipher is a special type of symmetric cipher. Both Feistel ciphers and substitution–permutation networks break the plain text into blocks and then encrypt each block. Stream ciphers take each bit of plain text and encrypt it with the keystream to produce the ciphertext stream. The combination is usually done via XOR operations.

There are two types of stream ciphers: synchronous and self-synchronizing. In a synchronous stream cipher (sometimes also referred to as a binary additive stream cipher), the key is a stream of bits produced by some pseudo-random number generator. The production of this keystream is completely independent of the plain text. The keystream is XOR'd with the plain text to produce the ciphertext. The reason this is called a synchronous cipher is that the sender and receiver need to be in synch. If digits are added or removed from the message during transmission (such as by an attacker) then that synchronization is lost and needs to be restored.

Self-synchronizing ciphers use parts of the current ciphertext to compute the rest of the keystream. In other words, the algorithm starts with an initial keystream, like an initialization vector, then creates the rest of the keystream based on the ciphertext already produced. These types of ciphers are sometimes called ciphertext autokey. The idea was patented in 1946, long before the advent of modern computer systems. With self-synchronizing ciphers if digits are added or removed in transit, the receiving end can still continue to decrypt.

LFSR

There are multiple ways to create the pseudo-random keystream. One of these is to use linear feedback shift registers. These are chosen because they are easy to implement in hardware. A linear feedback shift register (LFSR) is a shift register that uses its previous state to form the input for the current state. A shift register is simply a type of register found in digital circuits, basically a chain of flip flops that shifts the data stored in it by one bit, thus the name.

Usually a linear function is used to modify the data each round, most often an XOR is used. The LFSR is given an initial value, or seed. Then the register will shift that value and then XOR it with the previous value. Consider this, rather simplified example:

Input: 0101110
First it is shifted by the LFSR: 0010111

Then that value is XOR'd with the previous state so

0101110 XOR 0010111 = 0111001

Now this can be repeated with yet another shift and another XOR operation. You will see LSFRs used later when we discuss pseudo-random number generators.

RC4

RC4 is a widely known and used stream cipher, perhaps the most widely known. The RC stands for Ron's Cipher or Rivest cipher, it was designed by Ron Rivest. Ron Rivest is a name that is very familiar in cryptography. He is the R in RSA which we will explore in Chap. 10. RC4 is widely used in many security situations including WEP (Wired Equivalent Privacy) and TLS (Transport Layer Security). However, the algorithm was designed in 1987 and some experts have expressed concerns about its security. There has been speculation that it can be broken, and many people recommend no longer using it in TLS. However, it is the most widely known of stream ciphers and has a similar place in the history and study of stream ciphers as DES has in block ciphers.

RC4 Key Schedule Algorithm

The key scheduling algorithm is rather simple. We begin with an internal state that is denoted by a capital S. This state is a 256-byte array. While most of what you have seen so far involves bits, not bytes, this is not a typo. It is a 256-byte array. There are two indexes usually simply named i and j. These indexes are used to point to individual elements in the array. The key scheduling algorithm involves shuffling this array.

The first step in this algorithm involves simply initializing the state with what is termed the identity permutation. This simply means that the first element is initialized to 0, the second element to 1, the third to 0, and so on. Now obviously this is not very random at all, in fact it is the antithesis of random. So, the next step consists of shuffling. The shuffling involves iterating 256 times performing the following actions:

- compute $j = j + S[i] +$ key[i mod key length],
- swap $S[i]$ and $S[j]$,
- increment i.

After 256 iterations of this the array should be shuffled rather well. If you happen to have some programming experience, then the following pseudo code may assist you in understanding the shuffling:

for i from 0 to 255

   $S[i] := i$

end for loop
$j := 0$
for i from 0 to 255

   $j := (j + S[i] +$ key[i mod key length]) mod 256
   swap values of $S[i]$ and $S[j]$

end for loop

Now you may argue that this is too predictable, that it would generate the same key each time. And if the algorithm stopped here you would be correct. This is generating a state that will be used to create the keystream. We are not done yet.

The rest of the algorithm allows for the generation of a keystream of any size. The goal is to have a keystream that is the same size as the message you wish to encrypt.

The first step is to initialize the two pointers mentioned earlier, i and j, then generate the keystream one byte at a time. For each new byte the algorithm takes the following steps:

Step 1: Compute new value of i and j:

$i := (i + 1) \quad \% \quad 256$
$j := (j + S[i]) \quad \% \quad 256$

Step 2: Swap $S[i]$ and $S[j]$
Step 3: Retrieve the next byte of the keystream from the S array at the index

$S[i] + S[j] \quad \% \quad 256$

Again, if you have some background in programming, you may find it easier to understand this using pseudo code:

$i = 0$
$j = 0$

while generating the output:

$i := (i + 1) \quad \bmod \quad 256$
$j := (j + S[i]) \quad \bmod \quad 256$
swap values of $S[i]$ and $S[j]$
$K := S[(S[i] + S[j]) \quad \bmod \quad 256]$
output K

end while loop

Once you have generated the keystream you simply XOR it with the plain text to encrypt, or XOR it with the ciphertext to decrypt. One of the security issues with RC4 is that it does not use a nonce (recall this is a number only used once) along with the key. This means that if a key is being used repeatedly it can be compromised. More modern stream ciphers, such as eSTREAM, specify the use of a nonce.

In 1994 Ron Rivest published RC5 and then later, working with Matt Robshaw, Ray Sidney, and Yiqun Yin, he released RC6. However, both RC5 and RC6 are block ciphers, not stream ciphers. In fact, RC6 was designed for the AES competition.

FISH

This algorithm was published by the German engineering firm Siemens in 1993. The FISH (FIbonacci SHrinking) cipher is a software-based stream cipher using *Lagged Fibonacci* generator along with a concept borrowed from the shrinking generator ciphers.

A Lagged Fibonacci Generator, sometimes just called an LFG, is a particular type of pseudo-random number generator. It is based on the Fibonacci sequence (Blöcher and Dichtl 1993). Recall that the Fibonacci sequence is essentially.

$$F_n = F_{n-1} + F_{n-2}$$

This can be generalized to.

$$X_n = X_{n-l} + X_{n-k} (\text{mod } m) \quad \text{where } 0 < k < l$$

For most cryptographic applications the m is going to be some power of 2. Lagged Fibonacci generators have a maximum period of $(2^k - 1) * 2^{m-1}$. This brings us to the topic of the period of a pseudo-random number generator (PRNG). With any pseudo-random number generator, if you start with the same seed you will get the same sequence. The period of a PRNG is the maximum of the length of the repetition free sequence.

Ross Anderson's paper "On Fibonacci Keystream Generators" http://download. springer.com/static/pdf/347/chp%253A10.1007%252F3-540-60590-8_26.pdf? auth66=1427056723_46f58d9f89128bf6e3fc71a5fead8b99&ext=.pdf provides a very good discussion of Fibonacci-based key generation.

PIKE

This algorithm was published in a paper by Ross Anderson as an improvement on FISH. In that paper Anderson showed that Fish was vulnerable to known plaintext attacks. PIKE is both faster and stronger than FISH. The name PIKE is not an acronym, but rather a humorous play on the previous FISH, a pike being a type of fish.

eSTREAM

The eSTREAM project was a European search for new stream ciphers. The project ran from 2004 until 2008. It began with a list of 40 ciphers and used three phases to narrow that list to 7 ciphers. Four were meant for software implementations (HC-128, Rabbit, Salsa20/12, and SOSEMANUK) and three (Grain v1, MICKEY v2, and Trivium) for hardware. A brief description of some of these ciphers is provided here:

SNOW

SNOW 1.0 was submitted to the NESSIE project. It has since been supplanted by SNOW 2.0 and SNOW 3G. The SNOW cipher works on 32-bit words and can use either a 128- or 256-bit key. The cipher uses a Linear Feedback Shift Register along with a Finite State Machine (FSM).

Rabbit

Rabbit was designed by Martin Boesgaard, Mette Vetegrager, Thomas Pederson, Jesper Christiansen, and Ove Scavenis. It is a stream cipher that uses a 128-bit key along with a 64-bit initialization vector.

The cipher uses an internal state that is 513 bits. That consists of 8 variables each 32 bits in size, 8 counters that are also 32 bits in size, and one counter bit. The variables are denoted as $x_{j,i}$ meaning the state of the variable of subsystem j at iteration i. The counter variables are denoted by $c_{j,i}$. The entire algorithm is described in the original paper "The Stream Cipher Rabbit" that can be found online.

HC-128

HC-128 was invented by Hongjun Wu. It uses a 128-bit key with a 128-bit initialization vector. HC-128 consists of two secret tables, each one with 512 32-bit elements. At each step one element of a table is updated with non-linear feedback function. All the elements of the two tables will get updated every 1024 steps. At each step, one 32-bit output is generated from the non-linear output filtering function. The algorithm is described in detail in the paper "The Stream Cipher HC-128" which can be found online. In 2004, Wu published HC-256 which uses a 256-bit key and a 256-bit initialization vector.

MICKEY

MICKEY an acronym for Mutual Irregular Clocking KEY stream generator, was invented by Steve Babbage and Mathew Dodd. It uses an 80-bit key and a variable length initialization vector (up to 80 bits in length).

A5

A5/1 is a stream cipher that was used in GSM (Global System for Mobile Communications also known as 2g) cell phones. It was originally a secret but became public. A variation named A5/2 was developed specifically to be weaker for export purposes. KASUMI is used in A5/3. This algorithm is used in UMTS (Universal Mobile Telecommunications System also known as 3g). In January 2010, Orr Dunkelman, Nathan Keller, and Adi Shamir released a paper showing that they could break Kasumi with a related key attack. It should be noted that the attack was not effective against MISTY1, which we discussed in the previous chapter.

Phelix

This algorithm is interesting because the stream cipher has a built-in message authentication code for integrity. We will discuss message authentication codes in more detail in Chap. 9. The team of Bruce Schneier, Doug Whiting, Stefan Lucks, and Frédéric Muller created the algorithm and in 2004 submitted it to the eSTREAM contest. The primary operations used an xor, rotation, and addition modulo 2^{32}. The algorithm uses a 256-bit key with a 128-bit nonce.

Salsa20

Two related stream ciphers, Salsa20 and ChaCha, were developed by Daniel Bernstein. Salsa20 was published in 2005 and ChaCha in 2008. Both algorithms are based on a pseudo-random function that uses add-rotate-XOR operations (often called ARX operations). The core of the algorithm maps a 256-bit key and a 64-bit nonce to a 512-bit block of the keystream.

One-Time Pad

The one-time pad is the only true uncrackable encryption, if used properly. It should be clear that this is only true if used properly. This idea was first described in 1882, but then re-discovered and even patented in the early twentieth century. The first aspect of this idea is that a random key is used that is as long as the actual message. The reason this is so useful is that if the key is sufficiently random then there will be no period in the key. Periods in keys are used as part of cryptanalysis. The second aspect of this idea is actually in the name: the key is used for one single message then discarded and never used again. Should the encryption somehow be broken, and the key discovered (and this has never been done) it would cause minimal damage as that key will never be used again.

The patented version of this was invented in 1917 by Gilbert Vernam working at AT&T. It was patented in 1919 (U.S. Patent 1,310,719). This was called a Vernam cipher. It worked with tele-printer technology (the state of the art at that time). It combined each character of the message with a character on a paper tape key.

One-time pads are often described as being "information-theoretically secure." This is because the ciphertext provides no information about the original plain text. Claude Shannon, the father of information theory, provided that a one-time pad provided what he termed perfect secrecy. It should be obvious, however, that there are logistical issues with the one-time pad. Each message needs a new key. As we will see in Chap. 12, generating random numbers can be computationally intensive. Then we are left with the issue of key exchange. Imagine for a moment that secure website traffic was conducted with a one-time pad. That would require a key be generated and exchanged for each and every packet sent between the web browser and the server. The overhead would make communication impractical. For this reason, one-time pads are usually only used in highly sensitive communications wherein the need for security makes the cumbersome nature of key generation and exchange worth the effort.

Conclusions

In this chapter, you have seen substitution–permutation networks as well as stream ciphers. The most important algorithm to know well is AES/Rijndael. The other algorithms are interesting, but our focus was on AES because it is widely used around the world, and within the United States it is a national standard. There was some brief coverage of pseudo-random number generators, at least in relation to creating stream ciphers. However, the topic of PRNGs will be covered in much more detail in Chap. 12.

Test Your Knowledge

1. AES using a 192-bit key uses 12 rounds.
2. What happens in the rotate phase of the Rijndael Key Schedule? The rotate operation takes a 32-bit word (in hexadecimal) and rotates it eight bits to the left.
3. Serpent is a 32-round substitution–permutation matrix algorithm using key sizes of 128, 192, or 256 bits.
4. The FISH algorithm is a stream cipher developed by Seimans that uses the Fibonacci Lagged Sequence Generator for random numbers.
5. What are the two types of stream ciphers? synchronous and self-synchronizing.
6. SHARK uses a 64-bit block with a 128-bit key and operates in 6 rounds.
7. In Serpent there are 33 sub-keys or round keys, each 128 bits in size.

8. The RC4 key scheduling algorithm begins with a 256-byte array named the state.

9. Briefly describe the Rijndael shift rows step: each row of the state is shifted cyclically a certain number of steps. The first row is unchanged, the second row shifted one byte to the left, the third row 2 bytes to the left, and the fourth row three bytes to the left.

10. Which of the following steps does not occur in the final round of Rijndael:

    (a) SubBytes.
    (b) ShiftRows.
    (c) MixColumns.
    (d) AddRoundKey.

References

Barreto, P. S. L. M., & Rijmen, V. (2000). The Khazad legacy-level block cipher. Primitive submitted to NESSIE, 97, 106.

Biham, E., Anderson, R., & Knudsen, L. (1998, March). Serpent: A new block cipher proposal. In International workshop on fast software encryption (pp. 222–238). Springer, Berlin, Heidelberg.

Blöcher, U., & Dichtl, M. (1993, December). Fish: A fast software stream cipher. In International Workshop on Fast Software Encryption (pp. 41–44). Springer, Berlin, Heidelberg

Daemen, J., Knudsen, L., & Rijmen, V. (1997, January). The block cipher Square. In International Workshop on Fast Software Encryption (pp. 149–165). Springer, Berlin, Heidelberg.

Daemen, J., & Rijmen, V. (1998, September). The block cipher Rijndael. In International Conference on Smart Card Research and Advanced Applications (pp. 277–284). Springer, Berlin, Heidelberg.

Daemen, J., & Rijmen, V. (1999, March). The Rijndael block cipher: AES proposal. In First candidate conference (AeS1) (pp. 343–348).

Chapter 8
S-Box Design

Abstract Substitution boxes are central to symmetric ciphers. Many introductory cryptography books do not provide adequate coverage of this topic, in many cases no coverage at all. This is unfortunate, as s-boxes provide the primary source of nonlinearity in many symmetric ciphers. For this reason, it is important for the student of cryptography to have at least a basic knowldge of s-box design. In this chapter, we examine how s-boxes are designed, and we look at the DES s-boxes as well as the Rijndael s-box. This provides the reader with a working knowledge of s-box design.

Introduction

As you have seen in previous chapters, an s-box is an integral part of many ciphers. In many cases, it is key to the overall security of a given cipher. From DES to more modern ciphers, substitution boxes have provided a means of altering the ciphertext, and even transposing text, that goes far beyond simple XOR operations. An s-box is a lookup table wherein m number of input bits are replaced with n number of output bits.

Petr Tesar in his paper "A New Method for Generating High Non-Linearity S-Boxes" states: "All modern block and stream ciphers have one or more non-linear elements. S-box is one of the most used non-linear cornerstones of modern ciphers." Non-linearity is an important concept in cryptography. The goal of any cipher is to have the output look as much like a random number as possible, and still be something we can decrypt later to get back the original plaintext. Unfortunately, operations like XOR are linear. S-boxes provide a very good source of non-linearity in any block cipher.

Why Study S-Box Design?

Many cryptography textbooks give scant, if any coverage of s-box design. Usually the s-boxes used in DES or AES are explained in varying levels of detail, but that is the extent. And many texts won't go into any real depth on AES s-box design. So why devote an entire chapter to the study of s-boxes? Why not follow the de facto standard in cryptography texts and simply gloss over this topic? Put another way, why should you, the reader, devote time to studying this topic? There are actually three primary reasons you should study s-box design. Each is explained in detail in the following sections.

Critical to Block Ciphers

The first is one you should study s-box design is a reason you should already be aware of: that s-boxes form a major part of most block ciphers. It would be impossible to thoroughly study symmetric cryptography without knowing a bit about s-boxes. If you do not understand s-boxes, then there is a portion of most block ciphers that you must treat as simply a black-box, with no real understanding of what happens inside nor why. This fact should be readily apparent from the algorithms you studied in Chaps. 6 and 7. And as has already been pointed out in this book, s-boxes are the primary source for non-linearity in modern block ciphers.

Consider Feistel ciphers that you studied in Chap. 6. The XOR operation forms a part of every Feistel cipher. In most round functions there is an XOR with the round key, and of course there is a transposition of the two halves of the block each round. But the real substance of encrypting comes from the s-box. Without it, Feistel ciphers would be extremely weak, and simply not acceptable for modern use. In fact, without the s-boxes, many block ciphers would not be much better than combining a substitution cipher such as Caesar with a transposition cipher such as Rail-Fence and executing it several times.

Designing Ciphers

Secondly, should you ever be involved in the design of a symmetric cipher, you will need to design s-boxes. These are often key to the security of a symmetric cipher. It will be important to understand the principles of s-box design.

Typically designing your own cipher is an extremely bad idea. This is because you are unlikely to produce something that is sufficiently secure. It is most probable that in attempting such a task you will create a cipher that is, at best much weaker than currently used algorithms. However, someone must create the new ciphers that appear from time to time. So clearly some people do create new ciphers that are secure. However, the creators of real algorithms that are actually secure enough are people with a very substantial mathematical background and who have studied existing algorithms in quite some depth. My recommendation is that you first carefully study cryptography for several years, looking at existing ciphers in detail. Before considering developing your own cipher, you must thoroughly understand a wide range of existing ciphers. Consider the inventors of the algorithms you have seen so far in Chaps. 6 and 7. All had extensive related education, such as mathematics, and all have worked in the field of cryptography for many years. This is not a field where a novice is likely to make a substantive contribution. Becoming a cryptographer is a lengthy process. This book would just be the first step on that journey. Then, after careful and in-depth study, if you do feel compelled to create a new cipher, submit it to the peer review process so other experts in the field can evaluate your idea.

It is difficult to overemphasize the importance of the s-box in designing a block cipher. Anna Grocholewska-Czurylo, in her paper "Cryptographic properties of modified AES-like S-boxes," describes the importance of s-boxes as follows "S-box design is usually the most important task while designing a new cipher. This is because an S-box is the only non-linear element of the cipher upon which the whole cryptographic strength of the cipher depends. New methods of attacks are constantly being developed by researchers, so S-box design should always be one step ahead of those pursuits to ensure cipher's security." (Grocholewska-Czuryło 2011).

Altering S-Boxes

Finally, there are some organizations, primarily governments, who want the security of well-known algorithms such as Advanced Encryption Standard (AES), but wish to have an implementation that is private to their organization. One reason this may be desirable in some situations is that it provides an extra layer of security. Should an outside party obtain the symmetric key being used but apply it to intercepted ciphertext using the standard Rijndael cipher, the message will not be decrypted. The interest in this topic has increased as awareness of cryptographic backdoors has increased, particularly when such backdoors are used in random number generators used to generate keys for algorithms like AES. We will study cryptographic backdoors in detail in Chap. 18.

In 2013 Edward Snowden released classified documents that indicated that the United States National Security Administration (NSA) had placed backdoors in some random number generators. This prompted some concern as to the dependability of widely used random number generators. Some governments considered simply re-designing the s-boxes of AES so that their specific AES implementation was not standard, and thus even if the key was generated from a backdoor, one would still have to know the specifics of their AES implementation in order to compromise their security.

General Facts about S-Boxes

The core of most block ciphers (Blowfish, AES, DES, Serpent, GOST, etc.) is the s-box. A substitution box provides the source of non-linearity for a given cryptographic algorithm. Other facets of the algorithm are typically just various swapping mechanisms and exclusive or (XOR) operations. The XOR, in particular, provides no diffusion or confusion in the resulting ciphertext. In fact, the basis of differential cryptanalysis is the fact that XOR operation maintains the characteristics found in the plaintext to ciphertext. We will closely examine differential cryptanalysis in Chap. 17.

Types of S-Boxes

The first issue to discuss is that s-boxes can be divided into two types. These two types are substitution boxes and permutation boxes. A substitution box simply substitutes input bits for output bits. A permutation box transposes the bits. It is often the case that cryptologists simply refer to an "s-box" possibly meaning either type.

Let us first consider a simple 3-bit s-box. This s-box simply performs substitution. Each 3 bits of input are mapped with the 3 bits of output. This s-box is shown in Fig. 8.1. The first bit of input is on the left, the second two bits on the top. By matching those you will identify the output bits.

For example, an input of 110 would produce an output of 100. An input of 100 would produce an output of 101. This s-box is very simple and does not perform any transposition. The values of output were simply chosen at random. The other general type of s-box is the p-box or permutation box. A p-box is an s-box that transposes or permutes the input bits. It may, or may not, also perform substitution. You can see a p-box in Fig. 8.2.

Of course, in the process of permutation, the bit is also substituted. For example, if the least significant bit in the input is transposed with some other bit, then the least significant bit in the output is likely to have been changed. In the literature, you will often see the term s-box used to denote either a substitution box or a permutation box. This makes complete sense when you consider that, regardless of which type it

	00	11	10	01
1	101	011	100	111
0	010	000	001	110

Fig. 8.1 A three-bit s-box

Fig. 8.2 A p-box

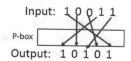

is, one inputs some bits and the output is different bits. So, for the remainder of this chapter, we simply use the term s-box to denote either a substitution or permutation box.

Whether it is a simple s-box or a p-box there are three sub-classifications: straight, compressed, and expansion. A straight s-box takes in a given number of bits and puts out the same number of bits. This is the design approach used with the Rijndael cipher. This is frankly the easiest and most common form of s-box.

A compression s-box puts out fewer bits than it takes in. A good example of this is the s-box used in DES. In the case of DES, each s-box takes in 6 bits but only outputs 4 bits. However, keep in mind that in the DES algorithm there is a bit expansion phase earlier in the round function. In that case, the 32 input bits are expanded by 16 bits to create 48 bits. So, when 8 inputs of 6 bits each are put into each DES s-box, and only 4 produced, the difference is 16 (8 X2). Therefore, the bits being dropped off are simply those that were previously added. You can see a compression s-box in Fig. 8.3.

The third type of s-box is similar compression s-box, it is the expansion s-box. The s-box puts out more bits than it takes in. One simple method whereby this can be accomplished is by simply duplicating some of the input bits. This is shown in Fig. 8.4.

There are significant issues associated with both compression and expansion s-boxes. The first issue is reversibility, or decryption. Since either type of s-box alters the total number of bits, reversing the process is difficult. One has to be very careful in the design of such an algorithm, or it is likely that decryption will not be possible. The second issue is a loss of information, particularly with compression s-boxes. In the case of DES, prior to the s-box, certain bits are replicated. Thus, what is lost in the compression step are duplicate bits and no information is lost. In

Fig. 8.3 A compression s-box

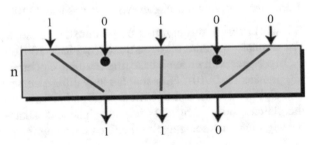

Fig. 8.4 An expansion s-box

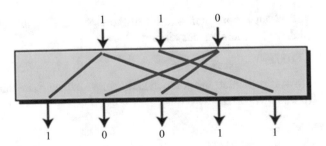

general, working with either compression or expansion s-boxes will introduce significant complexities in your s-box design. Therefore, straight s-boxes are far more common.

Design Considerations

Regardless of the type or category of s-box that is being created, there are certain features that any s-box must exhibit in order to be an effective s-box. It is not the case that you can simply put together any substitution scheme you wish and have a good s-box. It is not enough that it simply substitutes values, the s-box needs to provide confusion and diffusion (Adams and Taveres 2001). The efficacy of an s-box is usually measured by examining three separate criteria. Each of these features contributes to the security of the s-box. Those features are Strict Avalanche Criteria, Balance, and Bit Independence Criteria.

Strict Avalanche Criteria

Strict Avalanche Criteria (SAC) is an important feature of an s-box (Easttom 2018a). Remember from Chap. 3 that avalanche is a term that indicates that when one bit in the plaintext is changed, multiple bits in the resultant ciphertext are changed. Consider the following example:

We begin with a plaintext 10110011
Then after applying our cipher we have this text 11100100.

But what if, prior to encrypting the plaintext, we change just one bit of the plaintext. For example, the third bit from the left we have 10010011.

In a cipher with no avalanche, the resulting ciphertext would only change by one bit, perhaps 11100101. Note that the only difference between this ciphertext and the first ciphertext is the last or least significant bit. This shows that a change of 1 bit in the plaintext only changed 1 bit in the ciphertext. That is no avalanche. However, if our algorithm has some avalanche, then changing the plaintext from

10110011 to 10010011

will change more than one bit in the ciphertext. In this case, before the change in plaintext, remember our ciphertext was:

11100100

Now, if our cipher has some avalanche, we expect more than one bit in the ciphertext to change, perhaps two bits:

10100101

Notice the second and last bits are different. So, a change in one bit of the plaintext produced a change in two bits in the ciphertext. Ideally one would like to get more avalanche than this, as much as having a change in a single bit of plaintext change ½ the ciphertext bits. Without some level of avalanche, a cryptanalyst can examine changes in input and the corresponding changes in output and make predictions about the key. It is therefore critical that any cipher exhibit at least some avalanche.

In most block ciphers, the primary way to achieve avalanche is the use of the s-box. Strict Avalanche Criteria is one way of measuring this phenomenon. Strict Avalanche Criteria requires that for any input bit, the output bit should be changed with a probability of 0.5. In other words, if you change any given input bit, there is a 50/50 chance that the corresponding output bit will change. One measurement of strict avalanche criteria is the Hamming weight. Remember from Chap. 3 that the Hamming weight of a specific binary vector, denoted by hwt(x), is the number of ones in that vector. Therefore, if you have an input of 8 bits with 3 one's and an output of 4 one's, the Hamming weight of the output is 4. Simply measuring the Hamming weight of the input and the output and comparing them will give one indication of whether or not SAC is satisfied. This is a simple test that one should subject any s-box too.

Balance

Balance is also important in effective s-box design (Grocholewska-Czuryło 2011). Various papers give slightly different definitions of balance. However, in the current context perhaps the best definition is that each output symbol should appear an equal number of times when the input is varied over all possible values. Some sources address s-boxes with unequal input and output: We say an s-box with n input bits and m output bits, $m < n$, is balanced if every output occurs 2^{n-m} times. So, an s box of 6-bit input and 4-bit output would need each output bit to occur 4 times (2^{6-4}).

Bit Independence Criteria

The bit independence criterion (BIC) is the third criterion for good s-box design (Easttom 2018a). Bit independence criterion states that output bits j and k should change independently when any single input bit i is inverted, for all i, j, and k. The output bits change independently of each other. They are, of course, dependent upon the input bits.

Approaches to S-Box Design

There are currently a number of approaches to s-box design. The first method simply uses a pseudo-random number generator for each entry in the s-box (Easttom 2018a). The problem with this approach is that you will not be able to predict whether or not your s-box actually fulfills the three criteria we have outlined for an effective s-box. Instead, you will have to test extensively. A second approach is the human made. This was the method used in DES. In fact, the details of how the s-box for DES was designed are not public information. The actual s-boxes for DES are public; however, the methodology in designing them is not. These s-boxes were designed in cooperation with the National Security Agency. The final method uses some mathematical-based method to generate the values for the s-box. This is the method used in AES.

DES S-Box

As we discussed in Chap. 6, the National Security Agency (NSA) was involved in the creation of DES. Specifically, they were deeply involved in the s-box design. In fact, one of the IBM employees who worked on DES is quoted as saying "We sent the s-boxes off to Washington. They came back and were all different." This led many people to believe that there might be a cryptographic backdoor embedded in the DES s-boxes, which would allow the NSA to more easily break DES-encrypted communications. However, many years of study and analysis have not revealed any such backdoor.

The DES s-boxes convey a resistance to differential cryptanalysis, which we will study in Chap. 17. In fact, it has been discovered that even a small change to the DES s-box can significantly weaken its resistance to differential cryptanalysis. Differential cryptanalysis was unknown to the public at the time DES was invented. In fact, differential cryptanalysis was invented (at least publicly) by Eli Biham and Adi Shamir in the late 1980s. It is interesting to note that both Biham and Shamir noticed that DES is very resistant to differential cryptanalysis. It therefore seems most likely that the NSA was aware of differential cryptanalysis long before it was publicly known and created DES to be resistant to that attack.

The Actual S-Boxes for DES

While the design choices themselves have not been made public, we can derive some knowledge from studying the s-boxes. As far as is publicly known, the s-boxes are not derived from a mathematical formula, as the s-boxes in AES are. It seems that each substitution was specifically and manually chosen (De Meyer and Vaudenay 2017). Figures 8.5, 8.6, 8.7, 8.8, 8.9, 8.10, 8.11, and 8.12 show DES s-boxes 1 through 8.

	x0000x	x0001x	x0010x	x0011x	x0100x	x0101x	x0110x	x0111x	x1000x	x1001x	x1010x	x1011x	x1100x	x1101x	x1110x	x1111x
0yyyy0	14	4	13	1	2	15	11	8	3	10	6	12	5	9	0	7
0yyyy1	0	15	7	4	14	2	13	1	10	6	12	11	9	5	3	8
1yyyy0	4	1	14	8	13	6	2	11	15	12	9	7	3	10	5	0
1yyyy1	15	12	8	2	4	9	1	7	5	11	3	14	10	0	6	13

Fig. 8.5 DES s-box #1

	x0000x	x0001x	x0010x	x0011x	x0100x	x0101x	x0110x	x0111x	x1000x	x1001x	x1010x	x1011x	x1100x	x1101x	x1110x	x1111x
0yyyy0	15	1	8	14	6	11	3	4	9	7	2	13	12	0	5	10
0yyyy1	3	13	4	7	15	2	8	14	12	0	1	10	6	9	11	5
1yyyy0	0	14	7	11	10	4	13	1	5	8	12	6	9	3	2	15
1yyyy1	13	8	10	1	3	15	4	2	11	6	7	12	0	5	14	9

Fig. 8.6 DES s-box #2

	x0000x	x0001x	x0010x	x0011x	x0100x	x0101x	x0110x	x0111x	x1000x	x1001x	x1010x	x1011x	x1100x	x1101x	x1110x	x1111x
0yyyy0	10	0	9	14	6	3	15	5	1	13	12	7	11	4	2	8
0yyyy1	13	7	0	9	3	4	6	10	2	8	5	14	12	11	15	1
1yyyy0	13	6	4	9	8	15	3	0	11	1	2	12	5	10	14	7
1yyyy1	1	10	13	0	6	9	8	7	4	15	14	3	11	5	2	12

Fig. 8.7 DES s-box #3

	x0000x	x0001x	x0010x	x0011x	x0100x	x0101x	x0110x	x0111x	x1000x	x1001x	x1010x	x1011x	x1100x	x1101x	x1110x	x1111x
0yyyy0	7	13	14	3	0	6	9	10	1	2	8	5	11	12	4	15
0yyyy1	13	8	11	5	6	15	0	3	4	7	2	12	1	10	14	9
1yyyy0	10	6	9	0	12	11	7	13	15	1	3	14	5	2	8	4
1yyyy1	3	15	0	6	10	1	13	8	9	4	5	11	12	7	2	14

Fig. 8.8 DES s-box #4

	x0000x	x0001x	x0010x	x0011x	x0100x	x0101x	x0110x	x0111x	x1000x	x1001x	x1010x	x1011x	x1100x	x1101x	x1110x	x1111x
0yyyy0	2	12	4	1	7	10	11	6	8	5	3	15	13	0	14	9
0yyyy1	14	11	2	12	4	7	13	1	5	0	15	10	3	9	8	6
1yyyy0	4	2	1	11	10	13	7	8	15	9	12	5	6	3	0	14
1yyyy1	11	8	12	7	1	14	2	13	6	15	0	9	10	4	5	3

Fig. 8.9 DES s-box #5

	x0000x	x0001x	x0010x	x0011x	x0100x	x0101x	x0110x	x0111x	x1000x	x1001x	x1010x	x1011x	x1100x	x1101x	x1110x	x1111x
0yyyy0	12	1	10	15	9	2	6	8	0	13	3	4	14	7	5	11
0yyyy1	10	15	4	2	7	12	9	5	6	1	13	14	0	11	3	8
1yyyy0	9	14	15	5	2	8	12	3	7	0	4	10	1	13	11	6
1yyyy1	4	3	2	12	9	5	15	10	11	14	1	7	6	0	8	13

Fig. 8.10 DES s-box #6

	x0000x	x0001x	x0010x	x0011x	x0100x	x0101x	x0110x	x0111x	x1000x	x1001x	x1010x	x1011x	x1100x	x1101x	x1110x	x1111x
0yyyy0	4	11	2	14	15	0	8	13	3	12	9	7	5	10	6	1
0yyyy1	13	0	11	7	4	9	1	10	14	3	5	12	2	15	8	6
1yyyy0	1	4	11	13	12	3	7	14	10	15	6	8	0	5	9	2
1yyyy1	6	11	13	8	1	4	10	7	9	5	0	15	14	2	3	12

Fig. 8.11   DES s-box #7

	x0000x	x0001x	x0010x	x0011x	x0100x	x0101x	x0110x	x0111x	x1000x	x1001x	x1010x	x1011x	x1100x	x1101x	x1110x	x1111x
0yyyy0	13	2	8	4	6	15	11	1	10	9	3	14	5	0	12	7
0yyyy1	1	15	13	8	10	3	7	4	12	5	6	11	0	14	9	2
1yyyy0	7	11	4	1	9	12	14	2	0	6	10	13	15	3	5	8
1yyyy1	2	1	14	7	4	10	8	13	15	12	9	0	3	5	6	11

Fig. 8.12   DES s-box #8

As you should already know, the DES s-boxes are compression s-boxes. They take in 6 input bits and produce 4 output bits. If we begin by examining the first s-box, you can see how this is done. All possible combinations of the four middle bits of the input are listed on the top row of the s-box. All possible combinations of the outer two bits are listed on the far-left column. By matching the out bits on the left with the inner bits on the top, the output bits are found. Some substitutions change several bits. For example, in s-box 1, an input of all 0's "000000" produces "1110". However, others produce far less change, for example, again focusing on s-box 1 we see that an input of 001000 produces 0010. A simple shift of the 1 to the right.

Although most of the s-boxes provide a different substitution for any input, there is some overlap. For example, inputting 000110 in either s-box 2 or s-box 3 will produce 1110. It is also noteworthy that there are several cases wherein different inputs to an s-box produce the same output. For example, if you consider s-box 5 notice that an input of 000001 produces 1110. However, an input of 111110 also produces 1110.

There has been no public disclosure of why the specific design choices for DES s-boxes were made. As we have already mentioned, resistance to differential cryptanalysis appears to have played a significant role. However, another factor is the nature of these s-boxes as compression boxes. As mentioned earlier in this chapter, it is difficult to design an s-box that uses compression without losing data. In the case of DES, it is only possible because an earlier step in the algorithm expanded bits. At least some of the design choices in DES are related to providing the compression without losing data.

The Rijndael S-Box

Given the prominence of AES, the Rijndael s-box is a good candidate for analysis. In fact, this is probably the most important portion of this chapter. Before we can delve deeply into the s-boxes for Rijndael, we need to look at some of the mathematics behind the s-box design. The actual Rijndael s-box is shown in Fig. 8.13.

```
 | 0   1   2   3   4   5   6   7   8   9   a   b   c   d   e   f
---|--|--|--|--|--|--|--|--|--|--|--|--|--|--|--|--|
00 |63  7c  77  7b  f2  6b  6f  c5  30  01  67  2b  fe  d7  ab  76
10 |ca  82  c9  7d  fa  59  47  f0  ad  d4  a2  af  9c  a4  72  c0
20 |b7  fd  93  26  36  3f  f7  cc  34  a5  e5  f1  71  d8  31  15
30 |04  c7  23  c3  18  96  05  9a  07  12  80  e2  eb  27  b2  75
40 |09  83  2c  1a  1b  6e  5a  a0  52  3b  d6  b3  29  e3  2f  84
50 |53  d1  00  ed  20  fc  b1  5b  6a  cb  be  39  4a  4c  58  cf
60 |d0  ef  aa  fb  43  4d  33  85  45  f9  02  7f  50  3c  9f  a8
70 |51  a3  40  8f  92  9d  38  f5  bc  b6  da  21  10  ff  f3  d2
80 |cd  0c  13  ec  5f  97  44  17  c4  a7  7e  3d  64  5d  19  73
90 |60  81  4f  dc  22  2a  90  88  46  ee  b8  14  de  5e  0b  db
a0 |e0  32  3a  0a  49  06  24  5c  c2  d3  ac  62  91  95  e4  79
b0 |e7  c8  37  6d  8d  d5  4e  a9  6c  56  f4  ea  65  7a  ae  08
c0 |ba  78  25  2e  1c  a6  b4  c6  e8  dd  74  1f  4b  bd  8b  8a
d0 |70  3e  b5  66  48  03  f6  0e  61  35  57  b9  86  c1  1d  9e
e0 |e1  f8  98  11  69  d9  8e  94  9b  1e  87  e9  ce  55  28  df
f0 |8c  a1  89  0d  bf  e6  42  68  41  99  2d  0f  b0  54  bb  16
```

Fig. 8.13 Rijndael s-box

The Irreducible Polynomial

The Rijndael s-box is based on a specific irreducible polynomial in a specific Galois Field (Daemen and Rijmen 1999):

$$\mathrm{GF}\left(2^8\right) = \mathrm{GF}(2)[x]/\left(x^8 + x^4 + x^3 + x + 1\right)$$

In hexadecimal this is 11B, in binary it is 100011011.

What is an irreducible polynomial? An irreducible polynomial that cannot be factored into the product of two other polynomials. In other words, it cannot be reduced. This is in reference to a specific field; in the case of the irreducible polynomial we are considering it is in reference to the Galois Field $\mathrm{GF}(2^8)$. Put more formally: A polynomial is irreducible in $\mathrm{GF}(p)$ if it does not factor over $\mathrm{GF}(p)$. Otherwise it is reducible.

Why was this specific irreducible polynomial chosen? Does it have some special property that makes it more suitable for cryptography? Well to answer that question let us consider the actual words of the inventors of Rijndael "The polynomial m (x) ('11B') for the multiplication in $\mathrm{GF}(2^8)$ is the first one of the list of irreducible polynomials of degree 8" (Daemen and Rijmen 1999). In other words, they looked at a list of irreducible polynomials in a specific text and chose the first one. This is important to keep in mind. Any irreducible polynomial of the appropriate size can be used.

The text that Daemen and Rijmen consulted for their list of irreducible polynomials was "Introduction to finite fields and their applications," Cambridge University Press, 1986. You can check the same source that was cited by the inventors of Rijndael. Here are a few irreducible polynomials from that list (in binary form, you may place them in polynomial or hex form if you wish).

100101011
100111001
100111111
101001101
101011111
101110111
110001011

You may have noticed that all of these, and the one chosen for Rijndael have 9 digits. Why use degree 8 (9 digits) isn't that one too many? "Clearly, the result will be a binary polynomial of degree below 8. Unlike for addition, there is no simple operation at byte level." – page ¾ of the specification.

The reason an irreducible polynomial must be used, instead of just any polynomial (also called a primitive polynomial), is that we are trying to make a non-linear permutation function that has diffusion, spreading input bits to output bits in a non-linear way.

Multiplicative Inverse

In mathematics, the reciprocal, or multiplicative inverse, of a number x is the number which, when multiplied by x, yields 1. The multiplicative inverse for the real numbers, for example, is $1/x$. To avoid confusion by writing the inverse using set-specific notation, it is generally written as x^{-1}.

Multiplication in Galois Field, however, requires more tedious work. Suppose $f(p)$ and $g(p)$ are polynomials in $gf(pn)$ and let $m(p)$ be an irreducible polynomial (or a polynomial that cannot be factored) of degree at least n in $gf(pn)$. We want $m(p)$ to be a polynomial of degree at least n so that the product of two $f(p)$ and $g(p)$ does not exceed $11111111 = 255$ as the product needs to be stored as a byte. If $h(p)$ denotes the resulting product then.

$$h(p) = (\, f(p) * g(p)) \ (\mathrm{mod}\ m(p))$$

On the other hand, the multiplicative inverse of $f(p)$ is given by $a(p)$ such that

$$(\, f(p) * a(p)) \ (\mathrm{mod}\ m(p)) = 1$$

Note that calculating the product of two polynomials and the multiplicative inverse of a polynomial requires both reducing coefficients modulo p and reducing polynomials modulo $m(p)$. The reduced polynomial can be calculated easily with long division while the best way to compute the multiplicative inverse is by using Extended Euclidean Algorithm. The details on the calculations in $gf(2^8)$ are best explained in the following example.

Finite field multiplication is more difficult than addition and is achieved by multiplying the polynomials for the two elements concerned and collecting like powers of x in the result. Since each polynomial can have powers of x up to 7, the result can have powers of x up to 14 and will no longer fit within a single byte. This situation is handled by replacing the result with the remainder polynomial after division by a special eighth order irreducible polynomial, which, as you may recall for Rijndael, is:

$$m(x) = x8 + x4 + x3 + x + 1$$

The finite field element (00000010) is the polynomial x, which means that multiplying another element by this value increases all its powers of x by 1. This is equivalent to shifting its byte representation up by 1 bit so that the bit at position i moves to position $i + 1$. If the top bit is set prior to this move, it will overflow to create an $x8$ term, in which case the modular polynomial is added to cancel this additional bit, leaving a result that fits within a single byte.

For example, multiplying (11001000) by x, that is (00000010), the initial result is 1(10010000). The "overflow" bit is then removed by adding 1(00011011), the modular polynomial, using an exclusive-or operation to give a final result of (10001011). However, you need not calculate the multiplicative inverse manually, the table in 8–14 provides multiplicative inverses (Fig. 8.14).

Affine Transformation

This concept originates in graphics and is also used in transforming graphics. Moving pixels in one direction or another is very similar to moving a value in a matrix, so the concept gets applied to matrices (as in AES). In geometry, an affine transformation or affine map or an affinity (from the Latin, affinis, "connected with") between two vector spaces (strictly speaking, two affine spaces) consists of a linear transformation followed by a translation:

In general, an affine transform is composed of linear transformations (rotation, scaling, or shear) and a translation (or "shift"). Several linear transformations can be combined into a single one, so that the general formula given above is still applicable. For our purposes, it is just a word for a linear transformation.

	0	1	2	3	4	5	6	7	8	9	a	b	c	d	e	f
00	--	01	8d	f6	cb	52	7b	d1	e8	4f	29	c0	b0	e1	e5	c7
10	74	b4	aa	4b	99	2b	60	5f	58	3f	fd	cc	ff	40	ee	b2
20	3a	6e	5a	f1	55	4d	a8	c9	c1	0a	98	15	30	44	a2	c2
30	2c	45	92	6c	f3	39	66	42	f2	35	20	6f	77	bb	59	19
40	1d	fe	37	67	2d	31	f5	69	a7	64	ab	13	54	25	e9	09
50	ed	5c	05	ca	4c	24	87	bf	18	3e	22	f0	51	ec	61	17
60	16	5e	af	d3	49	a6	36	43	f4	47	91	df	33	93	21	3b
70	79	b7	97	85	10	b5	ba	3c	b6	70	d0	06	a1	fa	81	82
80	83	7e	7f	80	96	73	be	56	9b	9e	95	d9	f7	02	b9	a4
90	de	6a	32	6d	d8	8a	84	72	2a	14	9f	88	f9	dc	89	9a
a0	fb	7c	2e	c3	8f	b8	65	48	26	c8	12	4a	ce	e7	d2	62
b0	0c	e0	1f	ef	11	75	78	71	a5	8e	76	3d	bd	bc	86	57
c0	0b	28	2f	a3	da	d4	e4	0f	a9	27	53	04	1b	fc	ac	e6
d0	7a	07	ae	63	c5	db	e2	ea	94	8b	c4	d5	9d	f8	90	6b
e0	b1	0d	d6	eb	c6	0e	cf	ad	08	4e	d7	e3	5d	50	1e	b3
f0	5b	23	38	34	68	46	03	8c	dd	9c	7d	a0	cd	1a	41	1c

Fig. 8.14 Multiplicative inverses

Generating the S-Box

The Rijndael s-box can be generated with a series of shift operations, in fact, this is exactly how it is usually implemented in programming. These shifts essentially accomplish the same process as matrix multiplication.

The matrix multiplication can be calculated by the following algorithm:

1. Store the multiplicative inverse of the input number in two 8-bit unsigned temporary variables: s and x.
2. Rotate the value s one bit to the left; if the value of s had a high bit (eighth bit from the right) of one, make the low bit of s one; otherwise the low bit of s is zero.
3. Exclusive or the value of x with the value of s, storing the value in x.
4. For three more iterations, repeat steps two and three; steps two and three are done a total of four times.
5. The value of x will now have the result of the multiplication.

After the matrix multiplication is complete, exclusive or (XOR) the resultant value by the decimal number 99 (the hexadecimal number 0x63, the binary number 1100011, and the bit string 11000110 representing the number in least significant bit first notation). This value is termed the translation vector. This last operation, the final XOR, is meant to prevent any situation wherein the output is the same as the input. In other words, to prevent S-box(a) = a.

An Example

It may be helpful for you to see an actual example generating the s-box.

Take a number for $GF(2^8)$ lets pick 7. Looking at the multiplicative inverse table that gives us d1 in hex or 11010001 in binary.

Now we need to do four iterations of the process of affine transformation. We start by putting the multiplicative inverse into two variables s and x:

$s = 11010001$
$x = 11010001$

Now we simply rotate s(10001101) to the left $= 00011010$

- If the high bit is 1, make the low bit 1
- else low bit is 0.

Now in this case the high bit was 1 so we change the low bit

- thus, s is 00011011.

  xor that number with x so 00011011 xor 10001101 $= 10010110$
  $s = 00011011$; $x = 10010110$

Next rotate s (00011011) to the left $= 00110110$

If this still gives us 00110110
xor with x so 00110110 xor 10010110 $= 10100000$
$s = 00110110$; $x = 10100000$

Next rotate s(00110110) to the left $= 01101100$

if this still gives us 01101100
xor with x so 01101100 xor 10100000 $= 11001100$
$s = 01101100$; $x = 11001100$

Next rotate s(01101100) to the left $= 11011000$

if this still gives us 11011000
xor with x so 01101100 xor 11001100 $= 00010100$
$s = 11011000$; $x = 00010100$

Now x (00010100) gets xor'd with decimal 99 (hex x63 binary 1100011) $= 1110111$ or 77. Remember the output of the matrix multiplication (which we have accomplished via shift operations) is finally xor'd with the translation vector (decimal 99). This process allows one to create the Rijndael s-box, and it is in fact how that s-box is often created in code.

Changing the Rijndael S-Box

After studying this previous section, you should realize that there are three factors in generating the AES s-box. Those are the selection of the irreducible polynomial, in this case it was $P = x8 + x4 + x3 + x + 1$, which is 11B in hexadecimal notation, or 100011011 in binary numbers. As we mentioned previously, the creators of the Rijndael cipher stated clearly that this number was chosen simply because it was the first one of the list of irreducible polynomials of degree 8 in the reference book they chose (Daemen and Rijmen 1999). That means that one could choose other irreducible polynomials.

There are a total of 30 irreducible polynomials of degree 8 to choose from. This gives you 29 alternatives to the traditional s-box for AES, each with well-tested security. For more details on this alternative, you can look into Rabin's test for irreducibility. Das, Sanjoy, Subhrapratim, and Subhash demonstrated equally secure variation of the Rijndael, by changing the chosen irreducible polynomial. You can use any of the 30 possible irreducible polynomials, each of these is equally secure to the original Rijndael cipher s-box.

Altering the Rijndael s-box is only practical if you have the ability to ensure that all parties to encrypted communication will be using your modified s-box. If you simply modify the s-box on your end, then you would render communication with other parties impossible. Even though those other parties will have the same key, and the same algorithm (AES), they will be using standard s-boxes. This is why altering AES s-boxes is primarily an issue for government entities who want to have secure communication with a limited number of involved parties.

A second option, one that may be the simplest to implement, is to change the translation vector (the final number you xor with). Obviously, there are 255 possible variations. Rather than utilize 0x63, use any of the other possible variations for that final byte. While simple to implement, it may be more difficult to test. Some variations might adversely affect one of the three criteria we are attempting to maintain. In fact, selecting the wrong translation vector may lead to no change at all when applied to the product of the preceding matrix multiplication.

The third method is to change the affine transform. This can be more difficult to implement but safe if you simply alter parameters within the existing transform. Section 5.2 of Sinha and Arya paper discusses this in detail. According to Cui, Huang, Zhong, Chang, and Yang, the choice of affine transformation matrix or irreducible polynomial has no significant impact on the security of the resultant ciphertext.

Conclusions

The s-box (or in some cases s-boxes) is a critical part of most block ciphers. To fully understand any block cipher, it is necessary to have some understanding of s-box design. S-boxes fall into two main categories: substitution and permutation. Either of these can be divided into three subcategories: straight, compression, and expansion.

In addition to understanding s-boxes in general, the Rijndael s-box warrants particular attention. In addition to the standard Rijndael s-box commonly used, there are three relatively simple methods to alter the s-box used in the Rijndael cipher. This will lead to permutations of Rijndael that are equal, or very nearly equal, in security to the original Rijndael cipher. Using such permutations will essentially lead to a private version of the AES algorithm. This can be useful for certain governmental and military applications.

If you wish to delve into s-box design beyond the scope of this chapter, unfortunately there are not books on s-box design. As previously mentioned, many cryptography texts avoid this topic altogether, or provide only a cursory coverage. There are, however, a number of papers you may find useful in the reference section of this chapter.

Test Your Knowledge

1. In mathematics, the _____ of a number x is the number which, when multiplied by x, yields 1.
2. What is the irreducible polynomial used in standard AES?
3. What is the Strict Avalanche Criteria in s-box design?
4. What is the Bit independence Criteria in s-box design?
5. What is the value of the Rijndael translation vector?
6. How many irreducible polynomials are there for the generation of the Rijndael s-box.
7. What is the primary advantage that DES s-boxes convey on the DES cipher?
8. A _____ provides the source of non-linearity for a given cryptographic algorithm.
9. An s-box that transposes bits is called a _____ box.
10. What are the two concerns with using a compression s-box?

References

Adams, C. & Taveres, S. (2001). Good S-Boxes Are Easy To Find. Advances in Cryptology — CRYPTO' 89 Proceedings Lecture Notes in Computer Science, 435 pp. 612–615

Daemen, J., & Rijmen, V. (1999, March). The Rijndael block cipher: AES proposal. In First candidate conference (AeS1) (pp. 343–348).

De Meyer, L., & Vaudenay, S. (2017). DES S-box generator. Cryptologia, 41(2), 153-171.

Easttom, C. (2018a). A Generalized Methodology for Designing Non-Linear Elements in Symmetric Cryptographic Primitives. In Computing and Communication Workshop and Conference (CCWC), 2018 IEEE 8th Annual. IEEE.

Easttom, C. (2018b). An Examination of Inefficiencies in Key Dependent Variations of the Rijndael S-Box. In Electrical Engineering (ICEE), Iranian Conference on (pp. 1658–1663). IEEE

Grocholewska-Czuryło, A. (2011). Cryptographic properties of modified AES-like S-boxes. Annales Universitatis Mariae Curie-Skłodowska, sectio AI–Informatica, 11(2).

Kazlauskas, K., & Kazlauskas, J. (2009). Key-dependent S-box generation in AES block cipher system. Informatica, 20(1), 23–34.

Chapter 9
Cryptographic Hashes

Abstract Cryptographic hashes are utilized for many purposes, the most common is data integrity. The nature of a hash makes it an excellent tool for verifying message integrity. Hashes are also utilized to secure passwords. Hashes are algorithms that are one-way, have a fixed length output, and are collision resistant. This chapter provides an overview of how hashes are used, alternatives such as message authentication codes (MAC) and hash message authentication codes (HMAC). Specific algorithms will also be examined in enough detail for you to gain a working knowledge of cryptographic hashes.

Introduction

In the preceding eight chapters, we examined cryptography as a means of keeping information confidential. Cryptographic hashes are not concerned with protecting confidentiality, instead they are concerned with ensuring integrity. Computer security has a handful of bedrock concepts upon which it is built. The CIA triangle, consisting of Confidentiality, Integrity, and Availability is one such concept. Different technologies support different aspects of the CIA triangle. For example, backups, disaster recovery plans, and redundant equipment support availability. Certainly, encrypting a message so others cannot read it supports confidentiality. But cryptographic hashes are about supporting integrity.

You should be aware that as with previous chapters, some algorithms will be discussed in depth, others will simply be briefly described. The goal is not necessarily for you to memorize every major cryptographic hashing function, but rather to gain a general familiarity with cryptographic hashes. It is also important for you to understand how cryptographic hashing algorithms are used to support security.

In This Chapter, We Will Cover

What is a cryptographic hash?
Specific algorithms
Message authentication codes (MACs)
Applications of cryptographic hashes

What Is a Cryptographic Hash?

A cryptographic hash is a special type of algorithm. William Stallings describes a hash as follows:

1. H can be applied to a block of data of variable size.
2. H produces a fixed-length output.
3. H(X)is relatively easy to compute for any given x, making both hardware and software implementations practical. X is whatever you input into the hash.
4. For any given value h, it is computationally infeasible to find x such that
   H(x) = h. This is sometimes referred to in the literature as the one-way property.
5. For any given block x, it is computationally infeasible to find y ! = x such that
   H(y) = H(x). This is sometimes referred to as weak collision resistance.
6. It is computationally infeasible to find any pair x, y such that H(x) = H(y)
   This is sometimes referred to as strong collision resistance.

This is a very accurate definition but may be a bit technical for the novice. Allow me to explain the properties of a cryptographic hash in a manner that is a bit less technical, but no less true. In order to be a cryptographic hash function, an algorithm needs to have three properties. The first property is that the function is one way. That means it cannot be "unhashed." Now this may seem a bit odd at first. An algorithm that is not reversible? Not simply that it is difficult to reverse, but that it is literally impossible to reverse. Yes, that is exactly what I mean. Much like trying to take a scrambled egg and unscramble it and put it back in the eggshell, it is just not possible. When we examine specific hashing algorithms later in this chapter, the reason why a cryptographic hash is irreversible should become very clear.

The second property that any cryptographic has must have is that a variable length input produces a fixed length output. That means that no matter what size of input you have, you will get the same size output. Each particular cryptographic hash algorithm has a specific size output. For example, SHA-1 produces a 160-bit hash. It does not matter whether you input 1 byte or 1 terabyte, you get out 160 bits.

How do you get fixed length output regardless of the size of the input? Different algorithms will each use their own specific approach, but in general it involves compressing all the data into a block of a specific size. If the input is smaller than the block, then pad it. Consider the following example. This particular example is trivial and for demonstrative purposes only. It would not suffice as a secure cryptographic hash. We will call this trivial hashing algorithm or THA:

Step 1: if the input is less than 64 bits, then pad it with zeros until you achieve 64 bits. If it is greater than 64 bits, then divide it into 64-bit segments. Make sure the last segment is exactly 64 bits, even if you need to pad it with zeros.

Step 2. Divide each 64-bit block into two halves.

Step 3: XOR the left have of each block with the right half of the block.

Step 4: If there is more than one block, start at the first block XORing it with the next block. Continue this until you get to the last block. The output from the final XOR operation is your hash. If you had only one block, then take the result of XORing the left half with the right half and that is your hash.

Now I cannot stress enough this would not be a secure hashing algorithm. In fact, it likely would not be referred to as a cryptographic hashing algorithm. It is very easy to envision collisions occurring in this scenario, and quite easily in fact. However, this does illustrate a rather primitive way in which the input text can be condensed (or padded) to reach a specific size. Actuall hashing algorithms that we will explore later in this chapter are more complex. However this trivial psuedo-hash should give you a feel for the process.

Finally, the algorithm must be collision resistant. But what precisely does that mean? A collision occurs if two different inputs produce the same output. If you use SHA-1, then you have a 160-bit output. That means 2^{160} possible outputs. Clearly, you could have trillions of different inputs and never see a collision. It should be noted that the size of the output (also called a digest or message digest) is only one factor in collision resistance. The nature of the algorithm itself also has an impact on collision resistance.

How Are Cryptographic Hashes Used?

There are many ways that a cryptographic hash can be used. In this section, you will see some of the most common ways they are used. Each of these will depend on one or more of the three key properties of a cryptographic hash that we discussed in the previous section. This section should help you understand the importance of cryptographic hashes in computer security.

Message Integrity

One common use of hashing algorithms is in ensuring integrity of messages (Easttom 2019). It should be fairly obvious that messages can be altered in transit, either intentionally or accidentally. Hashing algorithms can be used to detect that such an alteration has occurred. Consider the simple example of an email message. If you put the body of the message into a hashing algorithm, let's just say SHA-1, the output is a 160-bit hash. That hash can be appended at the end of the message.

When the message is received, the recipient can re-calculate the cryptographic hash of the message and compare that result to the hash that was attached to the message. If the two do not match exactly, this indicates that there has been some alteration in the message and the message contents are no longer reliable.

Cryptographic hashes are also used in file integrity systems. For example, the very popular TripWire product (both the open source Linux version and the Windows version) creates a cryptographic hash of key files (as designated by the TripWire administrator). At any time, a hash of the current file can be compared to the previously computed cryptographic hash to determine if there has been any change in the file. This can detect anything from a simple edit of a file such as a spreadsheet to an executable that has been infected with a Trojan Horse.

It is likely that many readers already know what a Trojan Horse is, but in case you don't it is a program or file that has had malware attached to it. Often wrapper programs are used to tie a virus or spyware to a legitimate program. When the user executes the legitimate program, he or she does not realize that they also just launched the malware.

Password Storage

Cryptographic hashes also provide a level of security against insider threats. Consider the possibility that someone with access to a system, for example, a network administrator, has ill intent. Such a person might simply read a user's password from the database, then use that user's login credentials to accomplish some attack on the system. Then, should the attack become known, it is the end user who will be a suspect, not the administrator who actually perpetrated the breach. One way to avoid this is to store passwords in a cryptographic hash. When the user logs into the system, whatever password they typed in is hashed, then compared to the hash in the database. If it matches exactly, then the user is logged into the system.

Given that the database only stores a hash of the password, and hashes are not reversible, even a network administrator or database administrator cannot retrieve the password from the database. If someone attempted to type in the hash as a password, the system will hash whatever input is placed into the password field, thus yielding a different hash than is stored in the database. The storing of passwords as a hash is widely used and strongly recommended.

Hashing is in fact how Windows stores passwords. For example, if your password is "password," then Windows will first hash it producing something like:

8846F7EAEE8FB117AD06BDD830B7586C

Then store that in the SAM (Security Accounts Manager) file in the Windows System directory (Easttom 2017). When you log on, Windows cannot "un hash" your password. What Windows does is take whatever password you type in, hash it, then compare that result with what is in the SAM file. If they match (exactly), then you can login.

Password	NTLM Hash
password	8846F7EAEE8FB117AD06BDD830B7586C
ab3tt34p@ssw0rd	375E832BD79801DA4CE32A1458F237A6
anotherpassword	69AA7AD81335466852B12532895FA9A9
pa$$w0rd0!	03EDECCBCCA5A08A6581804ED776067D
january10	AA8E009EC11ED01079FACD6F9372D23D

Fig. 9.1 Rainbow table example

It is worth noting here, that there are methods for circumventing this security and retrieving passwords. A rainbow table is one such mechanism. A rainbow table is a table of pre-computed hashes. Windows uses the NTLMv2 hashing algorithm to store passwords. Imagine you make a table of all common 8-character passwords in one column, and the NTLMv2 hash of them in the second column. Then you repeat this for all common 9-character passwords. Then for all 10-character passwords. You can take this as far as you like, and your computing resources will support. Then if you can extract a hash from the target machine, you search the tables for a match. If you find a match in the second column, whatever is in the first column must be that person's password. It will look something like what is shown in Fig. 9.1.

If an attacker is able to get the windows SAM file, then he or she can take the hashes and search the rainbow table seeking a match. There are even rainbow tables available online:

https://crackstation.net/
http://rainbowtables.it64.com/

However, there are ways for the defender to fight back against rainbow table attacks. The most common method is to use salt. Salt is a term for bits added to the text, before it is hashed. This is transparent to the user, happens without his or her knowledge. Basically, what occurs is when the user first selects their password, something is added to it, then that value is hashed. As an example, if the user selects the password "password" the system first adds something to it, perhaps digits 11, then hashes "password11". The next time the user logs in, he or she types in "password", but the system knows to add the digits 11 before hashing the input and comparing to the stored hash. The addition of salt means that if an attacker utilizes a rainbow table, what he or she gets back won't be your actual password.

Forensic Integrity

When conducting a forensic examination of a hard drive, one of the first steps is to create an image of the drive and perform the analysis on the image. After creating an image, it is necessary to verify that the imaging process was accurate. You do this by creating a hash of the original and a hash of the image and comparing the two (Easttom 2017). This is where the "variable length input produces fixed length output" and "few or no collisions" comes in. If you have a terabyte drive and you image it, you want to make sure your image is an exact copy. The fact that the hash is only a fixed size, such as 160-bit for SHA-1, is very useful. It would be quite unwieldy if the hash itself was the size of the input. Comparing terabyte size hashes would be problematic at best.

If you make a hash of the original and the image, and compare them, you can verify that everything is exactly copied into the image. If it was not the case that a variable length input produced a fixed length output, then you would have hashes that were humongous and took forever to compute. Also, if two different inputs could produce the same output, then you could not verify the image was an exact copy.

Merkle-Damgard

Before we delve into specific hashing algorithms, it is worthwhile to examine a function that is key to many commonly used cryptographic hashes. A Merkle-Damgard function (also called a Merkle-Damgard construction) is a method for building hash functions (Backes et al. 2012). Merkle-Damgard functions form the basis for MD5, SHA1, SHA2, and other hashing algorithms. This function was first described in Ralph Merkle's doctoral dissertation published in 1979.

The function starts by applying some padding function to create an output that is of some particular size (256-bits, 512-bits, 1024-bits, etc.) (Backes et al. 2012). The specific size will vary from one algorithm to another, but 512-bits is a common size that many algorithms use. The function then processes blocks one at a time, combining the new block of input with the block from the previous round (recall our trivial hashing algorithm that takes output from one block and combines it with the next). Put another way if you have a 1024-bit message that you break into four 256-bit blocks. Block 1 will be processed, then its output is combined with block 2 before block 2 is processed. Then that output is combined with block 3 before it is processed. And finally, that output is combined with block 4 before that block is processed. Thus, Merkle-Damgard is often referred to as a compression function as it compresses all the message into a single output block. The algorithm will start with some initial value, or initialization vector that is specific to the implementation. The final message block is always padded to the appropriate size (256-bits, 512-bits, etc.) and includes a 64-bit integer that indicates the size of the original message.

Specific Algorithms

There are a variety of specific algorithms utilized to create cryptographic hashes. Remember that to be a viable cryptographic hash, an algorithm needs three properties:

1. Variable length input produces fixed length output.
2. Few or no collisions.
3. It is not reversible.

There are a variety of cryptographic hashing functions that meet these three criteria. In the United States, as well as much of the world, the MD5 and SHA algorithms are the most widely used, and thus will be prominently covered in this chapter.

Checksums

A much simpler algorithm than a cryptographic hash is a checksum. It can serve similar purposes, for example, checking for message integrity. The word checksum is often used as a generic term for the actual output of a specific algorithm. The algorithms are often called a checksum function or checksum algorithm.

Longitudinal Parity Check

One of the simplest checksum algorithms is the longitudinal parity check. It breaks data into segments (called words) with a fixed number of bits. Then the algorithm computes the XOR of all of these words, with the final result being a single word or checksum. Let us assume a text that says.

"Euler was a genius"

Then convert that to binary (first converting to ASCII codes, then converting that to Binary):

01000101 01110101 01101100 01100101 01110010 00100000 01110111
  01100001 01110011 00100000 01100001 00100000 01100111 01100101
  01101110 01101001 01110101 01110011.

The segments can be any size but let us assume a 2 byte (16-bit word). This text is now divided into nine words (shown here separated with brackets).

[01000101 01110101][01101100 01100101][01110010 00100000][01110111 01100001][01110011 00100000][01100001 00100000][01100111 01100101] [01101110 01101001][01110101 01110011].

The next step is to XOR the first word with the second:

01000101 01110101 XOR 01101100 01100101 = 0010100100010000.

Then that result is XOR'd with the next word.

0010100100010000 XOR 00100000 01110111 = 0000100101100111.

This process is continued with the result of the previous XOR, then XOR'd with the next word until a result is achieved. That result is called the longitudinal parity check. This type of checksum (as well as others) works well for error detection. The checksum of a message can be appended to a message. The recipient then recalculates the checksum and compares it. Checksums are usually quite fast, much faster than cryptographic hashes. However, there exists possibility of collisions (different inputs producing identical outputs).

Fletcher Checksum

The Fletcher checksum is an algorithm for computing a position-dependent checksum devised by John G. Fletcher at Lawrence Livermore Labs in the late 1970s. The objective of the Fletcher checksum was to provide error-detection properties approaching those of a cyclic redundancy check but with the lower computational effort associated with summation techniques. The Fletcher checksum works by first dividing the input into words and computing the modular sum of those words. Any modulus can be used for computing the modular sum. The checksum is appended to the original message. The recipient recalculates the checksum and compares it to the original.

MD5

This is a 128-bit hash that is specified by RFC 1321. It was designed by Ron Rivest in 1991 to replace an earlier hash function, MD4. MD 5 produces a 128-bit hash or digest (Easttom 2019). It has been found to not be as collision resistant as SHA. As you might suspect, MD5 is an improvement on an earlier algorithm, MD4. As early as 1996, a flaw as found in MD5 and by 2004, it was shown that MD5 was not collision resistant.

Keep in mind that the MD5 algorithm will ultimately produce a 128-bit output. The first step in the algorithm is to break the message to be encrypted into 512-bit segments. Each of those segments consists of sixteen 32-bit words. If the last segment is less than 512 bits, it is padded to reach 512 bits. The final 64 bits represents the length of the original message (Rivest 1992).

Since MD5 ultimately seeks to create a 128-bit output, it works on a 128-bit state that consists of four 32-bit words labeled A, B, C, and D. These 32-bit words are initialized to some fixed constant. The algorithm has four rounds consisting of a number of rotations and binary operations. There is a different round function for each round. Note that since MD5 is a Merkle-Damgard function, the following steps are done to each 512-bit block. If the message is more than 512 bits, then final output of a given block is then combined with the next block until the last block is reached (Rivest 1992).

Round 1: B is XORd with C. The output of that operation is OR'd with the output of the negation of B XORd with D.

Round 2: B is XORd with D. The output of that operation is OR'd with the output of C XORd with the negation of D.

Round 3: B is XORd with C, the output is XORd with D.

Round 4: C is XORd with the output of B ORd with the negation of D.

These round functions are often denoted as F, G, H, and I, with F being round 1, G being round 2, H being round 3, and I being round 4.

The algorithm is illustrated in Fig. 9.2.

The blocks A–D have already been explained. The F in Fig. 9.2 represents the round function for that given round (F, G, H, or I depending on the round in question). The message word M_i and K_i is a constant. The "<<<" denotes a binary left shift by s bits. After all rounds are done, the A, B, C, and D words contain the digest, or hash of the original message.

Fig. 9.2 The MD5 algorithm

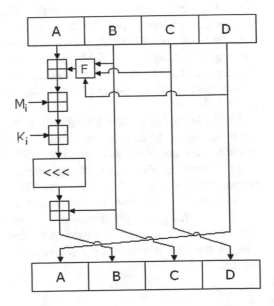

It should be noted that MD6 was published in 2008. It was submitted to the NIST SHA-3 competition but had speed issues and was not quite ready. In September 2011, a paper was publishes showing an improved MD6 was resistant to several common cryptanalytic attacks.

SHA

The Secure Hash Algorithm is perhaps the most widely used hash algorithm today. There are now several versions of SHA. SHA (all versions) is considered secure and collision free.

SHA-1: This is a 160-bit hash function, which resembles the earlier MD5 algorithm. This was designed by the National Security Agency (NSA) to be part of the Digital Signature Algorithm.

SHA-2: This is actually two similar hash functions, with different block sizes, known as SHA-256 and SHA-512. They differ in the word size; SHA-256 uses 32-byte (256 bits) words where SHA-512 uses 64-byte(512 bits) words. There are also truncated versions of each standardized, known as SHA-224 and SHA-384. These were also designed by the NSA.

SHA-3: This is the latest version of SHA. It was adopted in October of 2012.

SHA-1

Let us begin by taking a closer look at the SHA-1 algorithm. Much like MD5 there will be some padding. And like MD5, SHA-1 uses 512-bit blocks. The final block needs to be 448 bits followed by a 64-bit unsigned integer that represents the size of the original message. Since SHA-1 is also a Merkle-Damgard function, the following steps are done to each 512-bit block. If the message is more than 512 bits, then final output of a given block is then combined with the next block until the last block is reached.

Unlike MD5, SHA1 uses 5 blocks often denoted as h1, h2, h3, h4, and h5. These are initialized to some constant value. The message to be hashed is broken into characters and converted to binary format. Pad the message so it evenly breaks into 512-bit segments with the 64-bit original message length at the end. Each 512-bit segment is broken into 16 32-bit words. The first step after preparing the message is a process that will create 80 words from the 16 in the block. This is a loop that will be executed until a given condition is true. You start by taking the 16 words in the 512-bit block and putting those into the first 16 words of the 80-word array. The other words are generated by XORing previous words and shifting one bit to the left.

Fig. 9.3 The SHA1
algorithm

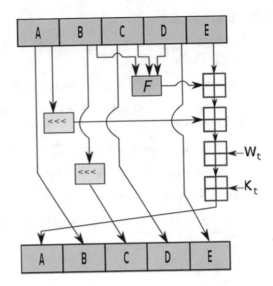

Then 80 times you loop through a process that

1. Calculates the SHA function and a constant K both based on the current round.
2. Set word $e = d$; word $d = c$; word $c = b$ after rotating b left 30 bits; word $b = 1$; word a = word A rotated left 5 bits + the SHA function+ $e + k$ + word $[i]$, where i is the current round.
3. Concatenate the final hash as a, b, c, d, and e. The reason we have 5 final words (each 32 bit) whereas MD5 only had 4 is that MD5 produces a 128-bit digest, SHA-1 produces a 160-bit digest.

You can see this process depicted in Fig. 9.3.

Notice the round function in Fig. 9.3. Like MD5, SHA-1 has a varying round function that changes slightly each round:

Variation 1: f(t;B,C,D) = (B AND C) OR ((NOT B) AND D) (0 <= t <= 19)
Variation 2: f(t;B,C,D) = B XOR C XOR D (20 <= t <= 39)
Variation 3: f(t;B,C,D) = (B AND C) OR (B AND D) OR (C AND D) (40 <= t <=59)
Variation 4: f(t;B,C,D) = B XOR C XOR D (60 <= t <= 79)

At this point, you may find both MD5 and SHA 1 to be a bit convoluted. That is to be expected. Remember the goal is to have a non-reversible algorithm. Thus, there are a number of convoluted steps designed simply to scramble the message so that it cannot be unscrambled.

If you have a programming background, or are simply familiar with algorithm analysis, you may find the following pseudo code helpful in understanding SHA-1.

For loop on k = 1 to L

(W(0),W(1),...,W(15)) = M[k] /* *Divide M[k] into 16 words* */
For t = 16 to 79 do:

W(t) = (W(t-3) XOR W(t-8) XOR W(t-14) XOR W(t-16)) <<< 1

A = H0, B = H1, C = H2, D = H3, E = H4
For t = 0 to 79 do:

TEMP = A<<<5 + f(t;B,C,D) + E + W(t) + K(t) E = D, D = C,

C = B<<<30, B = A, A = TEMP

End of for loop

H0 = H0 + A, H1 = H1 + B, H2 = H2 + C, H3 = H3 + D, H4 = H4 + E
End of for loop

SHA-2

SHA-2 is similar in structure and function to SHA-1, it is also a Merkle-Damgard construction. However, it has a variety of sizes. The most common are SHA-256 and SHA-512 but there is also SHA-224 and SHA-384.

SHA-256 uses 32-bit words and SHA-512 uses 64-bit words. In the United states, FIPS PUB 180-2 defines SHA-2 as a standard.

SHA-1 is one of the most widely used hashing algorithms in existence today. SHA-2 is widely used in PGP, SSL, TLS, SSH, IPSec, and other security protocols. Certain Linux distributions, such as Debian, use SHA-2 variations to authenticate software packages. Some implementations of Bitcoin use SHA-2 to verify transactions.

SHA-3

SHA-3 is an interesting algorithm in that it was not designed to replace SHA-2. There are no known significant flaws with SHA-2. There are issues with MD-5 and at least theoretical issues with SHA-1. SHA-3 was the result of a contest to find a new hashing algorithm. The actual algorithm was named Keccak designed by Guido Bertoni, Joan Daemen (of Rijndael cipher fame), Michaël Peeters, and Gilles Van Assche (Bertoni et al. 2013). The United States NIST (National Institute of Standards) published FIPS 202 standardizing the use of SHA-3.

Unlike SHA-1 and SHA-2 that use a Merkle-Damgard construction, SHA-3 uses a sponge construction (Dworkin 2015). A sponge construction (also known as a sponge function) is a type of algorithm. It uses an internal state and takes input of any size producing a specific sized output. This makes it a good fit for cryptographic hashes which need to take variable length input and produce a fixed length output.

Any sponge construction uses three components. The first is a state, usually called S that is of some fixed size. Then there is some function f that transforms the state, and finally a padding function to ensure the output is of the proper size.

The state is usually divided into two sections denoted R and C. The R represents the size bitrate and the C is the size of the capacity. The bitrate is simply the base size for the algorithm. Just as Merkle-Damgard constructions might use 512-bit blocks, Sponge constructions have a certain base size and the entire input must be a multiple of that size. The padding function is to ensure that the input is multiple of R.

The Sponge function consists of iterations of the following steps (Bertoni et al. 2013):

The state (S) is initialized to zero.

The input string is padded using the padding function.

R is then XOR'd with the first r-bit block of padded input. Remember R is the bit size being used as a base for this implementation.
S is replaced by $F(S)$. The function F is specific to a particular Sponge function.
R is then XOR'd with the next r-bit block of padded input.

This last two steps (s being replaced by $F(s)$ and R being XOR'd with the next block of input) continue until the end of the input. Once the end of input has been reached, it is time to produce the output. That is done using the following steps:

The R portion of the state memory is the first r bits of output.
If more output bits are desired, S is replaced by $F(S)$.
The R portion of the state memory is the next r bits of output.

There are variations on this process, and what has been presented here is simply a generic overview of the essentials of a Sponge construction. The details of the function F will vary with specific implementations. In SHA-3, the state consists of a 5×5 array of 64-bit words, 1600 bits total. The authors claim 12.5 cycles per byte on an Intel Core 2 CPU. In hardware implementations, it was notably faster than all other finalists. The permutations of SHA-3 that were approved by the NIST include 224, 256, 384, or 512-bit outputs.

RipeMD

RACE Integrity Primitives Evaluation Message Digest is a 160-bit hash algorithm developed by Hans Dobbertin, Antoon Bosselaers and Bart Preneel. There exist 128, 256 and 320-bit versions of this algorithm, called RIPEMD-128, RIPEMD-256, and RIPEMD-320, respectively. These all replace the original RIPEMD which was found to have collision issues. The larger bit sizes make this far more secure that MD5 or RIPEMD.

RIPEMD-160 was developed in Europe as part of RIPE project and is recommended by the German Security Agency. The authors of the algorithm describe RipeMD as follows "RIPEMD-160 is a fast-cryptographic hash function that is tuned towards software implementations on 32-bit architectures. It has evolved from the 256-bit extension of MD4, which was introduced in 1990 by Ron Rivest [20, 21]. Its main design feature are two different and independent parallel chains, the result of which are combined at the end of every application of the compression function." (Dobbertin et al. 1996).

The RipeMD-160 algorithm is slower than SHA-1, but perhaps more secure (at least according to some experts). The general algorithm is that it uses two parallel lines of processing. Each consists of 5 rounds and 16 steps. Note that since the 5 rounds are in parallel, some sources describe RipeMD as using 10 rounds. Either description is accurate, depending on your perspective.

As with many other hashing algorithms, RipeMD works with 512-bit segments of input text. It first must pad so that the final message block is 448 bits plus a 64-bit value that is the length of the original message. You may have noticed by now that this is a very common approach to creating hashing algorithms. Now the initial 5-word buffer (each word being 32 bits totally 160 bits) is initialized to a set value. The buffers are labeled A, B, C, D, and E (much as you have seen with previous algorithms). The message being hashed is processed in 512-bit segments. The final output is the last 160 bits left in the buffer when the algorithm has completed all rounds (Dobbertin et al. 1996).

Tiger

Tiger is designed using the Merkle-Damgård construction (sometimes call the Merkle-Damgård paradigm). Tiger produces a 192-bit digest. This cryptographic hash was invented by Ross Anderson and Eli Biham and published in 1995. The one-way compression function operates on 64-bit words, maintaining three words of state and processing eight words of data. There are 24 rounds, using a combination of operation mixing with XOR and addition/subtraction, rotates, and S-box lookups, and a fairly intricate key scheduling algorithm for deriving 24-round keys from the eight input words (Anderson and Biham 1996).

HAVAL

HAVAL is a cryptographic hash function. Unlike MD5, but like most other modern cryptographic hash functions, HAVAL can produce hashes of different lengths. HAVAL can produce hashes in lengths of 128 bits, 160 bits, 192 bits, 224 bits, and 256 bits. HAVAL also allows users to specify the number of rounds (3, 4, or 5) to be used to generate the hash. HAVAL was invented by Yuliang Zheng, Josef Pieprzyk, and Jennifer Seberry in 1992.

NTLM

Windows currently hashes local passwords using the NTLM v2 (New Technology Lan Manager version 2) algorithm. Given the ubiquitous nature of Windows, it would be an oversight not to discuss this hashing algorithm. Windows began with using Lan Manager. It was as simplistic algorithm and not a true cryptographic hash (Easttom 2019). It is described here:

- Limited to 14 characters.
- Input text is converted to upper case before hashing.
- Null padded to 14 bytes (if less than 14).
- Split into two 7-byte halves.
- These are used to create two separate DES Keys.
- Each of these are used to encrypt a constant ASCII string KGS!@#$% yielding 2 8-byte ciphertext strings. These are then concatenated to form a 16-byte LM hash, which is not actually a true hashing algorithm.

Then Microsoft moved on to NT Hash, which uses MD4 on the password (in Unicode), so it is an actual hash function. There is a client server exchange consisting of client sending a NEGOTIATE_MESSAGE, server responses with C HALLENGE _MESSAGE, and client responds with AUTHENTICATE _MES-SAGE. One or both of the two hashed password values that are stored on the server are used to authenticate the client. These two are the LM Hash and the NT Hash.

NTLM v1

Server authenticates client by sending a random 8-byte number as the challenge. The client performs an operation involving the challenge and a secret shared between client and server. The shared secret is one of the hashed passwords discussed in the previous screen The result is a 24-byte value that is sent to the server. The server verifies it is correct.

NTLMv2

The client sends two responses to the 8-byte challenge from the server. Each is a 16-byte HMAC hash of the server challenged and an HMAC of the user's password (the HMAC uses the MD5 hashing algorithm). The second response also includes the current time, an 8-byte random value, and the domain name.

Whirlpool

Whirlpool was invented by Paulo Barreto and Vincent Rijmen, who was one of the creators of the Rijndael cipher, which was chosen to become AES. Paulo Barreto is a Brazilian cryptographer who also worked on the Anubis and KHAZAD ciphers. He has worked on various projects with Vincent Rijmen and published on Elliptic Curve cryptography, which you will see in Chap. 11.

Whirlpool was chosen as part of the NESSIE project (mentioned in Chap. 7) and has been adopted by the International Organization for Standardization (ISO) as a standard. Whirlpool is interesting because it is a modified block cipher (Kitsos and Koufopavlou 2004). It is a common practice to create a hashing algorithm from a block cipher. In this case, the Square cipher (precursor to Rijndael) was used as the basis. Whirlpool produces a 512-bit digest and can take in inputs up to 2^{256} bits in size. The algorithm is no patented and can be used free of charge.

Skein

Skein is a cryptographic hash function and one out of five finalists in the NIST hash function competition. Entered as a candidate to become the SHA-3 standard, the successor of SHA-1 and SHA-2, it ultimately lost to NIST hash candidate Keccak. Skein was created by Bruce Schneier, Stefan Lucks, Niels Ferguson, Doug Whiting, Mihir Bellare, Tadayoshi Kohno, Jon Callas, and Jesse Walker. Skein is based on the Threefish block cipher compressed using Unique Block Iteration (UBI) chaining mode while leveraging an optional low-overhead argument-system for flexibility (Ferguson et al. 2010).

FSB

In cryptography, the Fast Syndrome-based hash Functions (FSB) are a family of cryptographic hash functions introduced in 2003 by Daniel Augot, Matthieu Finiasz, and Nicolas Sendrier. Danial Augot is a French researcher known for work in cryptography, as well as algebraic coding theory. Matthieu Finiasz has numerous publications covering a wide variety of cryptographic topics. Nicolas Sendrier is a French researcher with extensive publications in both cryptography and computer security.

FSB is distinctive in that it can at least to a certain extent be proven to be secure. It has been proven that breaking FSB is at least as difficult as solving a certain NP-complete problem known as Regular Syndrome Decoding. What this means, in practical terms, is that FSB is provably secure. You should recall a discussion of NP-complete problems from Chap. 5.

Often times, security and efficiency are conflicting goals. And it is often the case that provably secure algorithms are a bit slower. Therefore, as you might expect, FSB is slower than many traditional hash functions and uses quite a lot of memory. There have been various versions of FSB, one of which was submitted to the SHA-3 competition but was rejected.

Gost

This hash algorithm was initially defined in the Russian national standard GOST R 34.11-94 Information Technology—Cryptographic Information Security—Hash Function. This hash algorithm produces a fixed-length output of 256 bits. The input message is broken up into chunks of 256-bit blocks. If a block is less than 256 bits, then the message is padded by appending as many zeros to it as are required to bring the length of the message up to 256 bits. The remaining bits are filled up with a 256-bit integer arithmetic sum of all previously hashed blocks, and then a 256-bit integer representing the length of the original message, in bits, is produced. It is based on the GOST block cipher.

BLAKE

BLAKE is a cryptographic hash that is based on the ChaCha stream cipher. There are variations on the size digest produced such as BLAKE-256, BLAKE-512. BLAKE2 was announced in 2012, and BLAKE3 in 2020.

Grøstl

Grøstl is a hash function developed by Praveen Gauravaram, Lars Knudsen, Krystian Matusiewicz, Florian Mendel, Christian Rechberger, Martin Schläffer, and Søren. It uses an s-box, in fact the AES s-box. In fact, much of the algorithm is based on the Rijndael cipher.

SWIFFT

SWIFFT is actually a collection of hash functions. They are notable because they are provably secure. That literally means that they have been mathematically proven to be secure. The mathematics of the SWIFFT hashes is a bit more complex, it utilizes a fast Fourier Transform, and uses lattice-based mathematics. That is beyond the scope of this text to delve deeply into. But since the algorithms are provably secure, it is worthwhile to mention them.

MAC and HMAC

As we have previously discussed, hashing algorithms are often used to ensure message integrity. If a message is altered in transit, the recipient can compare the hash they received against the hash they computer and detect the error in transmission. But what about intentional alteration of messages? What happens if someone alters the message intentionally, deletes the original hash, and re-computes a new one? Unfortunately, a simple hashing algorithm cannot account for this scenario.

A Message Authentication Code (or MAC) is one way to detect intentionally alterations in a message. A MAC is also often called a keyed cryptographic hash function. That name should tell you how this works. One way to do this is the HMAC or Hashing Message Authentication Code. Let us assume you are using MD5 to verify message integrity. To detect an intercepting party intentionally altering a message, both the sender and recipient must previously exchange a key of the appropriate size (in this case 128 bits). The sender will hash the message, then XOR that hash with this key. The recipient will hash what they receive, and XOR that computed hash with the key. Then the two hashes are exchanged. Should an intercepting party simply re-compute the hash, they will not have the key to XOR that with (and may not even be aware that it should be XOR'd) and thus the hash the interceptor creates won't match the hash the recipient computes and the interference will be detected (Easttom 2019).

Another common way of accomplishing a MAC is called a CBC-MAC. For this a block cipher is used (any cipher will do) rather than a hash. The algorithm is used in CBC mode. Then only the final block is used for the Message Authentication Code. This can be done with any block cipher you choose. From a practical perspective both MAC's and HMAC's are more secure than hashes. The addition of a shared secret key improves the security. Whenever possible, using a MAC or HMAC is preferred overusing just a hash.

Key Derivation Functions

Related to hashes are key derivation functions. One widely used function is Password-Based Key Derivation Function 2(PBKDF2). It is part of the public key cryptography standards, specifically PKCS #5 v2.0. PBKDF2 uses an HMAC to input a password or pass phrase with a salt value and repeats that process several times to generate derived key (Easttom and Dulaney 2017). This is one way to generate cryptographic keys. The algorithm bcrypt is a password hashing function that is based on the Blowfish cipher. It incorporates sale and is quite resistant to brute force search attacks. There are implementations of bcrypt in many popular programming languages.

Conclusions

In this chapter, you have seen a broad overview of cryptographic hashes. The first important thing you should absolutely commit to memory are the three criteria for an algorithm to be a cryptographic hash. You should also be familiar with the practical applications of cryptographic hashes.

You have also been exposed to several cryptographic hashing algorithms, some in more detail than others. The SHA family of hashing algorithms is the most widely used, and thus is the most important for you to be familiar with. The other algorithms presented in this chapter are widely used and warrant some study. You may even with to take one or more of these other algorithms and study them in depth. You will find a great deal of information, including research papers, on the internet using any standard search tool.

Test Your Knowledge

1. _____ is a message authentication code that depends on a block cipher wherein only the final block is used.
2. The _____ works by fist dividing the input into words and computing the modular sum of those words.
3. A hashing algorithm that uses a modified block cipher and was chosen as part of the NESSIE project? _____.
4. The algorithm ultimately chosen as SHA-3?
5. What are the three properties all cryptographic hash functions must exhibit?
6. What is the compression function used as a basis for MD5, SHA1, and SHA2?
7. _____ was developed in Europe and is recommended by the German Security Agency.
8. Which algorithm discussed in this chapter was published in 1995 and produces a 192-bit digest?
9. _____ was one of the five finalist in the SHA-3 competition (but did not win) and is based on the Threefish block cipher.
10. Which algorithm introduced in this chapter has provable security?

References

Anderson, R., & Biham, E. (1996, February). Tiger: A fast new hash function. In International Workshop on Fast Software Encryption (pp. 89–97). Springer, Berlin, Heidelberg.

Backes, M., Barthe, G., Berg, M., Grégoire, B., Kunz, C., Skoruppa, M., & Béguelin, S. Z. (2012, June). Verified security of merkle-damgård. In 2012 IEEE 25th Computer Security Foundations Symposium (pp. 354–368). IEEE.

Bertoni, G., Daemen, J., Peeters, M., & Van Assche, G. (2013, May). Keccak. In Annual international conference on the theory and applications of cryptographic techniques (pp. 313–314). Springer, Berlin, Heidelberg.

Dobbertin, H., Bosselaers, A., & Preneel, B. (1996, February). RIPEMD-160: A strengthened version of RIPEMD. In International Workshop on Fast Software Encryption (pp. 71–82). Springer, Berlin, Heidelberg.

Dworkin, M. J. (2015). SHA-3 standard: Permutation-based hash and extendable-output functions (No. Federal Inf. Process. Stds.(NIST FIPS)-202).

Ferguson, N., Lucks, S., Schneier, B., Whiting, D., Bellare, M., Kohno, T., . . . & Walker, J. (2010). The Skein hash function family. Submission to NIST (round 3), 7(7.5), 3.

Easttom, C. (2017). System Forensics, Investigation, and Response, 3rd Edition. Burlington Massachusetts: Jones & Bartlett.

Easttom, C. (2019). Computer Security Fundamentals, 4th Edition. New York City, New York: Pearson Press

Easttom, C & Dulaney, E. (2017). CompTIA Security+ Study Guide: SY0-501. Hoboken, New Jersey: Sybex Press.

Kitsos, P., & Koufopavlou, O. (2004, May). Whirlpool hash function: architecture and vlsi implementation. In 2004 IEEE International Symposium on Circuits and Systems (IEEE Cat. No. 04CH37512) (Vol. 2, pp. II-893). IEEE.

Rivest, R. (1992). RFC1321: The MD5 message-digest algorithm.

Chapter 10
Asymmetric Algorithms

Abstract Asymmetric algorithms are essential for modern e-commerce, banking, and secure communications. It is not an overstatement to assert that much of the modern use of the interent would not occur without the security provided by asymmetric algorithms. Asymmetric algorithms overcome the issue of key exchange. Asymmetric algorithms are based on specific problems in number theory that are considered "hard," which generally means cannot be solved in polynomial time. This chapter explores common asymmetric algorithms. Some algorithms, those most widely used, will be covered in detail. Others will be briefly described.

Introduction

In this chapter, we discuss asymmetric cryptography. This includes a discussion of the general concepts and usage, as well as an in-depth discussion of the more common asymmetric algorithms. You will apply the mathematics reviewed in Chaps. 4 and 5. Where necessary, a brief reminder of the mathematical principles is provided. However, if those chapters were new or difficult to you, you may wish to briefly review Chaps. 4 and 5 before proceeding with this chapter. A basic grasp of number theory and algebraic concepts is necessary to follow the algorithms presented in this chapter.

In This Chapter, We Will Cover

What is asymmetric cryptography
RSA
Diffie-Hellman
ElGamal
MQV
Homomorphic encryption
Applications

© The Author(s), under exclusive license to Springer Nature Switzerland AG 2021   225
W. Easttom, *Modern Cryptography*, https://doi.org/10.1007/978-3-030-63115-4_10

What Is Asymmetric Cryptography?

Asymmetric cryptography, as the name suggests, is a form of cryptography wherein one key is used to encrypt a message, and a different (but related) key is used to decrypt. This concept often baffles those new to cryptography and students in network security courses. How can it be that a key used to encrypt will not also decrypt? This will be clearer to you once we examine a few algorithms and you see the actual mathematics involved. For now, set that issue to one side and simply accept that one key encrypts but cannot decrypt the message. Another key is used to decrypt.

The reason asymmetric cryptography is such a powerful concept is because symmetric cryptography (that you studied in Chaps. 6 and 7) has a serious problem. That problem is key exchange. Most cryptography books and papers prefer to use the fictitious characters Alice, Bob, and Eve to explain how asymmetric cryptography works, and I will continue that tradition. This stems from the original paper describing the RSA algorithm.

Let's assume Alice would like to send Bob a message. But Alice is concerned that Eve might eavesdrop (thus her name!) on the communication. Now let us further assume that we don't have asymmetric cryptography, that all you have available to you are the symmetric ciphers that you learned in Chaps. 6 and 7. And assume Bob and Alice do not live in the same location. How can they exchange a key so that they might encrypt messages? Any method you might think of has the very real chance of being compromised, short of a secure/trusted courier manually taking keys between the two parties. If a courier was needed to exchange keys every time secure communication was required, then we would not have online banking, e-commerce, or a host of other useful technologies.

With public key/asymmetric cryptography, Alice will get Bob's public key and use that to encrypt the message she sends to Bob. Now should Eve intercept the message and have access to Bob's public key, it is ok. That key won't decrypt the message. Only Bob's private key will, and this he safeguards. You can see this in Fig. 10.1.

If Bob wishes to respond to Alice, he reverses the process. He gets Alice's public key and encrypts a message to her, which only her private key will decrypt. Thus, asymmetric cryptography solves the problem of key exchange. It does not impede security if literally every person on the planet has both Bob and Alice's public keys. Those keys can only be used to encrypt messages to Bob and Alice (respectively) and cannot decrypt the messages. So as long as Bob and Alice keep their private keys secret, secure communication is achieved with no problems in key exchange.

This basic concept of one key being public and another being private is why asymmetric cryptography is often referred to as public key cryptography. Unfortunately, it is as far as many security courses go with explaining asymmetric cryptography. Of course, we will be delving into the actual algorithms.

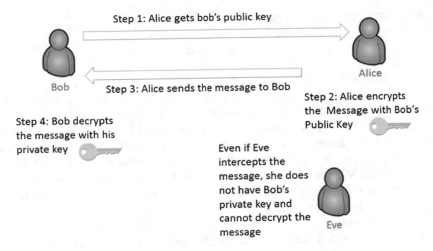

Fig. 10.1 Alice sends Bob a message with asymmetric cryptography

RSA

RSA may be the most widely used asymmetric algorithm. It is certainly one of the most well-known. It was first published in 1977 by Ron Rivest, Adi Shamir, and Leonard Adleman. RSA is perhaps the most widely known asymmetric algorithm. This is a public key method developed in 1977 by three mathematicians, Ron Rivest, Adi Shamir, and Len Adleman (Easttom and Dulaney 2017). The name RSA is derived from the first letter of each mathematician's last name. The algorithm is based on prime numbers and the difficulty of factoring a large number into its prime factors.

The three inventors of RSA are very well known and respected cryptographers. Ron Rivest has been a professor at MIT and invented several algorithms including RC2, RC4, RC5, MD2, MD4, MD5, and MD6. Adi Shamir is an Israeli cryptographer and one of the inventors of differential cryptanalysis. Leonard Adleman has made significant contributions to using DNA as a computational system.

As often happens in the history of cryptography, it turns out that there was a similar system developed earlier, but it was classified. In 1973, Clifford Cocks developed a similar system that remained classified until the late 1990s. Mr. Cocks worked for the English government, specifically the Government Communications Headquarters (GCHQ). He did go on to develop other, non-classified, cryptographic innovations. In 2001, he invented one of the early identity-based encryption methodologies, using aspects of number theory.

First to create the public and private key pair, you start by generating two large random primes, p and q, of approximately equal size (Hinek 2009). You will need to select two numbers so that when multiplied together the product will be the size you want (i.e., 2048 bits, 4096 bits, etc.).

Next, multiply p and q to get n.

$$\text{Let } n = pq$$

The next step is to multiply Euler's Totient for each of these primes.

If you have forgotten about Euler's totient from Chap. 4, this brief reminder should help you. Two numbers are considered co-prime if they have no common factors. For example, if the original number is 8, then 8 and 9 would be co-prime. Factors of 8 are 2 and 4, and 9's factor is 3. Euler asked a few questions about co-prime integers. Given a number X, how many numbers smaller than X are co-prime to X. Well, we call that number Euler's totient, or just the totient. It just so happens that for prime numbers, this is always the number minus 1. For example, 7 has 6 numbers that are co-prime to it.

Well when you multiply two primes together, you get a composite number. And there is no easy way to determine the Euler's totient of a composite number. Recall that Euler found that if you multiple any two prime numbers together, the Euler's totient of that product is the Euler's totient of each prime multiplied together. So, our next step is:

$$\text{Let } m = (p - 1)(q - 1)$$

So, m, is the Euler's totient of n.

Now we are going to select another number, we will call this number e. We want to pick e so that it is co-prime to m. Frequently, a prime number is chosen for e. That way, if e does not evenly divide m, then we are confident that e and m are co-prime, as e does not have any factors to consider. Many RSA implementations use $e = 2^{16} + 1 = 65,537$. This is considered large enough to be effective but small enough to still be fast.

At this point, you have almost completed generating the key. Now we just find a number d that when multiplied by e and modulo m would yield a 1 (note: modulo means to divide two numbers and return the remainder. For example, 8 modulo 3 would be 2). In other words:

$$\text{Find } d, \text{ such that } de\%m \equiv 1$$

Note, the % symbol is used in many programming languages to denote the modulus operator. You can view the % and mod as synonomous. Now you will publish e and n as the public key.

Keep d as the secret key. To encrypt, you simply take your message raised to the e power and modulo n.

$$\text{Ciphertext} = \text{Message}^e \bmod n$$

To decrypt, you take the ciphertext, raise it to the d power modulo n.

$$\text{Plaintext} = \text{Ciphertext}^d \bmod n$$

RSA Example 1

Let's look at an example that might help you understand. Of course, RSA would be done with very large integers. To make the math easy to follow, we will use small integers in this example (note this example is from Wikipedia):

Choose two distinct prime numbers, such as $p = 61$ and $q = 53$.

Compute $n = pq$ giving n $= 61 \cdot 53 = 3233$.

Compute the totient of the product as $\varphi(n) = (p - 1)(q - 1)$ giving $\varphi(3233) = (61 - 1)(53 - 1) = 3120$.

Choose any number $1 < e < 3120$ that is co-prime to 3120. Choosing a prime number for e leaves us only to check that e is not a divisor of 3120. Let $e = 17$.

Compute d, the modular multiplicative inverse of yielding $d = 2753$.

The public key is ($n = 3233$, $e = 17$). For a padded plaintext message m, the encryption function is m^{17} (mod 3233).

The private key is ($n = 3233$, $d = 2753$). For an encrypted ciphertext c, the decryption function is c^{2753} (mod 3233).

RSA Example 2

For those readers new to RSA, or new to cryptography in general, it might be helpful to see one more example, with even smaller numbers.

1. Select primes: $p = 17$ & $q = 11$.
2. Compute $n = pq = 17 \times 11 = 187$.
3. Compute $\phi(n) = (p - 1)(q - 1) = 16 \times 10 = 160$.
4. Select e: $gcd(e, 160) = 1$; choose $e = 7$.
5. Determine d: $de = 1 \mod 160$ and $d < 160$. Value is d $= 23$ since $23 \times 7 = 161 = 10 \times 160 + 1$.
6. Publish public key (7 and 187).
7. Keep secret private key 23.

Now let us apply this to encrypting something. For some reason, you have decided to encrypt the number 3. Thus, we will use the number 3 as the plaintext. Remember $e = 7$, $d = 23$, and n $= 187$. So, we will use the recipients public key which is e and n.

Ciphertext = Plaintext^e mod n or

Ciphertext = 3^7 mod 187

Ciphertext = 2187 mod 187

Ciphertext = 130

When the recipient receives this ciphertext, he or she will use their secret key to decrypt it as follows:

Plaintext = Ciphertext d mod n
Plaintext = 13,023 mod 187
Plaintext = 4.17539054134131116367045797e+48 mod 187
Plaintext = 3

As you can see it works, even if there is not really any particular reason to encrypt the number three. But are we absolutely sure that the public key won't decrypt the message? That is easy to put to the test. Remember the ciphertext is 130. Remember e = 7, d = 23, and n = 187. What if you tried the public key (e,n) again, instead of the private key (d,n)?

Plaintext = Ciphertext e mod n
Plaintext = 1307 mod 187
Plaintext = 627485170000000 mod 187
Plaintext = 37

You absolutely do not get back the original text. There is one exception to this. RSA cannot encrypt any plaintext larger than the modulus. If you do, then the public key will decrypt, as well as encrypt! This is why algorithms such as RSA are not used to encrypt drives and large input items. If you are struggling with the concepts of RSA key generation, I suggest you first work through these two examples. Since the complete process and the answers are provided for you, it will be easy for you to check your work. You may also want to do a few more examples. Start with any two prime numbers that are small enough to make the calculations feasible.

Factoring RSA Keys

You may be thinking, couldn't someone take the public key and use factoring to derive the private key? Well hypothetically yes. However, it turns out that factoring really large numbers into their prime factors is extremely difficult. The more technical description would be to state that it is computationally infeasible. There is no efficient algorithm for doing it. And when we say large numbers, RSA can use 1024, 2048, 4096, 8192 bit, and larger keys. Those are extremely large numbers. Let us look at an example. A 2048-bit number represented in decimal format:

5148324789325478963214778006950135669987541002514563021458614785-
 1483247893254789632147780069501514832478932547896321477800695̄01-
 35669987541002514563021458614785514832478932547896321477800695
 0132566631245886314458770233565889635023235865890014522̄1478
 5336547

In most modern implementations, at least as of this writing, 2048 bits is the smallest RSA key used. Reflect on the rather large number typed out above and contemplate attempting to factor that number. There are mathematical techniques that will improve the process, but nothing that makes factoring such large numbers a feasible endeavor. Of course, should anyone ever invent an efficient algorithm that will factor a large number into its prime factors, RSA would be dead. There certainly have been incidents where someone was able to factor a small RSA key. In 2009, Benjamin Moody factored a 512-bit RSA key in 73 days. In 2010 researchers were able to factor a 768-bit RSA key. Due to these advances in factorization, modern implementations of RSA are using larger key sizes. In Chap. 17, we will examine cryptanalysis techniques used on RSA.

This section is just a basic introduction to RSA. A great resource for delving deeper into RSA is the book *Cryptanalysis of RSA and Its Variants* by Hinek from Chapman and Hall/CRC. You also might it interesting to read the paper "Fast variants of RSA" by Boneh and Shacham. CryptoBytes Vol. 5, No.1.

The Rabin Cryptosystem

This algorithm was created in 1979 by Michael Rabin. Michael Rabin is an Israeli cryptographer and a recipient of the Turing Award. The Rabin cryptosystem can be thought of as an RSA cryptosystem in which the value of e and d are fixed (Hinek 2009).

$$e = 2 \text{ and } d = \tfrac{1}{2}$$

The encryption is $C \equiv P^2 \pmod{n}$ and the decryption is $P \equiv C^{1/2} \pmod{n}$.
Here is a very trivial example to show the idea.
Bob selects $p = 23$ and $q = 7$.
Bob calculates $n = p \times q = 161$.
Bob announces n publicly; he keeps p and q private.
Alice wants to send the plaintext message $M = 24$. Note that 161 and 24 are relatively prime; 24 is in the group selected, $Z_{161}{}^*$.

Encryption: $C = 24^2 \bmod 161 = 93$ and sends the ciphertext 93 to Bob.

This algorithm is not as widely used as RSA or Diffie Hellman but is presented to give you a general overview of alternative asymmetric algorithms.

Diffie–Hellman

While some security textbooks state that RSA was the first asymmetric algorithm, this is not accurate. In fact, Diffie–Hellman was the first publicly described asymmetric algorithm. This is a cryptographic protocol that allows two parties to establish a shared key over an insecure channel. In other words, Diffie–Hellman is often used

Both parties know p and g

Alice

1. Alice generates a
2. Alice's public value is
   g^a mod p
3. Alice computes g^{ab} =
   $(g^b)^a$ mod p,

Bob

1. Bob generates B
2. Bob's public value is g^b
   mod p
3. Bob computes g^{ba} =
   $(g^a)^b$ mod p.

Since $g^{ab} = g^{ba}$ they now have a shared secret key usually called k ($K = g^{ab} = g^{ba}$)

Fig. 10.2 Diffie–Hellman

to allow parties to exchange a symmetric key through some unsecure medium, such as the internet. It was developed by Whitfield Diffie and Martin Hellman in 1976.

As we already saw with RSA, one problem with working in cryptology is that much of the work is classified. You could labor away and create something wonderful...that you cannot tell anyone about. Then to make matters worse, years later someone else might develop something similar and released, getting all the credit. This is exactly the situation with Diffie Hellman. It turns out that a similar method had been developed a few years earlier by Malcolm J. Williamson of the British Intelligence Service, but it was classified (Easttom 2019).

The system has two parameters called p and g (Rescorla 1999). Parameter p is a prime number and parameter g (usually called a generator) is an integer less than p, with the following property: for every number n between 1 and $p-1$ inclusive, there is a power k of g such that $n = g^k$ mod p. Let us revisit our old friends Alice and Bob to illustrate this.

Alice generates a random private value a and Bob generates a random private value b. Both a and b are drawn from the set of integers.

They derive their public values using parameters p and g and their private values. Alice's public value is g^a mod p and Bob's public value is g^b mod p.

They exchange their public values.

Alice computes $g^{ab} = (g^b)^a$ mod p, and Bob computes $g^{ba} = (g^a)^b$ mod p. Since $g^{ab} = g^{ba} = k$, Alice and Bob now have a shared secret key k.

This process is shown in Fig. 10.2.

ElGamal

It is based on the Diffie–Hellman key exchange. It was first described by Taher Elgamal in 1984. ElGamal is based on the Diffie–Hellman key exchange algorithm described earlier in this chapter. It is used in some versions of PGP. The ElGamal algorithm has three components: the key generator, the encryption algorithm, and the decryption algorithm (Tsiounis and Yung 1998). We will keep with the format we have used so far, that of using Alice and Bob.

- Alice generates an efficient description of a multiplicative cyclic group G of order q with generator g. You should remember groups from Chap. 5. A cyclic group is a group that is generated by a single element, in this case that is the generator g. With a multiplicative cyclic group, each element can be written as some power of g.
- Next Alice chooses a random from x from a set of numbers $\{0,\ldots,q-1)$.
- Then Alice computes h = gx. Remember g is the generator for the group and x is a random number from within the group.
- h, G,q, and g are the public key, x is the private key.
- If Bob wants to encrypt a message m with the public key Alice generated, the following process is done:

  - Bob generates a random number y chosen from $\{0,..,q-1\}$. Y is often called an "ephemeral key."
  - Next Bob will calculate c1. That calculation is simple: $c1 = g^y$.
  - Next a shared secret $s = h^y$ is computed.
  - The message m is converted to m' of G.
  - Next Bob must calculate c2. That calculation is relatively easy: c2 = m' * s.
  - Bob can now send c1 and c2 = as the encrypted text.

- To decrypt a message m with the public key the first person generated, the following process is done:

  - The recipient calculates $s = c1^x$.
  - The then the recipient calculates $m' = c2 * s^{-1}$.
  - Finally, m' is converted back to the plaintext m.

The structure should look somewhat similar to Diffie–Hellman.

MQV

Like ElGamal, MQV (Menezes–Qu–Vanstone) is a protocol for key agreement that is based on Diffie–Hellman. It was first proposed by Menezes, Qu, and Vanstone in 1995, then modified in 1998. MQV is incorporated in the public-key standard IEEE P1363 (Easttom 2019). HQMV is an improved version. The specific algorithm is related to elliptic curves, so we will address those specifics in Chap. 11.

YAK

This algorithm is interesting because it includes authentication with key agreement. Like MQV and HMQV, it is authenticated, but it is also considered one of the simplest to implement.

With authenticated key exchange, Bob has a private key of b and a public key of:

$$Bpub = gb(modp)$$

Alice has a private key of a and a public key of:

$$Apub = ga(modp)$$

For the key exchange, Alice generates a secret x and sends Bob:

$$A = gx(modp)$$

For the key exchange, Bob generates a secret y and sends Alice:

$$B = gy(modp)$$

Alice computes:

$$K = (gbgy)x + a(modp)$$

Bob computes:

$$K = (gagx)y + b(modp)$$

These values should be the same, as:

$$K = (gbgy)x + a(modp) = g(b + y)(x + a)$$

Bob computes:

$$K = (gagx)y + b(modp) = g(a + x)(y + b)$$

This is how YAK works. This algorithm might be a bit complex for the novice. It is not critical that you fully understand this algorithm. It would be outstanding if you did, but as long as you understand the general framework of the algorithm, that would be sufficient for a novice.

Forward Secrecy

In cryptography, forward secrecy (FS), sometimes also called perfect forward secrecy (PFS), is a feature of specific key agreement protocols that gives assurances that session keys will not be compromised even if the private key of the server is compromised. The term "perfect forward secrecy" was coined by C. G. Günther in 1990. It has been used to describe a property of the Station-to-Station protocol. Definition: An encryption system has the property of forward secrecy if plaintext (decrypted) inspection of the data exchange that occurs during key agreement phase of session initiation does not reveal the key that was used to encrypt the remainder of the session.

Optimal Asymmetric Encryption Padding

OAEP (Optimal Asymmetric Encryption Padding) was introduced by Bellare and Rogaway and is standardized in RFC 2437. This process processes the plaintext prior to encryption with an asymmetric algorithm. When used with an algorithm such as RSA gives a cryptography scheme that is proven to be secure against a chosen plaintext attack.

OAEP satisfies the following two goals:

1. Add an element of randomness which can be used to convert a deterministic encryption scheme like RSA into a probabilistic scheme.
2. Prevent partial decryption of ciphertexts (or other information leakage) by ensuring that an adversary cannot recover any portion of the plaintext without being able to invert the trapdoor one-way permutation f.

Cramer–Shoup

The Cramer–Shoup system is an asymmetric key encryption algorithm. It was developed by Ronald Cramer and Victor Shoup in 1998, and it is an extension of the Elgamal cryptosystem. This was the first efficient algorithm proven to be secure against adaptive chosen ciphertext attack. This algorithm is not as widely used as RSA or Diffie-Hellman. However, it is a part of the history of asymmetric cryptography.

Applications

By this point, you should have some general understanding of several asymmetric algorithms, and a very good understanding of at least RSA and Diffie–Hellman. What is lacking, at this point, is a thorough discussion of how asymmetric algorithms are used. In this section, we will look at common applications for asymmetric algorithms.

Key Exchange

As has been previously stated in this book, symmetric algorithms are much faster than asymmetric and achieve the same security with far smaller keys. However, asymmetric algorithms overcome the issue of key exchange. Therefore, it is common for asymmetric algorithms to be used for exactly that purpose. For example, in SSL and TLS, an asymmetric algorithm (such as RSA) is used to exchange a symmetric key (such as AES). This is a common way to use asymmetric algorithms. We will explore SSL and TLS in more detail in later chapters.

Digital Signatures

I am sure you have heard the term, digital signature before, but do you know what they are and how they work? Remember that cryptographic algorithms are about protecting confidentiality, ensuring that only the intended recipient can read the message. Essentially, digital signatures take asymmetric cryptography and reverse it, so that they can protect integrity. Put another way, assume your boss sends you an email telling you that you have done such a great job, he thinks you should take next week off with pay. It would probably be a good thing to verify that this is legitimate, really sent from him, and not a prank. Well what a digital signature does is to take the senders private and encrypt either the entire message, or a portion (like the signature block). Now anyone with the sender's public key can decrypt that (Easttom 2019). So, let us return to Alice and Bob to see how this works.

Bob wants to send Alice a message and make certain she knows it's from him. So, he signs it with his private key. When Alice uses Bob's public key, the message decrypts, and she can read it. Now suppose that Bob did not really send this. Instead Eve sent it, pretending to be Bob. Well since Eve does not have Bob's private key, she had to use some other key to sign this message. When Alice tries to decrypt it (i.e., verify the signature) with Bob's public key, she will get back gibberish, nonsense.

A digital signature verifies that the sender really is who he or she claims to be. Digital signatures form an important part of security of messages. In essence to have total security for a message, one would execute the following steps:

1. Use a hash, MAC, or HMAC on the message and put the digest at the end of the message
2. Digitally sign the message, usually just a portion such as the hash or signature block, with your own private key
3. Encrypt the message with the recipient's public key

This process is depicted in Fig. 10.3.
The recipient reverses the process:

1. Decrypt the message with the recipient's private key
2. Verify the signature with the sender's public key
3. Re-calculate the hash, MAC, HMAC, and compare it to the one received to ensure no errors in transmission

Keep in mind, that as was stated in a previous chapter, the use of a MAC or HMAC is preferred over a hash. Figure 10.3 is just meant to demonstrate the concept of incorporating message integrity, confidentiality, and digital signatures.

There exists more than one type of digital signature. The type I just described is the most common and is referred to as a direct digital signature. A second type is the arbitrated digital signature. It is just like the process described above, with one signature. Instead of the sender digitally signing each message, a trusted third party digitally signs the message, attesting to the sender's identity.

A third type of digital signature exists, that is, the blind signature. The basic idea is that a Sender makes a Signer to sign a message m without knowing m; therefore, this is considered a blind signature. Blind signing can be achieved by a two-party protocol, between the sender and the Signer, that has the following properties.

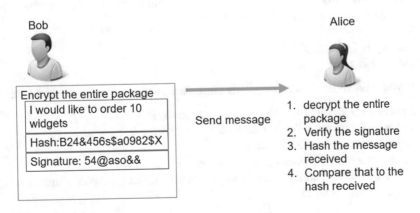

Fig. 10.3 message security

1. In order to sign (by a Signer) a message m, the Sender computes, using a blinding procedure, from m an m^* from which m cannot be obtained without knowing a secret, and sends m^* to the signer.
2. The Signer signs m^* to get a signature s_{m*} (of m*) and sends s_{m*} to the Sender. Signing is done in such a way that the Sender can afterward compute, using an unblinding procedure, from Signer's signature s_{m*} of m^*—the signer signature s_m of m.

This allows the arbiter to sign the message confirming that it was created on a given date by a specific sender, without knowing the contents of the message.

Digital Signature Algorithm

A digital signature can be done with any asymmetric algorithm. However, some algorithms have been created specifically for digitally signing messages. Digital Signature Algorithm is described in U.S. Patent 5,231,668, filed July 26, 1991, and attributed to David W. Kravitz. It was adopted by the US government in 1993 with FIPS 186. The actual algorithm functions as follows:

Choose a Hash function (traditionally this has been SHA1 but the stronger the hash the better).

Select a Key length L and N.

Note: The original Digital Signature Standard constrained L to be a multiple of 64 between 512 and 1024. Now lengths of 2048 are recommended. US government documents now specify L and N length pairs of (1024,160), (2048,224), (2048,256), and (3072,256).

Choose a prime number q that must be less than or equal to the hash output length.
Choose a prime number p such that $p-1$ is a multiple of q.
Choose g. This number g must be a number whose multiplicative order modulo p is q.
Choose a random number x, where $0 < x < q$.
Calculate $y = g^x \bmod p$.
Public key is (p, q, g, y).
Private key is x.

If you wish to use DSA to digitally sign, then the following steps are taken:

1. Let H be the hashing function and m the message:
2. Generate a random value for each message k where $0 < k < q$.
3. Calculate $r = (g^k \bmod p) \bmod q$.
4. Calculate $s = (k^{-1}(H(m) + x * r)) \bmod q$.
5. If r or s = zero, then re-calculate for a non-zero result (i.e., pick a different K).
6. The signature is (r, s).

DSA is a commonly used digital signature algorithm. In Chap. 11, we will see an elliptic curve variation of DSA.

Digital Certificates

A digital certificate is a digital document that contains information about the certificate holder and (if it is an X.509 certificate) a method to verify this information with a trusted third party. Digital certificates are how websites distribute their public keys to end users, and they are how web sites can be authenticated. The most common type of certificates are X.509. Before we go further into this topic, you should first know the contents of an X.509 certificate (Easttom and Dulaney 2017).

Version: What version of the X.509 standard is this certificate using.

Certificate holder's public key: This is one of the reasons we use digital certificates so that the recipient can get the certificate holders public key. If you visit a website that uses SSL/TLS (explained later in this chapter), then your browser gets the websites public key from the websites digital certificate.

Serial number: This identifies the specific certificate.

Certificate holder's distinguished name: Something to uniquely identify the certificate holder. This will often be an email address or domain name.

Certificate's validity period: How long is this certificate good for?

Unique name of certificate issuer: The items above identify the certificate, the certificate holder, and provide the certificate holder's public key. But how do you know this certificate really belongs to who it claims? How do you know it is not a fake? You verify the certificate with a trusted certificate issuer, such as Verisign.

Digital signature of issuer: To prove this certificate was issued by a trusted certificate issuer, that issuer signs the certificate. What they sign is usually the preceding information in the certificate.

Signature algorithm identifier: What digital signing algorithm did the certificate issuer use in this process?

For example, when you visit a website that uses SSL/TLS, your browser will first retrieve the websites certificate. Then it notes the unique name of the certificate issuer. The browser then retrieves that issuer's public key from the issuer, in order to verify the digital signature. If it is verified, then secure communications can proceed. You can see this process in Fig. 10.4.

It should be noted, that one usually no longer needs to retrieve the certificate issuer's public key. Most computers come with some sort of certificate store that has the digital certificate (and thus the public key) for the well-known certificate authorities. This is shown in Fig. 10.5.

There are some general terms associated with digital certificates that you should be familiar with. Those terms are briefly described here:

- PKI (public key infrastructure): It uses asymmetric key pairs and combines software, encryption, and services to provide a means of protecting security of business communication and transactions.
- PKCS (Public Key Cryptography Standards): It is put in place by RSA to ensure uniform Certificate management throughout the internet.

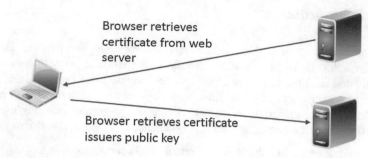

Browser retrieves
certificate from web
server

Browser retrieves certificate
issuers public key

Fig. 10.4 Retrieving a digital signature

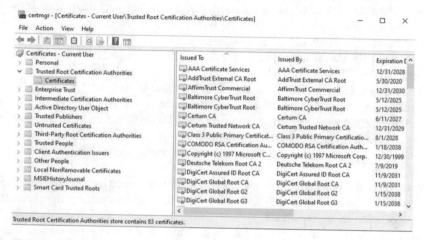

Fig. 10.5 Certificate Store

- CA (Certification Authority): It is an entity trusted by one or more users to issue and to manage certificates.
- RA (Registration Authority): It is used to take the burden off of a CA by handling verification prior to certificates being issued. RA acts as a proxy between user and CA. RA receives request, authenticates it, and forwards it to the CA.
- X.509: This is an international standard for the format and information contained in a digital certificate. X.509 is the most common type of digital certificate in the World. It is a digital document that contains a public key signed by the trusted third party, which is known as a Certificate Authority, or CA.
- CRL (Certificate Revocation List): This is a list of certificates issued by a CA that are no longer valid. CRLs are distributed in two main ways: PUSH model—CA automatically sends the CRL out at regular intervals. Pull model—The CRL is downloaded from the CA by those who want to see it to verify a certificate. End user is responsible.
- The Online Certificate Status Protocol (OCSP) is an internet protocol used for obtaining the revocation status of an X.509 digital certificate. It is described in

RFC 2560 and is on the internet standards track. It was created as an alternative to certificate revocation lists (CRL), specifically addressing certain problems associated with using CRLs in a public key infrastructure (PKI). The OCSP allows the authenticity of a certificate to be immediately verified.

While X.509 certificates are the most common type, they are not the only type. Usually websites will use an X.509 certificate, but for email some people use a PGP certificate. PGP or Pretty Good Privacy is a software that provides encryption, as well as integrity. It was created by Phil Zimmerman in 1991. PGP software defines its own certificate. It does not have certificate authorities that issue certificates, so there is no third-party verification of the certificate holder's identity. But PGP certificates can be used to exchange public keys. Here are some basic fields found in most PGP certificates:

PGP version number
Certificate holder's public key
Certificate holder's information
Digital signature of certificate owner
Certificate's validity period
Preferred symmetric encryption algorithm for the key

The critical issue to keep in mind with PGP certificates is that they do not include any trusted third-party verification. Therefore, they are not used in applications where such verification is important, such as e-commerce. However, PGP is often used to encrypt email.

SSL/TLS

Chapter 13 has a full coverage of SSL and TLS. This section is meant as a basic introduction to the topic. If you ever use a secure website, for example, to check your bank account, shop on Amazon.com, or any sort of e-commerce, then you have used SSL/TLS. SSL, or Secure Sockets Layer, is a technology employed to allow for transport-layer security via public-key encryption. Most websites now use TLS (the successor to SSL) but the term SSL stuck, so many people simply refer to SSL meaning either SSL or TLS. SSL/TLS is a protocol developed by Netscape for transmitting private documents via the internet. URLs that require an SSL connection start with *https:* instead of *http:*. SSL/TLS works by using X.509 certificates so the browser can get the web server's public key, then that public key is used to exchange a symmetric key. There have been several versions as of this writing (Easttom, 2019):

Unreleased v1 (Netscape).
Version 2 was released in 1995 but had many flaws
Version 3 was released in 1996 RFC 6101.
Standard TLS1.0 RFC 2246 was released in 1999.

TLS 1.1 was defined in RFC 4346 in April 2006.

TLS 1.2 was defined in RFC 5246 in August 2008. It is based on the earlier TLS 1.1 spec.

TLS 1.3 as of July 2014, TLS 1.3.

The process of establishing an SSL/TLS connection is actually somewhat straight forward and is described in the following steps:

1. When the client browser first encounters a website that is indicating the use of SSL/TLS, the client sends the server the client's SSL version number, cipher settings (i.e., what algorithms the client is capable of), and some session-specific data.

2. The server responds to the client with similar information from the server: the server's SSL version number, cipher settings, and some session-specific data. The server also sends its X.509 certificate. If mutual authentication is being used, or if the client is requesting a server resource that requires client authentication, the server requests the client's certificate.

3. The client browser first uses the X.509 certificate from the server to authenticate the server. If the server cannot be authenticated, the user is warned of the problem and informed that an encrypted and authenticated connection cannot be established. If the server can be successfully authenticated, the client proceeds to the next step, using the server's public key which the client retrieved from the X.509 certificate.

4. Using all data generated in the handshake thus far, the client creates the pre-master secret for the session. Then the client encrypts this pre-master secret with the server's public key and sends the encrypted pre-master secret to the server.

5. (optional step) If the server has requested client authentication, the client will also send its own X.509 certificate so that the server can authenticate the client. The server attempts to authenticate the client. If the client cannot be authenticated, the session ends. If the client can be successfully authenticated, the server uses its private key to decrypt the pre-master secret, and then performs a series of steps to generate the master secret. These are the exact steps the client will use on the pre-master secret to generate the same master secret on the client side.

6. Both the client and the server use the master secret to generate the session keys, which are symmetric keys (using whatever algorithm the client and server agreed upon). All communication between the client and the server after this point will be encrypted with that session key.

7. The client sends a message to the server informing it that future messages from the client will be encrypted with the session key. It then sends a message indicating that the client portion of the handshake is finished. The server responds with a similar message, telling the client that it all future messages from the server will be encrypted, and the server portion of the handshake is complete.

This handshake process may seem a bit complex, but it serves several purposes. First, it allows the client to authenticate the server and get the server's public key. It then allows the client and server both to generate the same symmetric key, then use that key to encrypt all communication between the two parties. This provides a very robust means of secure communication.

Homomorphic Encryption

Homomorphic encryption is about allowing one to perform mathematical operations on data that is still encrypted. In other words, analysis can be conducted on the ciphertext itself, without the need to first decipher the text. Before we delve into how this is accomplished, it may be useful to first understand why it is done. There are situations where you may desire a third party to calculate some value regarding data, without exposing them to the actual plaintext data. In such situations, homomorphic encryption is the solution.

Homomorphic encryption plays an important part in cloud computing, allowing companies to store encrypted data in a public cloud and still use the cloud providers analytic tools. The cloud provider can analyze aspects of the data without decrypting the data. The Pallier Cryptosystem is a homomorphic cryptography system. It is an asymmetric algorithm invented by Pascal Paillier in 1999. It is one of the modern homomorphic cryptographic algorithms.

Conclusions

This chapter gave you an overview of asymmetric cryptography. The most important algorithm for you to understand from this chapter is RSA. That is why you were shown two different examples of RSA. It is imperative that you fully understand RSA before proceeding. The next most important algorithm in this chapter is Diffie-Hellman. The other algorithms are interesting, and if you proceed further into cryptography you will, undoubtedly, delve deeper into those algorithms.

The chapter also gave an introduction to applications of cryptography. Digital certificates and SSL/TLS are so commonly used; it is very important that you have a strong understanding of these applications before proceeding. Homomorphic encryption is a relatively new topic, and at this point you need only have a general understanding of what it is.

Test Your Knowledge

1. An algorithm based on Diffie–Hellman first described in 1984 that is named after its inventor. _____.
2. U.S. Patent 5,231,668 filed July 26, 1991, and attributed to David W. Kravitz. It was adopted by the US government in 1993 with FIPS 186. _____.
3. _____ can be thought of as an RSA cryptosystem in which the value of e and d are fixed. e = 2 and d = ½
4. Explain the basic setup of Diffie–Hellman (the basic math including key generation).
5. Explain RSA key generation:
6. What is the formula for encrypting with RSA?
7. _____ was introduced by Bellare and Rogaway and is standardized in RFC 2437.
8. What does PKCS stand for?
9. What is the most widely used digital certificate standard?

X.509 certificates contain the digital signature of who.

References

Easttom, C. (2019). Computer Security Fundamentals, 4th Edition. New York City, New York: Pearson Press.

Easttom, C. & Dulaney, E. (2017). CompTIA Security+ Study Guide: SY0-501. Hoboken, New Jersey: Sybex Press.

Hinek, M. (2009). Cryptanalysis of RSA and its variants. England: Chapman and Hall.

Jonsson, J., & Kaliski, B. (2003). Public-key cryptography standards (PKCS)# 1: RSA cryptography specifications version 2.1 (pp. 1–68). RFC 3447, February.

Mao, W. (2011). Modern cryptography: Theory and practice. Upper Saddle River, New Jersey: Prentice Hall.

Rescorla, E. (1999). Diffie-hellman key agreement method. RFC 2631, June.

Tsiounis, Y., & Yung, M. (1998, February). On the security of ElGamal based encryption. In International Workshop on Public Key Cryptography (pp. 117–134). Springer, Berlin, Heidelberg.

Chapter 11
Elliptic Curve Cryptography

Abstract Elliptic curve cryptography is an important class of algorithms. There are currently implementations of elliptic curve being used in digital certificates and for key exchange. This class of algorithms provides robust security but with a substantially smaller key than RSA. In this chapter, we explore the basics of elliptic curve cryptography. The mathematics of elliptic curve cryptography is, however, more difficult than RSA. Thus, a seperate chapter is devoted to just this algorithm.

Introduction

This particular algorithm may be the most mathematically challenging that you will encounter in this book, or in any other introductory cryptography book. If you feel uncertain as to your mathematical acumen, you may wish to review Chaps. 4 and 5 (with particular attention to 5). Throughout the chapter, there are brief reminders as to key mathematical concepts to help you follow along. If your goal is a career related to cryptography, then you will, at some point, need to master this material. However, for those readers attempting to get a general overview of cryptography, whose primary focus is on computer/network security, then it is perfectly acceptable if you finish this chapter with just a broad overview of elliptic curve cryptography. Since this topic is often difficult for many readers, a chapter (albeit a short one) has been devoted to just this topic. Furthermore, in some cases, key concepts are explained more than once with slightly different wording to try and aid your understanding.

The reason why elliptic curve cryptography (commonly termed ECC) is more difficult for many people to learn is that fewer people have any prior exposure to the underlying mathematics. When one compares ECC to RSA, this difference becomes quite clear. Most people were exposed to prime numbers, factoring numbers, raising a number to a certain power, and basic arithmetic in primary and secondary school. But far fewer people are exposed to elliptic curves and discrete logarithms in school.

In This Chapter, We Will Cover

The basic math of elliptic curves
Elliptic curve groups as a basis for cryptography
ECC variations

General Overview

Elliptic curves have been studied, apart from cryptographic applications, for well over a century. As with other asymmetric algorithms, the mathematics has been a part of number theory and algebra, long before being applied to cryptography. As you saw in Chap. 10, many asymmetric algorithms depend on algebraic groups. There are multiple means to form finite groups. Elliptic curves can be used to form groups, and thus are appropriate for cryptographic purposes. There are two types of elliptic curve groups. The two most common (and the ones used in cryptography) are elliptic curve groups based on F_p where p is prime and those based on $F2^m$ (Rabah 2006). F, as you will see in this chapter, is the field being used. F is used because we are describing a field. Elliptic curve cryptography is an approach to public-key cryptography, based on elliptic curves over finite fields.

Remember that a field is an algebraic system consisting of a set, an identity element for each operation, two operations, and their respective inverse operations. A finite field, also called a Galois field, is a field with a finite number elements. That number is called the order of the field. Elliptic curves used for cryptographic purposes were first described in 1985 by Victor Miller (IBM) and Neil Koblitz (University of Washington). The security of elliptic curve cryptography is based on the fact that finding the discrete logarithm of a random elliptic curve element with respect to a publicly known base point is difficult to the point of being impractical to do (Rabah 2006).

Neal Koblitz is a mathematics professor at the University of Washington and a very well-known cryptographic researcher. In addition to his work on elliptic curve cryptography, he has published extensively in mathematics and cryptography. Victor Miller is a mathematician with the Institute for Defense Analysis in Princeton. He has worked on compression algorithms, combinatorics, and various subtopics in the field of cryptography.

First, we need to discuss what an elliptic curve is. An elliptic curve is the set of points that satisfy a specific mathematical equation. The equation for an elliptic curve looks something like this (Hankerson et al. 2006):

$$y^2 = x^3 + Ax + B$$

You can see this equation graphed in Fig. 11.1.

Fig. 11.1 The graph of an elliptic curve

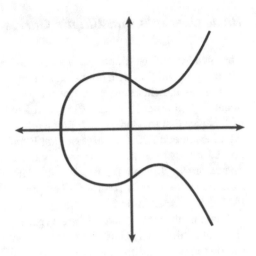

There are other ways to represent an elliptic curve, but Fig. 11.1 is the most common, and perhaps the easiest to understand. Another way to describe an elliptic curve is that it is simply the set of points that satisfy an equation that has two variables in the second degree and one variable in the third degree. The first thing you should notice from the graph in figure is the horizontal symmetry. Any point on the curve can be reflected about the x-axis without changing the shape of the curve.

Professor Lawrence Washington of the University of Maryland describes an elliptic curve a bit more formally, as follows: "an elliptic curve E is the graph of an equation of the form where A and B are constants. This will be referred to as the Weierstrass equation for an elliptic curve. We will need to specify what set A, B, x, and y belong to. Usually, they will be taken to be elements of a field, for example, the real numbers R, the complex numbers C, the rational numbers Q, one of the finite fields Fp (=Zp) for a prime p, or one of the finite fields Fq, where q=pk with k1."

This chapter only provides you with an overview of elliptic curve cryptography. Washington's book is an excellent place to get more detail. There have been two editions of his book, I have both and highly recommend them. However, the book does assume a level of mathematical sophistication that the current book you are reading does not.

The operation used with the elliptic curve is addition (remember the definition of a group requires a set along with an operation). Thus, elliptic curves form additive groups.

Recall from earlier in this book, a group is an algebraic system consisting of a set, an identity element, one operation, and its inverse operation. An Abelian Group or commutative group has an additional axiom a + b = b + a if the operation is addition and ab = ba if the operation is multiplication. A cyclic group is a group that has elements that are all powers of one of its elements.

Basic Operations on Elliptic Curves

The members of the elliptic curve field are integer points on the elliptic curve. You can perform addition with points on an elliptic curve. Throughout this chapter, as well as most of the literature on elliptic curve, we consider two points P and Q. The negative of a point $P = (xP, yP)$ is its reflection in the x-axis: the point $-P$ is $(xP, -yP)$. Notice that for each point P on an elliptic curve, the point $-P$ is also on the curve (Hankerson et al. 2006). Suppose that P and Q are two distinct points on an elliptic curve, and assume that P is not merely the inverse of Q. To add the points P and Q, a line is drawn through the two points. This line will intersect the elliptic curve in exactly one more point, call $-R$. The point $-R$ is reflected in the x-axis to the point R. The law for addition in an elliptic curve group is $P + Q = R$ (Bos et al. 2014).

The line through P and $-P$ is a vertical line which does not intersect the elliptic curve at a third point; thus, the points P and $-P$ cannot be added as previously. It is for this reason that the elliptic curve group includes the point at infinity O. By definition, $P + (-P) = O$. As a result of this equation, $P + O = P$ in the elliptic curve group. O is called the additive identity of the elliptic curve group; all elliptic curves have an additive identity. See Fig. 11.2.

To add a point P to itself, a tangent line to the curve is drawn at the point P. If yP is not 0, then the tangent line intersects the elliptic curve at exactly one other point, $-R$. $-R$ is reflected in the x-axis to R. This operation is called doubling the point P and can be seen in Fig. 11.3.

Fig. 11.2 P + (−P)

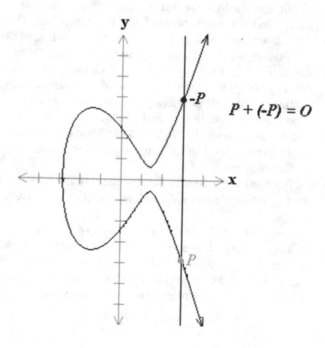

$P + (-P) = O$

The tangent from P is always vertical if yP = 0.

If a point P is such that yP = 0, then the tangent line to the elliptic curve at P is vertical and does not intersect the elliptic curve at any other point. By definition, 2P = O for such a point P.

Recall that the field Fp uses the numbers from 0 to p − 1, and computations end by taking the remainder on division by p (i.e., the modulus operations). For example, in F_{23}, the field is composed of integers from 0 to 22, and any operation within this field will result in an integer also between 0 and 22.

An elliptic curve with the underlying field of Fp can be formed by choosing the variables a and b within the field of F_p. The elliptic curve includes all points (x, y) which satisfy the elliptic curve equation modulo p (where x and y are numbers in F_p).

For example: $y^2 \bmod p = x^3 + ax + b \bmod p$ has an underlying field of Fp if a and b are in Fp.

If $x^3 + ax + b$ contains no repeating factors, then the elliptic curve can be used to form a group. An elliptic curve group over Fp consists of the points on the

Fig. 11.3 Doubling the P

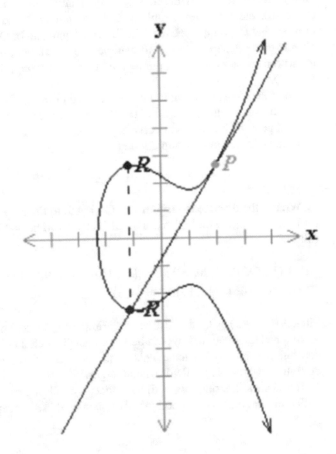

corresponding elliptic curve, together with a special point O called the point at infinity. There are finitely many points on such an elliptic curve.

At the foundation of every asymmetric cryptosystem is a hard-mathematical problem that is computationally infeasible to solve. The discrete logarithm problem is the basis for the security of many cryptosystems including the elliptic curve cryptosystem. More specifically, the ECC relies upon the difficulty of the elliptic curve discrete logarithm problem (ECDLP).

Recall that we examined two geometrically defined operations over certain elliptic curve groups. These two operations were point addition and point doubling. By selecting a point in an elliptic curve group, one can double it to obtain the point 2P. After that, one can add the point P to the point 2P to obtain the point 3P. The determination of a point nP in this manner is referred to as scalar multiplication of a point. The ECDLP is based on the intractability of scalar multiplication products.

In the multiplicative group Zp*, the discrete logarithm problem is: given elements r and q of the group, and a prime p, find a number k such that $r = qk \bmod p$. If the elliptic curve groups is described using multiplicative notation, then the elliptic curve discrete logarithm problem is: given points P and Q in the group, find a number that $Pk = Q$; k is called the discrete logarithm of Q to the base P. When the elliptic curve group is described using additive notation, the elliptic curve discrete logarithm problem is: given points P and Q in the group, find a number k such that $Pk = Q$.

The following is a common example, used in many textbooks, papers, and web pages. It uses rather small numbers that you can easily work with but makes the general point of how elliptic curve cryptography works.

In the elliptic curve group defined by

$$y2 = x3 + 9x + 17 \, over \, F23$$

What is the discrete logarithm k of $Q = (4,5)$ to the base $P = (16,5)$? One way to find k is to compute multiples of P until Q is found. The first few multiples of P are as follows:

$$P = (16, 5) \, 2P = (20, 20) \, 3P = (14, 14) \, 4P = (19, 20) \, 5P = (13, 10) \, 6P = (7, 3)$$
$$7P = (8, 7) \, 8P = (12, 17) \, 9P = (4, 5).$$

Since $9P = (4,5) = Q$, the discrete logarithm of Q to the base P is $k = 9$.

In a real application, k would be large enough such that it would be infeasible to determine k in this manner. This is the essence of elliptic curve cryptography, obviously we use larger fields, much larger.

If you are still struggling with the concepts of elliptic curves, there is an excellent online tutorial that does a good job of explaining these concepts. You might find that

reviewing that tutorial then re-reading this section aids you in understanding http://
arstechnica.com/security/2013/10/a-relatively-easy-to-understand-primer-on-ellip
tic-curve-cryptography/

The Algorithm

In the preceding section, we discussed elliptic curves in general, and in this section,
we discuss specific applications to cryptography. In cryptographic applications, the
elliptic curve along with some distinguished point at infinity is used along with the
group operation of the elliptic group theory form an Abelian group, with the point at
infinity as identity element. The structure of the group is inherited from the divisor
group of the underlying algebraic variety.

For cryptographic purposes, it is necessary to restrict ourselves to numbers in a
fixed range. The restriction is that only integers are allowed. You must also select a
maximum value. If one selects the maximum to be a prime number, the elliptic curve
is called a "prime curve" and can be used for cryptographic applications. To use
ECC, all parties must agree on all the elements defining the elliptic curve, that is, the
domain parameters of the scheme. The field is defined by p in the prime case and the
pair of m and f in the binary case. To focus on the prime case (the simpler), the field
is defined by some prime number p. The elliptic curve is defined by the constants A
and B used in the equation (Bos et al. 2014):

$$y^2 = x^3 + Ax + B$$

Finally, the cyclic subgroup is defined by its generator (i.e., base point) G.

The elliptic curve is the points defined by the above equation with an extra point
O which is called the point at infinity, and keep in mind that point O is the identity
element for this group.

Given a point $P = (x,y)$ and a positive integer n, we define $[n]P = P + P + ... + P$
(n times). The **order** of a point $P = (x,y)$ is the smallest positive integer n such that
$[n]P = O$.

We denote by $< P >$ the **group generated by P**. In other words, $< P > = \{O, P,
P + P, P + P + P,...\}$.

Now that you have a group, defined by an elliptic curve, the security of elliptic
curve cryptography is provided by the elliptic curve discrete logarithm problem.

The elliptic curve discrete logarithm problem (ECDLP) is this: given an elliptic
curve C defined over Fq and two points P, Q that are elements of the curve C, find an
integer x such that $Q = xP$.

Discrete logarithm review: To understand discrete logarithms, keep in mind the definition of a logarithm. It is the number to which some base must be raised to get another number. Discrete logarithms ask this same question but do so in regard to a finite group. Put more formally a discrete logarithm is some integer k that solves the equation $x^k = y$, where both x and y are elements of a finite group. A discrete logarithm is, essentially, a logarithm within some finite group.

In the preceding section we saw basic math, specifically addition, within an elliptic curve. You can choose any two points on an elliptic curve and produce another point. The process is not particularly complicated. You first begin with two points P1 and P2. These are defined as

$$P1 = (x1, y1)$$
$$P2 = (x2, y2)$$

The curve itself is symbolized as E given by the equation you already have seen y2 = x3 + AX + B.

Now to define a new point, we will call P3 and draw a line (which we will call L) through points P1 and P2. You can see that no matter where P1 and P2 are, that this line L will intersect E (the elliptic curve) at a third point P3. You can see this in Fig. 11.4.

Fig. 11.4 Line intersecting curve E

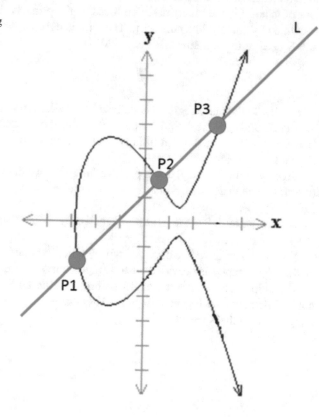

You can reflect that across the x-axis (simply change the sign of the y coordinate) and you have P3.

As you already know, an elliptic curve is the set of points that satisfy a specific mathematical equation. The equation for an elliptic curve looks something like this:

$$y^2 = x^3 + ax + b$$

And some point at infinity. This means that choosing different values for a and b changes the elliptic curve.

The size of the elliptic curve determines the difficulty of the finding the discrete logarithm, and thus the security of the implementation. The level of security afforded by an RSA-based system with a large modulus can be achieved with a much smaller elliptic curve group. This is one strong reason why elliptic curve cryptography has generated so much interest.

The U.S. National Security Agency has endorsed ECC algorithms and allows their use for protecting information classified up to top secret with 384-bit keys (Easttom 2019). This is important because that is much smaller than RSA keys. ECC achieves a level of security at least equal to RSA, but with key sizes almost as small as symmetric algorithms such as AES.

ECC Variations

As you can see from the previous sections, elliptic curves form groups, and those groups can be used just as any other algebraic group. The practical implication of this is that one can adapt various algorithms to elliptic curve groups. There are many permutations of elliptic curve cryptography including:

Elliptic curve Diffie–Hellman (used for key exchange)
Elliptic curve digital signature algorithm (ECDSA)
Elliptic curve MQV key agreement protocol
Elliptic Curve Integrated Encryption Scheme (ECIES)
In this section, we take a closer look at two of these.

ECC Diffie–Hellman

Diffie–Hellman, which you studied in Chap. 10, is the oldest key exchange protocol. It is a natural to modify for elliptic curves. Elliptic curve Diffie–Hellman (often simply called ECDH) is a key exchange or key agreement protocol used to establish a shared secret over an insecure medium (Easttom 2019). That shared secret is then used either directly or as the basis to derive another key. In the case of ECDH, the public private key pairs are based on elliptic curves.

- **Public:** Elliptic curve and point (x,y) on curve.
- **Secret:** Alice's A and Bob's B.
- Alice computes A(B(x,y)).
- Bob computes B(A(x,y)).
- These are the same since AB = BA.
- **Public:** Curve $y^2 = x^3 + 7x + b$ (mod 37) and point (2,5) $\Rightarrow$ b = 3.
- **Alice's secret:** A = 4.
- **Bob's secret:** B = 7.
- Alice sends Bob: 4(2,5) = (7,32).
- Bob sends Alice: 7(2,5) = (18,35).
- Alice computes: 4(18,35) = (22,1).
- Bob computes: 7(7,32) = (22,1).

For more details, consult NIST document 800-56A Revision 2 http://csrc.nist.gov/publications/nistpubs/800-56A/SP800-56A_Revision1_Mar08-2007.pdf

ECC DSA

The digital signature algorithm was invented specifically for digitally signing messages. Of course, one can use any asymmetric algorithm to sign a message, but the Digital Signature Algorithm was designed for that purpose. As you might expect, there is an elliptic curve variation on this algorithm.

To illustrate how this works, we will consider the fictitious Bob and Alice. First, the two parties must agree on some parameters. The curve, denoted as E; the base point/generator of the elliptic curve, denoted as G; and the order of G (an integer), denoted by n. Now to sign a message, Alice takes the following steps (Hankerson et al. 2006):

Select a random integer k that is less than n (i.e., K > 1; k < n).
Compute kG = (x_1, y_1) and r = x_1 mod n. If r = 0, then go to step 1.
Compute k^{-1} mod n.
Compute e = SHA-1(m). Most digital signature algorithms use a hash; in this case, the hash is usually SHA-1. So, this is stating that you compute the SHA-1 hash of the message.
Compute s = $k^{-1}\{e + d_A. r\}$ mod n.
If s = 0, then go to step 1. In other words, you keep repeating until s! = 0. This is not usually time consuming and could happen on the first attempt.
Alice's signature for the message m is (r, s).

In order for Bob to verify Alice's signature (r,s), he will execute the following steps:

Verify that r and s are integers in [1,n−1].
Compute e = SHA-1(m).

Compute $w = s^{-1} \bmod n$.
Compute $u_1 = ew \bmod n$ and $u_2 = rw \bmod n$.
Compute $(x_1, y_1) = u_1 G + u_2 Q_A$.
Compute $v = x_1 \bmod n$.
Accept the signature if and only if $v = r$.

This is very much like the traditional digital signature algorithm, except that it is using elliptic curve groups. ECC DSA is quite secure. There have been a few reported breaches, but those were based on faulty implementations, not on the algorithm itself being insecure. For example, in March 2011, researchers published a paper with the IACR (International Association of Cryptological Research) demonstrating that it is possible to retrieve a TLS private key of a server using OpenSSL using a timing attack. OpenSSL authenticates with elliptic curves DSA over a binary field. However, the vulnerability was fixed in a subsequent release of OpenSSL and was an implementation issue, not a flaw in the algorithm.

Conclusions

In this chapter, we have examined the use of elliptic curve groups for cryptography. This may be the most mathematically challenging chapter in this entire book. For most security applications, you need only be aware of the general description of the various elliptic curve algorithms. For those readers interested in pursuing cryptography in more depth, this chapter provides just an introduction to this topic. Various resources have been suggested in this chapter that will give you more detail and depth on ECC.

Test Your Knowledge

1. Which of the following equations is most related to elliptic curve cryptography?

   (a) $M^e \% n$
   (b) $P = C^d \% n$
   (c) $C^e \% n$
   (d) $y^2 = x^3 + Ax + B$

2. Which ECC variation requires the use of SHA?
3. What is an algebraic group?
4. What is a discrete logarithm?
5. What kind of key does ECC DH produce?

References

Bos, J. W., Halderman, J. A., Heninger, N., Moore, J., Naehrig, M., & Wustrow, E. (2014, March). Elliptic curve cryptography in practice. In International Conference on Financial Cryptography and Data Security (pp. 157–175). Springer, Berlin, Heidelberg.

Easttom, C. (2019). Computer Security Fundamentals, 4th Edition. New York City, New York: Pearson Press

Easttom, C. & Dulaney, E. (2017). CompTIA Security+ Study Guide: SY0-501. Hoboken, New Jersey: Sybex Press.

Hankerson, D., Menezes, A. J., & Vanstone, S. (2006). Guide to elliptic curve cryptography. Springer Science & Business Media.

Rabah, K. (2006). Elliptic curve cryptography over binary finite field GF. Information Technology Journal, 5 (1) 204–229. Retrieved from http://docsdrive.com/pdfs/ansinet/itj/2006/204-229.pdf.

Chapter 12
Random Number Generators

Abstract Random number generators are a critical component of crypto systems. They are used to generate symmetric keys, to provide initialization vectors, and for other purposes. However, it is critical to utilize a random number generator algorithm that is sufficiently random for cryptographic purposes. In this chapter, we explore the criteria for cryptographically secure random number generators, as well as several specific algorithms.

Introduction

Random numbers are a key part of cryptography. When you generate a key for a symmetric algorithm such as AES, Blowfish, or GOST, you need that key to be very random. Random numbers are also required for initialization vectors used with a variety of algorithms. A true totally random number is not possible to generate from a computer algorithm. It is only possible to generate a truly random number using other means including hardware, but not by using a software algorithm. Certain naturally occurring phenomena such as radioactive decay can form the basis for true random number generation. However, this is not a practical source of random numbers for use in cryptographic applications.

What is normally used in cryptography are algorithms called pseudorandom number generators. Pseudorandom number generators (PRNGs) are algorithms that can create long runs of numbers with good random properties but eventually the sequence repeats. Generating random numbers is not new. There are many ancient methods that have been around for many centuries, and we will explore some of these in this chapter. These are usually not sufficient for cryptographic purposes.

There are three types of pseudorandom number generators:

- Table lookup generators: Literally a table of precomputed pseudorandom numbers is compiled, and numbers are extracted from it as needed.
- Hardware generators: Some hardware process, perhaps packet loss on the network card, or fluctuations from a chip are used to produce pseudorandom numbers.
- Algorithmic (software) generators: This is the type most commonly used in cryptography, and what we will focus our attention on in this chapters.

In This Chapter, We Will Cover

Properties of a good PRNG
Different PRNG algorithms
How to test for randomness

What Makes a Good PRNG?

Now that we have discussed the fact that a true random number generator is impractical, we need to address the issue of what makes a good random number generator. Put another way, what makes a particular pseudorandom number generator good enough? To phrase that even more succinctly, is the output of a given algorithm random enough?

There are some specific properties one looks for in a PRNG. There are also some specific tests for randomness, as well as some standards one can use. In this section, we look at all three of these items. This information should inform your evaluation of specific algorithms we explore later in this chapter.

Desirable Properties of Pseudorandom Numbers

Before you consider actual algorithms and attempt to evaluate the efficacy of each for cryptographic purposes, it is important to understand what one is looking for in a cryptographic algorithm. Any pseudorandom number generator is going to generate a sequence of numbers. That sequence should have certain properties:

Uncorrelated sequences: This simply means that the sequences are not correlated. One cannot take a given stretch of numbers (say 16 bits) and use that to predict subsequent bits. There, quite literally, is no correlation between one section of output bits and another.

Long period: Ideally the series of digits (usually bits) should never have any repeating pattern. However, the reality is that there will eventually be some repetition. The distance (in digits or bits) between repetitions is the period of that sequence of numbers. The longer the period the better. Put another way: we accept that there will be repeated sequences, but those should be as far apart as possible.

Uniformity: It is most often that pseudorandom numbers are represented in binary format. There should be an equal number of 1s and 0s, though not distributed in any discernable pattern. The sequence of random numbers should be uniform and unbiased. If you have significantly more (or significantly less) 1s than 0s, then the output is biased.

Computational indistinguishability: Any subsection of numbers taken from the output of a given PRNG should not be distinguishable from any other subset of numbers in polynomial time by any efficient procedure. The two sequences are indistinguishable. That does not, however, mean they are identical. It means there is no efficient way to determine specific differences.

The third category is the one most often used in cryptography. It does not produce a truly random number but rather a pseudorandom number.

Tests of Randomness

How do you know if a given pseudorandom number generator is random enough? There are a variety of tests that can be applied to the output of any algorithm to determine the degree of randomness. More specifically, if that algorithm is suitable for cryptographic purposes. We will start with relatively simple tests that can easily be executed. Then, we will give an overview of more sophisticated statistical tests.

1-D Test

The 1-D test is a frequency test. It is a rather simple test, and essentially used a first pass. In other words, simply passing the 1-D test does not mean a given algorithm is suitable for cryptographic purposes. However, if a given PRNG fails the 1-D test, there is no need for further testing. Imagine a number line stretching from 0 to 1, with decimal points in between. Use the random number generator to plot random points on this line. First divide the line into a number of "bins." These can be of any size; in the graph below, there are four bins, each size 0.25.

```
|---------|---------|---------|---------|
        0.25      .50       .75      1.0
```

Now as random numbers (between 0 and 1.0) are generated, count how many fit into each bin. Essentially if the bins fill evenly that is a good sign that you have random dispersal. If there is a significant preference for one bin over another, then the PRNG is not sufficiently random. That PRNG has a bias, and further testing is not required to determine that it is not useful for cryptographic purposes.

Equidistribution

This test is also sometimes called the Monobit frequency test. This test just seeks to verify that the arithmetic mean of the sequence approaches 0.5. This can be applied to the entire output of a PRNG or to a given segment. Obviously, the larger the segment, the more meaningful the test. The further the arithmetic mean is from 0.5, the more of a bias the algorithm is displaying, and therefore the less suitable this PRNG is for cryptographic purposes. As the name suggests, the test is to determine if there is an equal distribution of 1s and 0s throughout the output of the PRNG. This test is described by the National Institute of Standards as follows:

> The focus of the test is the proportion of zeroes and ones for the entire sequence. The purpose of this test is to determine whether the number of ones and zeros in a sequence are approximately the same as would be expected for a truly random sequence. The test assesses the closeness of the fraction of ones to zero's that is, the number of ones and zeroes in a sequence should be about the same. All subsequent tests depend on the passing of this test; there is no evidence to indicate that the tested sequence is nonrandom.

This test (also sometimes just called the frequency test) has similarities to the 1-D test in that the purpose is to find out if there is bias in the output of a given PRNG. The focus of the test is the proportion of zeroes and ones for the entire sequence.

Runs Test

A run is an uninterrupted sequence of identical bits.

The focus of this test is the total number of runs in the sequence.

A run of length k consists of exactly k identical bits and is bounded before and after with a bit of the opposite value.

The purpose of the runs test is to determine whether the number of runs of ones and zeros of various lengths is as expected for a random sequence.

Determines whether the oscillation between such zeros and ones is too fast or too slow.

Fast oscillation occurs when there are a lot of changes, for example, 010101010 oscillates with every bit.

(input) E = 1100100100001111110110101010001000100001011010001100001000 1101001100010011000110011000101000101111000

(input) n = 100

(output) P-value = 0.500798

A p-value is a statistical test. A small p-value (typically ≤ 0.05) indicates strong evidence against the null hypothesis, so you reject the null hypothesis. A large

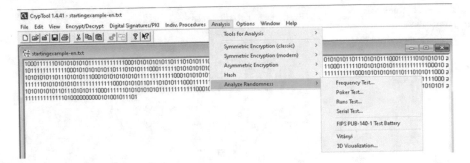

Fig. 12.1 CryptTool randomness analysis

p-value (> 0.05) indicates weak evidence against the null hypothesis, so you fail to reject the null hypothesis. You can use CryptTool to apply the Runs test to a sequence of numbers. You can find the various randomness tests by looking in the drop-down menu under Analysis and then *Analyze Randomness*, as shown in Fig. 12.1.

In this case, there is a seemingly random series of 1s and 0s, and to apply the runs test to this, you will first be prompted to select some test parameters, as shown in Fig. 12.2. For this example, simply use the default parameters.

The test results, shown in Fig. 12.3, are relatively simple to interpret. And in this case, the sequence of numbers did not pass a key part of the Runs test. If subsequent tests also cast doubt on the particular algorithm used, then it should definitely be rejected.

Test for Longest Run of 1s

This test is very similar to the Runs test. It is simply testing for the longest uninterrupted series of 1s within a given output for a given PRNG. The purpose of this test is to determine whether the length of the longest run of ones within the tested sequence is consistent with the length of the longest run of ones that would be expected in a random sequence.

Poker Test

The poker test for PRNGs is based on the frequency in which certain digits are repeated in a series of numbers. It is best understood by considering a trivial example, a three-digit number. In a three-digit number, there are only three possibilities. The first possibility is that the individual digits can be all different. The second possibility is that all three are different, and the third is that there is one pair of digits with one digit that does not match the other two.

Fig. 12.2 Runs test
parameters

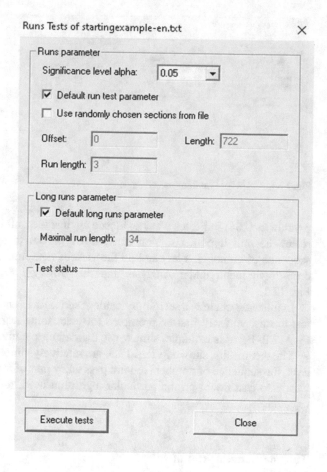

The tests actually assume sequences of five numbers, as there are five cards in a hand of poker. The actual five numbers are analyzed to determine if any given sequence appears more frequently than the other possible sequences. This is a rather primitive test. When using CryptTool to execute this test, you will need to select a few simple parameters. For our purposes, use the default settings. You can see the results in Fig. 12.4.

Statistical Tests

There are many statistical tests that can be applied to a sequence of numbers to determine how random that sequence is. Our purpose is not to evaluate all of these, but to give you some insight into how you might apply common statistical tests to determining how random a given PRNG's output is.

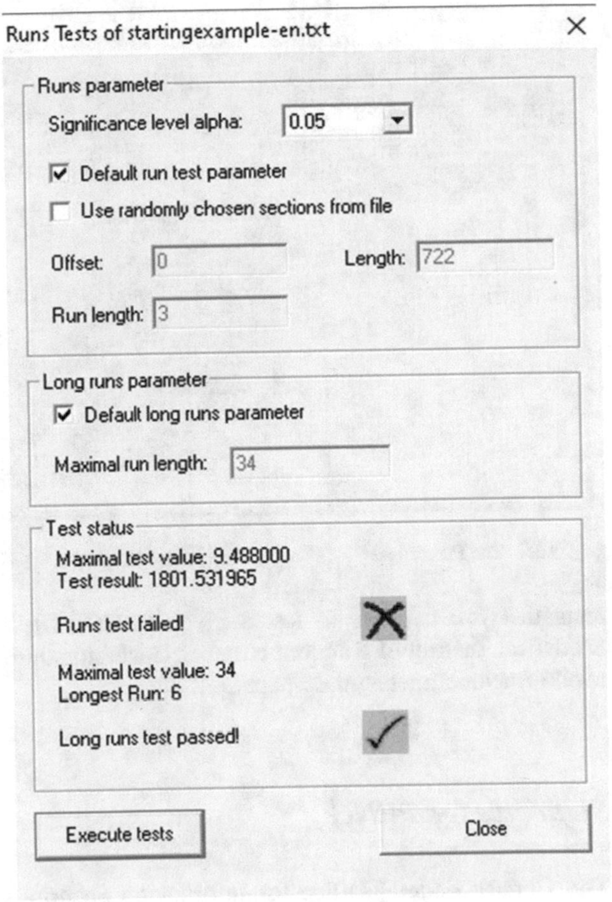

Fig. 12.3 Runs test results

Chi Squared

The chi-squared test is a common statistical test often used to test a sampling distribution. It is often used to compare observed data with data we would expect to obtain according to a specific hypothesis. It is not the purpose of this book or this chapter to teach statistics, so if you are not familiar with this test, it is recommended that you consult any elementary statistics textbook. However, the application of this test to PRNGs is fairly straight forward.

Test results are usually reported as a chi-square measure. A chi-square measure of $+-2$ is probably random noise, $+-3$ probably means the generator is biased, and $+-4$ almost certainly means the generator is biased. The lower the value, the more random the number sequence, the higher the value the less random. The United States National Institute of Standards (NIST) recommends a number of statistical

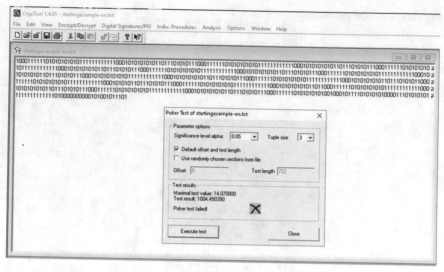

Fig. 12.4 The Poker test

tests that you can use to test a given PRNG. These tests are documented in Statistical Testing of Random Number Generators http://csrc.nist.gov/groups/ST/toolkit/rng/documents/nissc-paper.pdf

Standards for PRNG

The German Federal Office for Information Security (BSI) has established four criteria for quality of random number generators:

K1 A sequence of random numbers with a low probability of containing identical consecutive elements.

K2 A sequence of numbers which is indistinguishable from "true random" numbers according to specified statistical tests.

K3 It should be impossible for any attacker to calculate, or otherwise guess, from any given subsequence, any previous or future values in the sequence.

K4 It should be impossible for an attacker to calculate, or guess from an inner state of the generator, any previous numbers in the sequence or any previous inner generator states.

To be suitable for cryptography, any PRNG should meet K3 and K4 standards.

Specific Algorithms

In this section, we explore specific algorithms. We start with older methods that do produce a pseudorandom number, but perhaps not sufficient for cryptographic purposes. Just as we started in Chaps. 1 and 2 with historical ciphers, we start here with historical PRNGs.

Mid-Square

This is a very old algorithm. It was first described by a Franciscan friar in the thirteenth century. It was then re-introduced by John Von Neumann. It is a rather simple method and easy to follow. The following is a step-by-step description of the mid-square method:

1. Start with an initial seed (e.g., a four-digit integer).
2. Square the number.
3. Take the middle four digits.
4. This value becomes the new seed. Divide the number by 10,000. This becomes the random number. Go step 2. The following is a concrete example:

(1) $x_0 = 1234$
(2) x_1: $1234^2 = 01522756 \rightarrow x_1 = 5227$, $R_1 = 0.5227$
(3) x_2: $5227^2 = 27321529 \rightarrow x_2 = 3215$, $R_2 = 0.3215$
(4) x_3: $3215^2 = 10336225 \rightarrow x_3 = 3362$, $R_3 = 0.3362$

The process is repeated indefinitely, generating a new random number each time. The middle four bits of each output are the seed for the next iteration of the algorithm.

There are some definite limitations to this algorithm. The initial starting value is very important. If you start with all zeros, then you will continue with all zeros. However, if you start with leading zeros followed by some number, the subsequent values produced by the algorithm will be reducing to zero. Certain seed numbers are known to generate short periods of repeating cycles. In fact, the best that you can get from any mid-square implementation of n-digit numbers is a period of 8^n. And some have far shorter periods. This algorithm makes an excellent introduction to pseudorandom number generators, and it is easy to understand and frankly quite easy to code should you be interested in implementing this in a program. However, it is not adequate for modern cryptographic purposes.

Linear Congruential Generator

A linear congruential generator is an algorithm that depends on a linear equation to generate a pseudorandom sequence (Hallgren 1994). More specifically, it depends on a piecewise linear congruence equation. A piecewise linear function is essentially just a linear function that produces different results depending on input. In other words, it functions in pieces. Here is a simple example:

$$f(x) = \begin{pmatrix} X+3 & \text{if } X > 0 \\ X^2 & \text{if } X < 0 \\ X & \text{if } X = 0 \end{pmatrix}$$

This function takes X, and performs one of three operations, depending on the value of X.

There are a variety of linear congruential generators (also known as LCGs) but the most common form is

$$X_{n+1} = (aX_n + C) \bmod m$$

a is the multiplier
c is the increment
m is the modulus,
and X_0 is the initial value of X.

The period of a general LCG is at most m, so the choice of m is quite critical. Some linear congruential generators have a smaller period than m.

Not all LCGs are created equally. Some are quite good, others depend heavily on initial conditions, and still others are very bad. RANDU was an LCG used in the 1970s. It has a rather poor design and has widely been considered so bad at generating pseudorandom numbers, that it caused some to question the efficacy of LCGs in general.

There are a number of LCGs built into the random number generators of various libraries. Each with a value of m is sufficient for most random number purposes. The following table provides details on some of these implementations.

LCGs implemented in code

	m (modulus)	a (multiplier)	c (increment)
Borland C/C++	2^{32}	22695477	1
Microsoft Visual/Quick C/C++	2^{32}	214013	2531011
Java's java.util.Random	2^{48}	25214903917	11

LCGs are fast and easy to code. And as you can see many programming tools, libraries, and languages, have LCGs built in so you need not even code the algorithm itself. However, it must be kept in mind that LCGs are not considered random enough for many cryptographic applications. They are a great place to start learning about PRNGs, and for less rigorous applications might be used in some limited cryptographic scenarios, but in general are not sufficient for secure cryptography

Note: One of the major issues with cryptography is the implementation. Many common hard drive and file encryption products utilize sound algorithms such as AES and Blowfish but utilize substandard PRNGs to generate the cipher key for the chosen algorithm. Many programmers are not even aware that the library they are using may not have an adequate PRNG

CryptTool can also be used to generate random numbers. You can find a few common random number generators under *Individual Procedures*, *Tools*, and *Generate Random Numbers*. This is shown in Fig. 12.5.

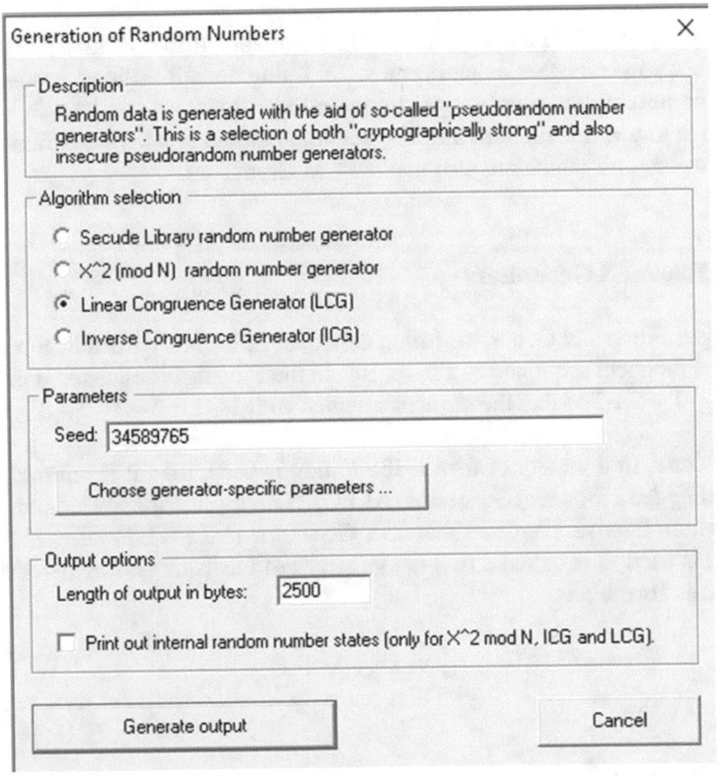

Fig. 12.5 CryptTool random number generation

```
                    Random data generated with the LCG generator - length 12000 bytes
00000000  81 0C EF 89 EB 9B 9C 77 1D 6A 76 85 B9 BD 64 51 96 57 B1 88 FB CE 75 F2 2F   A......w.jv...dQ.V....u./
00000019  18 03 1B AC 93 CA 12 97 1E 34 63 8C 18 93 56 94 DE 9F A7 EE 1B D5 AE AF 5B   .........4c...V..........[
00000032  B9 B9 21 06 37 60 17 B5 98 BD 9E A8 FD 34 0F 50 F9 DF CD C6 D6 B3 1E 62 F4   ..!.7`.......4.P.......b.
0000004B  FF 59 E4 55 B3 E5 7C EC CB C5 BF EE BB 59 F4 FC 20 0E 8A 6D 2D A9 35 D2 9E   .Y.U..|......Y.....m-.5..
00000064  0B C5 B5 04 9C 19 C0 12 C1 FC F3 9A 0F 76 D0 A2 97 4D F1 87 59 EB B9 EA 2B   .Y.U.........v...M..Y..+
0000007D  D6 11 3E D2 82 9A 95 D4 69 53 29 DE A0 DE AF 99 C6 74 54 6F D3 07 EB D3 C9   ..>.....iS)......tTo.....
00000096  72 84 42 1E 12 ED AE C8 3C 14 7B E1 4E 27 8B 04 DD 61 E0 97 D1 D2 0F 8E 85   r.B.....<.{.N'..a........
000000AF  F1 F4 C2 55 59 F8 BF 5B E6 5C 06 CD FD 16 74 52 74 3A 22 F5 27 CE 0A F4 87   ...UY..[.\....tRt:"......
000000C8  8E 5C 8A EE BD 4F 67 01 8F 0E C2 10 D5 DC 94 FB F6 C4 A9 0C FB 3E CD 86 38   ..Og............>..8
000000E1  44 B3 7E A6 05 C5 D5 7F 0B 9A 77 6E D4 9C F8 18 80 21 C8 16 85 39 06 74 45   D.~.......wn...!...9.tE
000000FA  00 F5 2C 52 23 0F 7S 02 FF 51 00 7C 17 4D F8 F7 C9 CC 31 73 E1 0C EE 6F 50   ..R#..s..Q.|.M..1s...oP
00000113  60 9D 4C C3 98 66 2F 93 77 5E 1A 1E 43 0E B6 60 E6 D1 E0 98 B1 67 CA 92 FC   `.L..f/.w^..C....`.....g..
0000012C  75 15 1F 47 DE 52 AD 28 DD 64 4F D6 0D F6 C2 9B A0 C2 09 B2 2C 66 03 C6 36   u..G.R.(.dO............6
00000145  84 63 40 48 C8 65 9A 2F 39 1F F9 68 35 BD EE 81 F4 77 C8 B8 5D 07 42 21 5F   .c@H.e./9..h5....w..].B!_
0000015E  52 66 80 83 27 4A C8 8B B9 10 E7 5D C9 F3 E6 15 59 D6 C9 A9 6F E8 DE 22 1E   Rf..'J.....]....Y...o.."
00000177  E8 BE 86 78 C0 A4 F2 9E 6A 2C DF F9 FE 02 18 63 E9 AB 9A 99 91 27 85 CC DD   ...x....j,.....c.......
00000190  AD 40 FF 5F AC C7 37 20 99 A6 7A 84 0E 97 E1 C6 5F 2D DC 81 46 DA 68 62 31   .@._..7 ..z..._-...F.hb1
000001A9  B7 FE 03 F5 09 27 6E E9 91 2C 72 F9 77 8F 0B 02 24 80 E0 8C 83 1F 9C 55 59   ......'n..,r.w...$.....UY
000001C2  69 E5 FF D5 03 1F 8B 97 D3 64 43 45 0F 88 0E 72 74 8E 37 14 3D 6B 30 41 D6   i.........dCE...rt.7.=k0A.
000001DB  11 77 82 45 8C BA FB F9 E5 EF 08 DB 26 E8 1C BE 32 7E 20 A9 E2 99 E1 2B   .w.E........&...2~ .....+
000001F4  93 47 E6 F9 1C E5 95 E9 9C 99 FF 0B 89 4D C9 20 05 28 10 07 44 84 E7 22 9B   .G...........M.. .(...D.."
0000020D  A4 3E 85 1C 23 1A 8B D8 64 2F 77 B8 18 AB 0E 74 2F 3E EC 93 EE 1F 4B E9 89   .>..#...d/w....t/>....K..
00000226  DA 0F 1F 46 97 E1 87 8E BD 21 FC 62 55 90 DF C5 58 55 3B CE 26 76 8C B5 E0   ...F.....!.bU...XU;.&v.
0000023F  2C 6A 8D 5B 49 13 F4 93 64 CF 0C 2A 9B 03 44 72 C9 DD E8 98 F7 67 42 36 1D   .F...![I..d..*.Dr....gB6.
00000258  56 C2 58 B8 30 6F C4 CF 73 E5 D6 AF 40 6A B8 35 4B F6 9C A6 90 1B 88 70 4A   V.X.0o..s...@j.5K.....pJ
00000271  6F 60 E2 4B 1F 3F 18 97 F1 D5 F7 34 E9 61 00 AB 01 FF 3A BC 20 E3 6D 4A A8   o`.K.?....4.a.....m.J.
0000028A  DD 5C DE 22 43 03 6C 3D A4 AA CA AD D6 1A 1C 4B 96 87 FE 6E C8 75 8F 2E 78   .\."C.l=........K..n.u..x
000002A3  28 16 97 22 EC 99 35 37 85 11 37 84 0D FA 3D 60 14 9A D7 31 79 9F E0 00 1C   (..".57..7..=`...1y..
000002BC  66 11 F3 6A 40 7A 61 0A A7 E4 40 3C 8A 3D 30 29 EF 08 7E 48 6E 6D 19 A3 CE   f..j@za..@<.=0)..~Hm...
000002D5  4B A4 E2 FF D0 4B 36 4E D9 79 9B B9 8D FB CE 66 DD 4B E2 2F BB 5A 58 7D DC   K..K6N.y....f.K./ZX).
000002EE  1B 3A FD DA 48 51 FF FE 5C 27 75 EA 4E 9A 70 B6 B3 2E 74 BF 44 16 E0 82 05   K....HQ..\'u..N.p...t.D.
00000307  75 14 B0 FF ED EB 3D 8C 6C 5E 6B BA 61 CF 3F D7 F3 AF 8C 53 50 F8 A5 FC C0   u....=.l^k.a.?...SP.
00000320  E3 25 C8 98 B6 1D 28 EC FC E2 D0 BE 24 BC 0B EB 28 87 9F 56 F0 94 FF 0C 93   .%....(....$...(..V.....
00000339  E7 B3 F5 D4 01 9F 17 C7 F6 CC 06 79 6C 25 7A 2A 0E 63 26 FB A5 CE AA 4E C7   ...........yl%z*.c&....N.
00000352  2B AF 15 66 BF F1 32 21 0B 70 1B C8 9B 1E 90 55 A5 9E 50 D5 85 01 93 60 4C   +..f..2!.p.....U..P...`L
0000036B  1A 88 3A 11 6E 2D 43 42 6A BD BF DC EB DE D4 C8 28 4D 33 C7 BB 64 DE A5   ..:.n-CBj.......(M3.d..
00000384  69 A5 91 CD 01 14 6E 8E 9C 98 04 4A 51 44 D7 04 0F 4B DD BC 36 7E 7F 2D D8   i....n...JQD..K..6~.-.
0000039D  76 81 07 E1 45 D8 FA 41 9C 02 18 CC 47 98 A9 47 8F 21 D0 16 4A 8C 0C 78 48   v..E..A..G..G.!.J.xH
000003B6  8D 02 5A 2F 71 84 E9 9C 04 1A 00 E1 3E 10 AB 7E C6 26 00 DD A6 0F 0A 54 D9   ..Z/q...........>..~.&.....T.
```

Fig. 12.6 CryptTool random numbers from LCG

As you can see, LCG is one of the PRNGs available. It starts with a preset seed of 314,149 but you can change that if you wish. Using default settings, produces 1200 bytes of pseudorandom numbers as shown in Fig. 12.6.

Feel free to test this random number with the randomness tests mentioned earlier in this chapter, and which are also available in CryptTool.

Lagged Fibonacci Generators

The Lagged Fibonacci Generator (often called an LFG) is a particular type of LCG using the Fibonacci sequence as a base. Recall the Fibonacci sequence is essentially $Sn = Sn - 1 + Sn-2$ so that the sequence begins with $1 - 1 - 2 - 3 - 5 - 8 - 13 - 21 - 35 -$ etc.

LFGs come in a variety of forms. If addition is used, then it is referred to as an Additive Lagged Fibonacci Generator (ALFG). If multiplication is used, it is referred to as a Multiplicative Lagged Fibonacci Generator (MLFG). Finally, if the XOR operation is used, it is called a two-tap generalized feedback shift register (GFS). The basic formula is

$$y = x^k + x^j + 1$$

Or

$$y = x^k \times x^j + 1 \,(\text{multiplicative LCG})$$

or

$$y = x^k \text{XOR} x^j + 1 \,(\text{GFS})$$

The indices j, k are the lags of the generator.

When the modulus is prime, a maximum period of $m^k - 1$ is possible. It is, however, more common to use a modulus which is a power of two, $m = 2^p$, with p being some prime number. In this case, the maximum periods are:

For an additive LFG : $\left(2^k - 1\right)2^{p-1}$

For a multiplicative LFG : $\left(2^k - 1\right)2^{p-3}$

Any LFG must be seeded with the initial k elements of the sequence.

Lehmer Algorithm

This PRNG is named after D. H. Lehmer, sometimes also referred to as the Park–Miller random number generator, after S. K. Park and K. W. Miller. It is the classic example of a Linear Congruential Generator. This algorithm produces a sequence of numbers $\{X_n\}$ using a standard LCG format:

$$X_{n+1} = (a \times X_n + c) \bmod m$$

Here is a relatively simple example:

Given a = 7, c = 0, m = 32, $X_0 = 1$
$X_1 = (7 \times 1 + 0) \bmod 32 = 7$
$X_2 = (7 \times 7 + 0) \bmod 32 = 49/32;\ Q = 1,\ R = 17$
$X_3 = (7 \times 17 + 0) \bmod 32 = 119/32;\ Q = 3,\ R = 23$
$X_4 = (7 \times 23 + 0) \bmod 32 = 161/32,\ Q = 5,\ R = 1$

However, with the small m you would expect a repetition at some point and indeed you find one:

$X_5 = (7 \times 1 + 0) \bmod 32 = 7/32;\ Q = 0,\ R = 7$
$X_6 =$ obviously we are at the same point as X_2 again

The sequence repeats with a period of 4 and is clearly unacceptable. Let us consider a few things that can be done to make this acceptable. For a range of m numbers $0 < m \le 2^m$, the function should generate all the numbers up to 2^m before repeating. So clearly a large value of m is important. Even a good algorithm with poorly chosen inputs will produce bad results. This is a substantial issue. Normally, these sorts of algorithms are used with very large values of m. As one example, something on the order of $m = 2^{48}$ would likely be sufficient for most applications.

Mersenne Twister

The original is not suitable for cryptographic purposes but permutations of it are. This PRNG was invented by Makoto Matsumoto and Takuji Nishimura. It has a very large period, $2^{19937}-1$, which is greater than the many other generators (Matsumoto and Nishimura 1998). Its name derives from the fact that its period length is chosen to be a Mersenne prime. The most commonly used version of the Mersenne Twister algorithm is based on the Mersenne prime $2^{19937}-1$ and uses a 32-bit word. It is often called MT 19937. There is a 64-bit word version called MT 199937-64. The Mersenne Twister is widely used and is implemented in PHP, MATLAB, Microsoft Visual C++, and Ruby.

The algorithm itself can be a bit complex to describe, and the following pseudo code might help those readers with even a rudimentary programming background to understand this algorithm:

```
// Create an array that is 624 bytes in size to store the state of the
// generator
int [0..623] MERSENNE //initial array named MERSENNE
int index = 0      // index

// Initialize the generator from a seed
function init_generator (int seed)
 {
index := 0
   MERSENNE [0] := seed
for i from 1 to 623 { // loop over each element
    MERSENNE [i] := lowest 32 bits of (1812433253 * (MERSENNE [i-1] xor
(right shift by 30 bits (MERSENNE [i-1]))) + i)
   }
 }

// Extract a pseudorandom number based on the index-th value,
// calling generate_numbers() every 624 numbers
function extract_number () {
if index == 0
   {
generate_numbers ()
   }

int y := MERSENNE [index]
y := y xor (right shift by 11 bits (y))
y := y xor (left shift by 7 bits (y) and (2636928640))
y := y xor (left shift by 15 bits (y) and (4022730752))
y := y xor (right shift by 18 bits (y))

index := (index + 1) mod 624
return y
 }
```

```
// Generate an array of 624 numbers
function generate_numbers()
{
forifrom 0 to 623
 {
int y:=(MERSENNE [i] and 0x80000000) // 32nd bit of MERSENNE [i]
     + (MERSENNE [(i+1) mod 624] and0x7fffffff // bits 0-30

  MERSENNE [i]:=MERSENNE [(i + 397) mod 624] xor (right shift by 1 bit (y))
if (y mod 2) != 0 { // y is odd
      MERSENNE[i] := MERSENNE[i] xor (2567483615)
    }
  }
}
```

Blum–Blum–Shub

Proposed in 1986 by Lenore Blum, Manuel Blum, and Michael Shub.
    The format of Blum–Blum–Shub is as follows:

$$X_{n+1} = X_n{}^2 \operatorname{Mod} M$$

M = pq is the product of two large primes p and q (this should remind you of the RSA algorithm). At each step of the algorithm, some output is derived from Xn + 1. The main difficulty of predicting the output of Blum–Blum–Shub lies in the difficulty of "quadratic residuosity" problem (Ding 1997). That mathematical problem, put simply is: given a composite number n, find whether x is a perfect square modulo n. It has been proven that this is as hard as breaking the RSA public-key cryptosystem which involves the factoring of a large composite. This makes Blum–Blum–Shub a quite effective PRNG.

Yarrow

This algorithm was invented by Bruce Schneier, John Kelsey, and Niels Ferguson. Like all of Bruce Schneier's inventions, this algorithm is unpatented and open source. Yarrow is no longer recommended by its inventors and has been supplanted by Fortuna. However, it is still an excellent algorithm to study because it is relatively simple to understand and implement and does a good job of generating sufficiently random numbers. The general structure of Yarrow is relatively simple to understand. Yarrow has four parts (Kelsey et al. 1999):

1. An entropy accumulator. This basically collects semirandom samples from various sources and accumulates them in two pools.
2. A generation mechanism. This generates the PRNG outputs.
3. Reseed mechanism to periodically re-seed the key with new entries from the entropy pools.
4. A reseed control that determines when reseeding should occur.

The two pools are called the fast pool and slow pool. The slow pool provides very conservative reseeds of the key, and the fast pool is, as the name suggest, used frequently to reseed the key. Both pools contain a hash of all inputs to that point in time. Both pools use SHA-1 so they produce a 160-bit hash output. Put more simply, each pool is fed some semirandom source, and that source is then put into a hash. As new data are fed to the pool, the hash is updated. This way there is a constantly changing hash value that could be used as a key (Kelsey et al. 1999).

The SHA-1 outputs are used to create keys for 3-DES. The outputs from 3-DES are the pseudorandom numbers. Periodically, the reseed mechanism goes back to the entropy accumulator to get a new SHA-1 hash and so that a new key is used with the 3-DES algorithm. Essentially the algorithm consists of accumulating semirandom input, hash that output, and use the hash as a seed/key for a block cipher (in this case 3-DES).

One reason this algorithm is worthy of study is that the same concepts could be easily modified. Allow me to illustrate with a simple but effective variation:

You begin with a poor PRNG such as mid-square. You use that to generate a pool of semirandom numbers. You can seed the PRNG with any value, even current date/ time stamp.

Each number in the pool is subjected to a hashing algorithm of your choice (SHA-1, RipeMD, etc.)

From that pool, you select two hashes. One will be the seed, the other will be the input or plain text value subjected to a cipher.

Then, use a block cipher of your choice (Blowfish, AES, Serpent, etc.) in Cipher Block Chaining Mode. The output of that cipher is your random number.

This is just provided as an example of how one can take existing cryptographic functions and combine them to produce numbers that should be sufficiently random for cryptographic purposes.

Fortuna

Fortuna is actually a group of PRNGs and has many options for whoever implements the algorithm. It has three main components (not this is quite similar to Yarrow):

1. A generator which is seeded and will produce pseudorandom data.
2. The entropy accumulator that collects random data from various sources and uses that to reseed the generator.
3. The seed file which has initial seed values.

The algorithm uses a generator that is based on any good block cipher (McEvoy et al. 2006). This can be DES, AES, Twofish, etc. The algorithm is run in counter mode. The generator is just a block cipher in counter mode. The counter mode generates a random stream of data, which will be the output of Fortuna. Because it is possible that a sequence would eventually repeat, it is recommended that the key used for the block cipher be periodically replaced. Note that counter mode is usually used to turn a block cipher into a stream cipher. It generates the next keystream block by encrypting successive values of the counter.

After each number is generated, the algorithm generates a fresh new key that is used for the next PRNG. This is done so that if an attacker were to compromise the PRNG and learn the state of the algorithm when generating a given PRNG, that this will not compromise previous or subsequent numbers generated by this algorithm (McEvoy et al. 2006).

Reseeding the algorithm is done with some arbitrary string. A hashing algorithm is often used for this purpose because it produces a somewhat random number itself, making it an excellent seed for Fortuna.

DUAL_EC_DRBG

Dual Elliptic Curve Deterministic Random Bit Generator became well known outside of cryptographic circles, when Edward Snowden revealed that the algorithm contained a backdoor inserted by the National Security Agency. The algorithm itself is based on elliptic curve mathematics and was standardized in NIST SP 800-90A. According to John Kelsey, the possibility of the backdoor by carefully chosen P and Q values was brought up at a meeting of the ANSI X9.82 committee. As a result, a way was specified for implementers to choose their own P and Q values.

In 2007, Bruce Schneier publishes an article with the title "Did NSA Put a Secret Backdoor in New Encryption Standard?" in *Wired* Magazine. His article is based on an earlier presentation by Dan Shumow and Niels Ferguson. Given that

DUAL_EC_DRBG is clearly not secure, the details of the algorithm are not important, however the story of this algorithm is. It illustrates an important point about the relationship between cryptography and network security. Most network security professionals only learn the most elementary facts about cryptography. Major industry security certifications such as the Certified Information Systems Security Professional (CISSP from ISC2), Certified Advanced Security Practitioner (CASP from the Computer Technology Industry Association), and even many university security textbooks give only a surface view of cryptography. This is dramatically illustrated by the story of DUAL_EC_DRBG. When Edward Snowden revealed, in 2013, that there was a backdoor in this algorithm, the network security community was stunned. However, the cryptography community had been discussing this possibility for many years. This is in fact, one of the major purposes of this book you are reading now. To present the world of cryptography in a manner that even general security practitioners can understand.

The Marsaglia CD ROM

As was mentioned earlier in this chapter, there are tables of pseudorandom numbers that can be used as sources for random numbers. George Marsaglia produced a CD-ROM containing 600 megabytes of random numbers. These were produced using various, dependable pseudorandom number generators, but were then combined with bytes from a variety of random sources or *semi-random* sources to produce quality random numbers.

The theory behind combining random numbers to create new random numbers can be described as follows. Suppose X and Y are independent random bytes, and at least one of them is uniformly distributed over the values 0–255 (the range found in a single byte when expressed in decimal format). Then both the bitwise exclusive-or of X and Y, and X + Y mod 256, are uniformly distributed over 0–255. If both X and Y are approximately uniformly distributed, then the combination will be more closely uniformly distributed. In the Marsaglia CD-ROM, the idea is to get the excellent properties of the pseudorandom number generator but to further "randomize" the numbers, by disrupting any remaining patterns with the combination operation.

Improving PRNGs

There are a number of ways to improve any given PRNG algorithm. One might think that simply creating new PRNG algorithms would be the ideal way to improve randomness. However, that approach can be difficult. One then needs to subject the new algorithm to lengthy peer review to ensure that a new problem has not been introduced. Furthermore, the more complex an algorithm, the more computationally intensive it will be. So, in some cases, it is desirable to take an existing algorithm and simply improve it. There are a few simple methods that can be applied to any PRNG to improve randomness.

Shuffling

One of the simplest methods for improving any PRNG is to shuffle the output. Consider the following example:

Start with an array of 100 bytes (you can actually use any size, but we will use 100 for this example). Then, fill that array with the output of the PRNG algorithm you are trying to improve. When a random number is required, take any two random numbers from the array and combine them. This combination, as described earlier in reference to specific PRNGs, will increase the randomness. You can make it even more random by combining nonsequential elements of the array.

Cryptographic Hash

This method is sometimes used to take a PRNG that is not cryptographically secure, something like a simple mid-square method, and attempt to generate a reasonably secure random number. The methodology is quite simple. You use the PRNG output as the input to a well-known cryptographic hash such as SHA-1. The output should be reasonably random.

Conclusions

In this chapter, you have been introduced to pseudorandom number generators. We have looked at specific algorithms, the desirable properties of a PRNG, and specific tests for randomness. It cannot be overstated how important PRNGs are in the creation of cryptographic keys for symmetric algorithms, as well as in the creation of initialization vectors. Even if one uses a secure algorithm, if the PRNG used to create the key is not sufficiently random, it will weaken the cryptographic implementation.

Test Your Knowledge

1. K3 of the German standard for PRNG states?
2. What is the basic formula for a Linear Congruential Generator?
3. Briefly describe the 1-D test?
4. What is "shuffling"?
5. Provide a general overview of Yarrow (the major steps).

References

Ding, C. (1997). Blum-Blum-Shub generator. Electronics Letters, 33(8), 677.

Hallgren, S. (1994). Linear congruential generators over elliptic curves. Carnegie-Mellon University. Department of Computer Science.

Kelsey, J., Schneier, B., & Ferguson, N. (1999, August). Yarrow-160: Notes on the design and analysis of the yarrow cryptographic pseudorandom number generator. In International Workshop on Selected Areas in Cryptography (pp. 13–33). Springer, Berlin, Heidelberg.

Matsumoto, M., & Nishimura, T. (1998). Mersenne twister: a 623-dimensionally equidistributed uniform pseudo-random number generator. ACM Transactions on Modeling and Computer Simulation (TOMACS), 8(1), 3–30

McEvoy, R., Curran, J., Cotter, P., & Murphy, C. (2006). Fortuna: cryptographically secure pseudorandom number generation in software and hardware.

Chapter 13
SSL/TLS

Abstract Secure Sockets Layer and the later version Transport Security Layer are the primary methods for security internet traffic. These protocols are utilized in e-commerce, Voice over IP (VoIP), and many other applications. One could claim that the modern uses we make of the internet, would not occur without SSL/TLS. The SSL/TLS protocols in turn depend on digital certificates. This chapter will provide you working knowledge of both digital certificates and SSL/TLS.

Introduction

The preceding 12 chapters have covered the algorithms used in cryptography including symmetric key cryptography, asymmetric key cryptography, cryptographic hashes, and generating random numbers. In this chapter, as well as the next two chapters, we will be focusing on practical applications of cryptography. Chapter 14 will discuss virtual private networks and Chap. 15 will cover military applications of cryptography. In this chapter, our primary focus is on SSL/TLS (Secure Sockets Layer/Transport Security Layer). However, in order to cover this topic adequately, we will first cover digital signatures, digital certificates, and the Public Key Infrastructure, all of which are needed for SSL/TLS.

The importance of applied cryptography cannot be overstated. Without modern protocols such as SSL/TLS modern e-commerce, online banking and similar technologies would not exist. While it is certainly important to understand the mathematics behind cryptography and the actual algorithms being implemented, it is also important to understand how all of this is utilized for secure communications.

In This Chapter We Will Cover

Digital Signatures
Digital Certificates

The Public Key Infrastructure (PKI)
SSL/TLS

Digital Signatures

We briefly touched on digital signatures when we discussed asymmetric cryptography in Chap. 10. The concept is to simply take some asymmetric algorithm and to reverse the process. Some piece of data is encrypted with the sender's public key. Anyone can access the sender's public key and decrypt/verify that message. There are primarily two types of digital signatures, direct and arbitrated. Each has its own strengths and weaknesses and is thus used in different situations. The question is not which type of signature is better, but rather which is better for a specific situation.

Direct Signature

A direct signature only involves the two parties (i.e., the sender/signer and the receiver). The signature is usually done in one of two ways. The first is to simply encrypt the entire message with the sender's private key (Easttom 2019). The second is to create a cryptographic hash, message authentication code, or HMAC that is then encrypted with the sender's private key. In either case the recipient will obtain the sender's public key to decrypt/verify the message.

With a direct signature there is no third-party verification of the sender's identity. This type of signature is often used in email communications. Verification is assumed based on (a) the sender's email address and (b) the fact that the recipient used the alleged sender's public key to verify the message and it worked. If the email was spoofed (i.e., it was not really the purported sender, but rather someone else faking their email address) then the message signature was actually done with some key other than the purported sender, and thus verification would fail. The basic process is shown in Fig. 13.1.

A typical process that was described earlier in this section involves the following steps:

1. Write the message
2. Create a hash of the message
3. Sign the hash with the senders private key
4. Encrypt the message with the recipients public key
5. Send the message

This multi-step process ensures message integrity with the hash, verifies the identity of the sender with the digital signature, and provides message confidentiality with the encryption. You can see this diagrammed in Fig. 13.2.

Alice

Bob

- Alice composes a message
- Alice creates a hash of the message
- Alice encrypts that hash with her private key
- Alice sends message to Bob

- Bob receives message
- Bob retrieves Alice's public key and decrypts the signature
- Bob then recomputes the hash of the message and compares it to the hash he received

Fig. 13.1 Direct digital signature

Fig. 13.2 Typical signature process

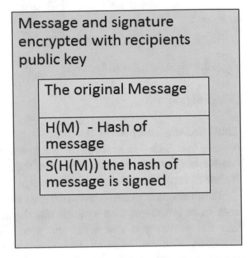

Message and signature encrypted with recipients public key

The original Message

H(M) - Hash of message

S(H(M)) the hash of message is signed

It should be noted in the diagram shown in Fig. 13.2. H(M) means the cryptographic hash function of the message M. H is the hashing function and M is the input or message. The S(H(M)) has similar meaning with S denoting the digital signature of the hash of the message M. And the hash could be replaced with a MAC or HMAC as needed.

Arbitrated Digital Signature

An arbitrated digital signature works similarly to a direct signature, however, the entity signing the message is not the sender, but rather a third party that is trusted by both sender and receiver (Easttom and Dulaney 2017). The rest of the process is the same. This sort of digital signature is often used to provide even more assurance that the sender is indeed, who they claim to be. A common implementation of this process is shown in Fig. 13.3.

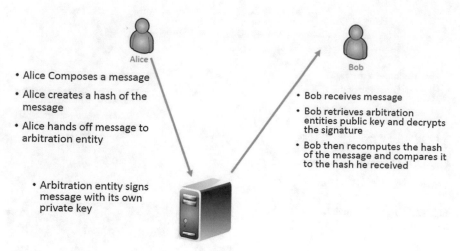

Fig. 13.3 Arbitrated Digital Signature

Blind Signatures

The blind signature takes the process of arbitrated digital signatures and adds a nuance to it. The signer simply signs a message verifying that the sender did indeed send it, but the signer has no idea what the content of the message is. This allows the signer to verify the origin of the message but not the content.

Think about it like this. Assume you observe a person write a note and put that note in an envelope, then seal the envelope. You then sign the envelope verifying that the person you saw write the note, did indeed write it and place it in the envelope. You can attest to the identity of the sender, but you have no idea what the contents are. A blind signature is a great option when the privacy of the data being sent is of paramount importance, for example, in voting situations or when using private digital currency. Blind signatures combine the enhance validation of the sender via arbitrated digital signatures but protect the privacy of the data.

Digital Certificates

In Chaps. 10 and 11 we discussed asymmetric key cryptography, also known as public key cryptography. The strength of asymmetric cryptography is that there is no issue with key exchange. It is perfectly acceptable for anyone, and indeed everyone, to have access to the public keys of both sender and receiver. However, this does not answer the question of how such public keys are disseminated. We discussed digital certificates briefly in a previous chapter. We will repeat some of that information here and expand upon it. The first step in that process is a digital certificate. A digital certificate is simply a digital document that contains some information about the certificate holder and contains that certificate holder's public key. There are two major types of digital certificates: X.509 and PGP.

X.509

X.509 is an international standard for the format and information contained in a digital certificate (Easttom and Dulaney 2017). X.509 is the most common type of digital certificate in the World. It is a digital document that contains a public key signed by the trusted third party which is known as a Certificate Authority or CA. The X.509 standard was first released in 1988. It has been revised since then, with the most recent version being X.509 v3, specified in RFC 5280. This system supports not only getting information about the certificate holder but verifying that information with a trusted third party. This is key to secure protocols such as SSL and TLS, as we will see later in this chapter.

The required content of an X.509 certificate is:

Version: What version of X.509 is being used. Today that is most likely going to be version 3.

Certificate holder's public key: This is the public key of the certificate holder, essentially this is how public keys are disseminated.

Serial number: This is a unique identifier that identifies this certificate.

Certificate holder's distinguished name: A distinguished or unique name for the certificate holder. Usually a URL for a web site or an email address.

Certificate's validity period: Most certificates are issued for 1 year, but the exact validity period is reflected in this field.

Unique name of certificate issuer: This identifies the trusted third party who issued the certificate. Public certificate authorities include Thawte, Verisign, GoDaddy, and others.

Digital signature of issuer: How do you know that this certificate was really issued by the certificate authority it claims to have been issued by? That is by having a digital signature.

Signature algorithm identifier: In order to verify the signer's digital signature, you will need the signer's public key and what algorithm they used.

There are other optional fields, but these are the required fields. Notice that the last three items are all about verification. One of the benefits of the X.509 digital certificate is the mechanism for verifying the certificate holder (Easttom 2019). This is a key to secure communications, not just encrypting the transmissions, but verifying the identity of the parties involved. An overview of the process is provided in Fig. 13.4.

Keep in mind that the certificate issuer's public certificate is likely already on the client machine, in their certificate store. Thus, Alice's browser can retrieve that from Alice's computer without the need to communicate with the certificate issuer's server. This leads to a significant question: why would you trust the certificate issuer to verify the identity of the certificate holder? Certificates are issued by certificate

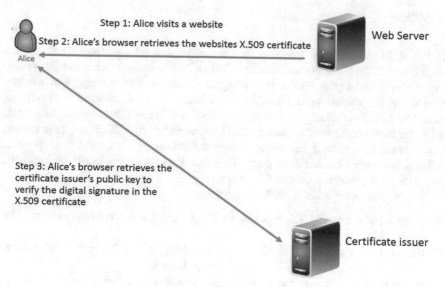

Fig. 13.4 Verifying X.509 certificates

authorities. These are entities that are trusted by many different third parties. The major certificate authorities include Symantec, GoDaddy, DigiCert, Global Sign, and Comodo. These certificate authorities (often abbreviated CA) are trusted vendors. It should also be noted that normally today the browser need not go to the certificate issuer to get the CA's public key. Most computers have a certificate store that contains the certificates (and thus public key's) for the major certificate authorities. A typical Windows 10 certificate store is shown in Fig. 13.5.

Companies and other organizations sometimes publish their own digital certificates, but those are only useful within the organizational network. Since the company's certificate authority is not trusted by other networks, then the certificates issued by that CA won't be trusted by other networks. There are varying levels of digital certificates. Each involves a different level of verification of the certificate holder's identity. The more verification used, the more expensive the certificate:

Class 1 for individuals, intended for email
Class 2 for organizations, for which proof of identity is required
Class 3 for servers and software signing, for which independent verification and
    checking of identity and authority is done by the issuing certificate authority
Class 4 for online business transactions between companies
Class 5 for private organizations or governmental security

Most websites will simply use a class 2 digital certificate.

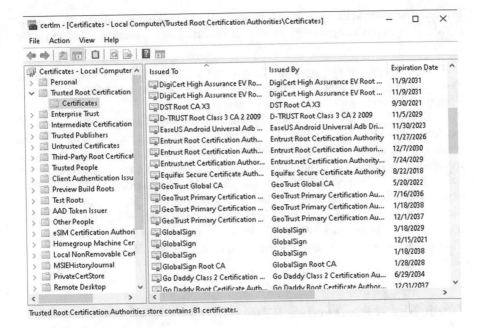

Fig. 13.5 Windows 10 certificate store

PGP

The other type of certificate is the PGP or Pretty Good Privacy certificate. Pretty Good Privacy is a methodology that combines cryptographic hashes, asymmetric cryptography, and symmetric cryptography to create secure communications. Each public key is tied to a username and/or email address, so PGP is often used for email encryption. PGP operates by using certificates but uses self-signed digital certificates. That means there is no certificate authority and thus no third-party verification of a sender's identity. This is why PGP certificates are not used for web pages/e-commerce. A PGP certificate includes:

PGP version number: This is just what version of PGP is being used.

Certificate holder's public key: The public key of the certificate holder. One of the purposes of using digital certificates is to disseminate public keys.

Certificate holder's information: This can be a bit more extensive that the X.509 certificate. It is information to identify who this certificate belongs to.

Digital signature of certificate owner: Since there is no CA, the certificate owner signs the certificate.

Certificate's validity period: How long the certificate is valid for.

Preferred symmetric encryption algorithm for the key: Once asymmetric cryptography is in use, what type of symmetric algorithm does the certificate holder prefer to use?

PGP was first created by Phil Zimmerman in 1991. The first version used a symmetric algorithm that Mr. Zimmerman had created himself. Later versions of PGP use a variety of well-known symmetric algorithms.

It should be noted that Phillip Zimmerman is well known in the world of cryptography. His PGP product is widely used to encrypt email, and he has also worked on voice over IP (VoIP) encryption protocols. He holds a bachelor's degree in computer science. The original release of PGP generated a criminal investigation because it enabled encryption of strengths that the U.S. government did not allow for export (see Chap. 15 for more on export laws). The case was dropped in 1996. Mr. Zimmerman has received significant recognition for his work including being inducted into the Internet Hall of Fame, being named one of the "Top 50 Tech Visionaries" by PC World and receiving the Louis Brandeis Award.

Public Key Infrastructure X.509

The public key infrastructure (often just called PKI) is essentially the infrastructure needed to create and distribute digital certificates (Easttom 2019). Since digital certificates are the means by which public keys for asymmetric algorithms are disseminated, the PKI is a key part of any implementation of asymmetric cryptography.

One role of the PKI is to bind public keys with some user's identity via a certificate authority. In other words, it is not adequate to simply have public keys widely available. There needs to be some mechanism to validate that a specific public key is associated with a specific user. With PKI this is done via a CA that validates the identity of the user.

There are several parts to the PKI. Each certificate issuer must be trusted by the other certificate issuers for the certificates to be interchangeable. Consider the process of visiting an online banking site. The sites have a digital certificate issued by some certificate authority. That certificate authority needs to be one that you and the bank both trust. Then later, perhaps you visit an e-commerce website. This website might use an entirely different certificate authority, but it must also be one that you trust.

The Certificate Authority is responsible for issuing and managing certificates. This includes revoking certificates. Revoking certificates is accomplished in one of two ways:

CRL (Certificate Revocation List) is simply a list of certificates that have been revoked. A certificate can be revoked for many reasons, as we already mentioned. There are two ways these lists are distributed: a) Push model: CA automatically sends the CRL out a regular intervals; b) Pull model: The CRL is downloaded from the CA by those who want to see it to verify a certificate. Neither model provides instant, real time updates.

Status Checking: Given that CRL's are not real0time, OCSP was invented. "Online Certificate Status Checking Protocol" is a real time protocol that can be used to verify if a certificate is still valid or not. OCSP is described in RFC 6960. OCSP uses HTTP to communicate messages. It is supported in Internet Explorer 7 and in Mozilla Firefox 3. Safari also supports OCSP.

The CA is often assisted by a Registration Authority (RA). The RA is responsible for verifying the person/entity requesting a digital certificate. Once that identity has been verified, the RA informs the CA that a certificate can be used.

The *Public Key Infrastructure X.509 (PKIX)* is the working group formed by the IETF to develop standards and models for the public key infrastructure. Among other things, this working group is responsible for updates to the X.509 standard.

The *Public Key Cryptography Standards (PKCS)* is a set of voluntary standards created by RSA and along with several other companies including Microsoft and Apple. As of this writing, there are 15 published PKCS standards:

PKCS #1: RSA Cryptography Standard.
PKCS #2: Incorporated in PKCS #1.
PKCS #3: Diffie-Hellman Key Agreement Standard.
PKCS #4: Incorporated in PKCS #1.
PKCS #5: Password-Based Cryptography Standard.
PKCS #6: Extended-Certificate Syntax Standard.
PKCS #7: Cryptographic Message Syntax Standard.
PKCS #8: Private-Key Information Syntax Standard.
PKCS #9: Selected Attribute Types.
PKCS #10: Certification Request Syntax Standard.
PKCS #11: Cryptographic Token Interface Standard.
PKCS #12: Personal Information Exchange Syntax Standard.
PKCS #13: Elliptic Curve Cryptography Standard.
PKCS #14: Pseudorandom Number Generators.
PKCS #15: Cryptographic Token Information Format Standard.

These standards are formed by the working group which involves experts from around the world, each contributing input to the standard.

SSL and TLS

SSL is the Secure Sockets Layer protocol. It has been supplanted by TLS (Transport Layer Security). Both protocols utilize X.509 certificates both for the exchange of public keys and to authenticate. Many references (books, magazines, courses, etc.) refer to SSL when in fact it is most likely that TLS is being used today. I will use the convention of referring to SSL/TLS when the specific version is not important.

History

When the web first began, no one considered security. HTTP, Hyper Text Transfer Protocol, is inherently quite insecure. It did not take long for computer scientists to realize that security was needed in order to use the web for sensitive communications like financial data. Netscape invented the SSL (Secure Sockets Layer) protocol, beginning with version 1.0 (Oppliger 2016). It was never released due to significant security flaws. However, version 2.0 was released in 1995 and began to be widely used. But security flaws were found with it, and it was subsequently supplanted with SSL version 3.0 in 1996. Version 3.0 was not just a minor improvement over past versions, it was a complete overhaul. It was published as an RFC, RFC 6101.

TLS 1.0 was released in 1999. It was essentially an upgrade to SSL 3.0. However, it was not compatible with SSL 3.0. TLS 1.0 also added support for GOST hashing algorithm as an option for message authentication and integrity. Previous versions had only supported MD5 and SHA-1 as hashing message authentication codes.

TLS 1.0 was eventually supplanted by TLS 1.1 released in April 2006. It had a number of specific cryptographic improvements including improved initialization vectors as well as supporting cipher block chaining for AES.

In August of 2008, TLS 1.2 was released as RFC 5246. It had many improvements over previous versions. Among them were:

Replacing MD5 and SHAQ with SHA 256
Support for advanced modes of AES encryption

TLS 1.3 was defined in RFC 8446 in August 2018. This new version offers a number of improvements for security. Those improvements include:

- Separating key agreement and authentication algorithms from the cipher suites
- Removing support for weak and lesser-used named elliptic curves
- Removing support for MD5 and SHA-224 cryptographic hash functions
- Requiring digital signatures even when a previous configuration is used
- Integrating HKDF and the semi-ephemeral DH proposal
- Replacing resumption with PSK and tickets
- Supporting 1-RTT handshakes and initial support for 0-RTT
- Mandating perfect forward secrecy, by means of using ephemeral keys during the (EC)DH key agreement
- Dropping support for many insecure or obsolete features including compression, renegotiation, non-AEAD ciphers, non-PFS key exchange (among which static RSA and static DH key exchanges), custom DHE groups, EC point format negotiation, Change Cipher Spec protocol, Hello message UNIX time, and the length field AD input to AEAD cipher
- Prohibiting SSL or RC4 negotiation for backwards compatibility
- Integrating use of session hash
- Deprecating use of the record layer version number and freezing the number for improved backwards compatibility

- Adding the ChaCha20 stream cipher with the Poly1305 message authentication code.
- Adding the Ed25519 and Ed448 digital signature algorithms
- Adding the x25519 and x448 key exchange protocols

You may recall that the ChaCha cipher was discussed in Chap. 7. The Ed25519 algorithm and the x25519 algorithms were discussed briefly in Chaps. 10 and 11. Perhaps the most important improvement in TLS 1.3 is that it removed support for various older and weaker cryptographic primitives. This includes weaker hash functions and weaker elliptic curves.

The Handshake Step By Step

The process of establishing an SSL/TLS connection is rather complex. The specific steps are described here (Davies 2011):

1. Communication begins with the client sending a Hello message. That message contains the client's SSL version number, cipher settings (i.e., what algorithms can the client support), session-specific data, and other information that the server needs to communicate with the client using SSL.
2. The server responds with a server Hello message. That message contains the server's SSL version number, cipher settings (i.e., what algorithms the server can support), session-specific data, and other information that the client needs to communicate with the server over SSL. The server also sends the client the server's X.509 certificate. The client can use this to authenticate the server and then use the server's public key. In some optional configurations, client authentication is required. In that case, part of the server Hello message is a request for the client's certificate.

   It should be noted that client authentication is not generally used in e-commerce as it would require each and every client to have an X.509 certificate from a well-known and trusted certificate authority. I suspect that most of you reading this book do not have such a certificate. If e-commerce sites did request such a certificate, it might reduce online fraud, but would also add an extra burden and cost to consumers. Consumers would have to purchase a certificate, at an average cost of $19.95 per year.
3. Now the client uses the server's X.509 certificate to authenticate the server. It does this by retrieving the public key of the certificate authority who issued this X.509 certificate and using that to verify the CA's digital signature on the X.509 certificate. Assuming authentication works, the client can now proceed with confidence that the server is indeed, who they claim to be.
4. Using all data generated in the handshake thus far, the client creates the pre-master secret for the session, encrypts it with the server's public key (obtained from the server's certificate, sent in step 2), and then sends the pre-master secret to the server.

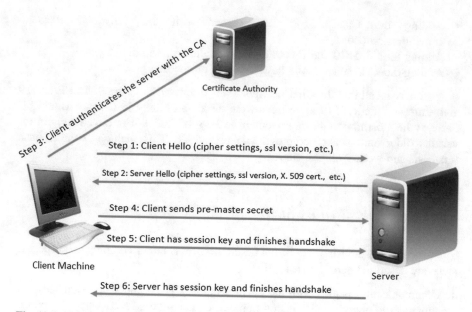

Fig. 13.6 SSL/TLS Handshake

5. If the server is configured to require client authentication, then at this point the server requires the client to send to the server, the client's X.509 certificate. The server will use this to attempt to authenticate the client.

6. If client authentication is required and the client cannot be authenticated, the session ends. If the client can be successfully authenticated, the server uses its private key to decrypt the pre-master secret that the client sent to it.

7. Both the client and the server use the pre-master secret that was sent from the client to the server, to generate the session keys. The session keys are symmetric keys and use whatever algorithm the client and server have agreed upon in steps 1 and 2 of the handshake process.

8. Once the client has completed generating the symmetric key from the pre-master secret, the client sends a message to the server stating that future messages from the client will be encrypted with the session key. It then sends an encrypted message indicating that the client portion of the handshake is finished.

9. Once the server has completed generating the symmetric key from the pre-master secret, the server sends a message to the client informing it that future messages from the server will be encrypted with the session key. The server then sends an encrypted message indicating that the server portion of the handshake is finished.

You can see this entire process in Fig. 13.6. What is shown in Fig. 13.6 is just the basic handshake process, without client authentication. Keep in mind, that as was discussed earlier in this chapter, the client machine need not communicate with the CA in order to validate the certificate of the server.

Fig. 13.7 SSL/TLS
indicator

The SSL/TLS process is very important. If you have ever purchased a product online, checked your bank statement online, or used any secure website, then you have already participated in this process. It is the basis for secure web traffic. Once the handshake is completed you will notice that your browser URL changes from http to https, indicating hypertext transfer protocol that is secured by SSL/TLS. Many browsers also display a small visual indicator that the site communications are now secure, as shown in Fig. 13.7.

Microsoft Edge displays the indicator at the end of the URL text filed, Mozilla Firefox and Google Chrome display the indicator just before the URL, but both use a padlock indicator.

Handshake Initiation

The hello messages from client to server were discussed in the handshake description given above. The details are provided here (Davies 2011; Oppliger 2016).

The Client Hello consists of the following elements:

Protocol version: This indicates if this is SSL version 3, TLS version 1.1, etc.

Random Number: This is a 32-byte random number. The first four bytes are the time of the day in seconds, the next 28 bits are just a random number. This is used to prevent replay attacks.

Session ID: This is a 32-byte number that is used to identify a given SSL/TLS session.

Compression algorithm: If compression is used in transmission, the specific algorithm is provided here.

Cipher Suite: This is a list of the cryptographic algorithms the client is capable. Often this will be common symmetric ciphers such as Blowfish, AES, etc. It may also include hashing or message authentication code algorithms the client is capable of in order to allow for message integrity.

The server hello is quite similar. It consists of the following elements:

Protocol Version: This indicates if this is SSL version 3, TLS version 1.1, TLS version 1.3, etc.

Random Number: This is a 32-byte random number. The first four bytes are the time of the day in seconds, the next 28 bits are just a random number. This is used to prevent replay attacks.

Session ID: This is a 32-byte number that is used to identify a given SSL/TLS session.

Compression algorithm: The server selects one of the algorithms the client has indicated it can support.

Cipher Suite: The server selects one of the algorithms the client has indicated it can support.

During the handshake, and throughout the SSL/TLS communication process, there are a number of specific error messages that both client and server send to one another. The most critical of these messages is shown in Table 13.1.

Applications of SSL/TLS

As we have already mentioned, secure web communications depend on SSL/TLS. If you use the web for e-commerce, online banking, secure social media, or other secure communications, then you are using SSL/TLS. Clearly, this is a critical technology.

Table 13.1 SSL/TLS messages

Message	Description
Unexpected_message	The message sent by the other party (client or server) is inappropriate and cannot be processed
Bad_record_mac	Incorrect message authentication code. This indicates that message integrity may be compromised
Decryption_failed	For some reason the party sending this message was unable to decrypt TLSCipher text correctly
Handshake_failure	Unacceptable security parameters, the handshake cannot be completed
Bad_certificate	There is a problem with the X.509 certificate that was sent
Unsupported_certificate	Certificate is unsupported. Either the type or format of the certificate cannot be supported
Certificate_revoked	Certificate has been revoked
Certificate_expired	Certificate has expired
Certificate_unknown	Certificate is unknown. This often happens with self-signed certificates
Unknown_CA	CA unknown. This also happens with self-signed as well as domain certificates
Access_denied	The other party is refusing to perform SSL/TLS handshake
Protocol_version	Protocol version not supported by both parties
Insufficient_security	Security requirements not met. The minimum-security level of one party exceeds the maximum level of the other party. This is not a common error message

OpenSSL

OpenSSL is a very popular open source implementation of SSL/TLS. It is written in C and is widely used for SSL/TLS. The OpenSSL project was founded in 1998 and is very widely used today with literally thousands of web servers running OpenSSL. OpenSSL can support a wide range of cryptographic algorithms. For symmetric ciphers it can support: DES, AES, Blowfish, 3DES, GOST, CAST 128, IDEA, RC2, RC4, RC5, Camelia, and SEED. For cryptographic hashes OpenSSL can support MD2, MD4, MD5, SHA-1, SHA-2, RIPEMD-160, GOST, and MDC-2. For asymmetric algorithms OpenSSL can support Diffie Hellman, RSA, DSA, and Elliptic Curve Cryptography.

Of course, no discussion of OpenSSL would be complete without discussing the infamous Heart Bleed bug. Heartbleed comes down to a very simple flaw, a buffer over-read. Essentially, the bug involved improper bounds checking that would allow an attacker to read more data than it should be able to. The flaw was implemented in the OpenSSL source code repository on December 31, 2011 and was released with OpenSSL version 1.0.1 on March 14, 2012.

The process worked as follows: A computer at one end of the connection could send a heartbeat request message. This is simply a request to ensure the other end is still active. This included a 16-bit integer, and the receiving computer would send back the same payload (16 bits). The bug allowed the attacker to set any return length size desired, up to 64 kilobytes of data. With this flaw, the attacker would request whatever was in active memory up to 64 kilobytes. In the case of servers, this 64 KB of data could include other users key, credit card information, or other sensitive data. This has since been patched, and OpenSSL continues to be widely used.

VoIP

Voice over IP is very widely used today. The ability to place phone calls using IP networks is quickly replacing traditional phone calls, at least for some people. Voice over IP depends primarily on two protocols. SIP or Session Initiation Protocol will establish a session/phone call. Then Real-time Transport Protocol or RTP is used to actually transmit the data. In many cases sRTP (Secure Real Time Transport Protocol) is used to secure the data. The sRTP protocol uses AES encryption. However, that still leaves the initial SIP communication insecure. There are implementations of VoIP that utilize an SSL/TLS connection established first, then both the SIP and RTP communication is conducted over the SSL/TLS connection. This allows for complete security of the entire communication session.

Email

Email is usually sent using the Simple Mail Transfer Protocol (SMTP) and received using either POP3 (Post Office Protocol version 3) or IMAP (Internet Message Access Protocol). All of these protocols are quite effective at transferring email but are simply not secure. In order to secure these protocols, SSL/TLS is added. SMTPS (SMTP using SSL/TLS) uses port 465 and can send email over an encrypted channel. IMAPS (IMAP using SSL/TLS) uses port 993 and can retrieve email from the email server over an encrypted channel.

Web Servers

Obviously, SSL/TLS is used to encrypt web traffic. In fact, that may be one of the most common applications of the SSL/TLS protocols. The two most common web servers are Apache web server and Microsoft Internet Information Server (IIS), let's look at configuring SSL/TLS for each of these commonly used web servers.

Apache Web Server

The Apache web server ships by default with many Linux distributions. You can also download Apache for either Linux or Windows. It is an open source product, and a free download. This, combined with its ease of use, has made it a very popular web server.

Depending on the version of Apache being used, the configuration file (httpd. conf) will be found in one of the following two locations:

/etc/apache2
/etc/httpd

You can configure your web server to accept secure connections, and in some cases to accept only secure connections simply by changing the httpd.conf file. For example, if you want the server to only accept SSL version 3:

SSLProtocol -all +SSLv3
SSLCipherSuite SSLv3:+HIGH:+MEDIUM:+LOW:+EXP

Figure 13.8 shows an excerpt from an example httpd.conf file. This excerpt comes from the Apache.org website.

From visiting the Apache.org website you can find other possible configurations. As you can see it is a rather trivial matter to enable SSL/TLS for Apache.

```
httpd.conf
# allow all ciphers for the initial handshake,
# so export browsers can upgrade via SGC facility
SSLCipherSuite ALL:!ADH:RC4+RSA:+HIGH:+MEDIUM:+LOW:+SSLv2:+EXP:+eNULL

<Directory /usr/local/apache2/htdocs>
# but finally deny all browsers which haven't upgraded
SSLRequire %{SSL_CIPHER_USEKEYSIZE} >= 128
</Directory>
```

Fig. 13.8 Configure SSL/TLS in Apache

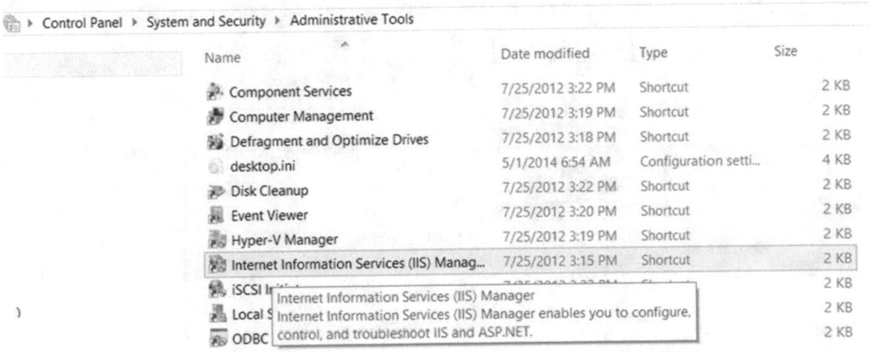

Fig. 13.9 Finding IIS in administrative tools

IIS

Microsoft Internet Information Services (IIS) is configured entirely from GUI (Graphical User Interface) methods. Unlike Apache, there is no configuration file to edit. However, the steps are relatively simple, and, in this section, you will see a step-by-step guide. The first step is to navigate to the Control Panel then Administrative Tools. This process will be the same in every version of Windows client (Windows 8, 8.1, 10, etc.). Note that it is slightly different in Windows server versions (Server 2012, Server 2016, Server 2019, etc.). Under administrative tools you should see Internet Information Services, as shown in Fig. 13.9.

If you do not see this option, you need to add it. In Windows Client operating systems (Windows 7, 8, etc.), you do this by finding Programs in the Control Panel then turn on Windows Features, as shown in Fig. 13.10.

IIS is a part of all versions of Windows and can be turned on/added to any Windows server or client. For Servers you simply use the Add Role function rather than turn on features. Once you have IIS turned on, you can use the IIS manager in

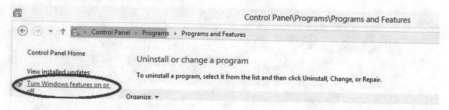

Fig. 13.10 Turning on Windows features

Fig. 13.11 IIS server settings

the Administrative Tools to configure IIS. There are really two segments to the IIS management. The first is for IIS server wide, and the second is for specific sites. As you go through these steps, if you do not see an option mentioned, it is most likely that you are looking at the wrong aspect of IIS manager. You can see the Server settings circled in Fig. 13.11.

Within those Server Settings you need to select Server Certificates as shown in Fig. 13.12.

On the right-hand sign of the Server Certificates screen, you can create a self-signed certificate, create a certificate request, create a domain certificate, or import an existing certificate. You can see this in Fig. 13.13.

In order to test this, you may wish to just create a self-signed certificate. In order to apply this certificate to a specific site, you need to navigate from the Server settings to a specific site settings and select SSL Settings as shown in Fig. 13.14. Notice that Windows refers to SSL when in fact you are most likely using TLS with any modern version of Windows.

Those settings allow you to decide if you wish to require SSL. This will mean that a client can only connect to this website using SSL/TLS, as opposed to the SSL/TLS being optional. If you choose this, then clients cannot connect to your website using http:// but rather must use https://. You can also select how to respond to client certificates. These settings are shown in Fig. 13.15.

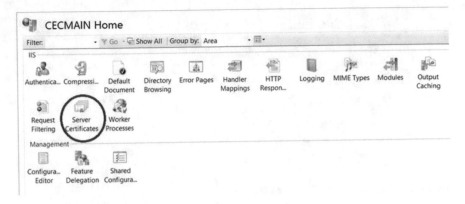

Fig. 13.12 IIS server certificates

Fig. 13.13 Creating or
importing certificates

Fig. 13.14 Site SSL settings

SSL Settings

This page lets you modify the SSL settings for the content of a website or application.

☑ Require SSL

Client certificates:

   ● Ignore
   ○ Accept
   ○ Require

Fig. 13.15 Handling SSL/TLS connections

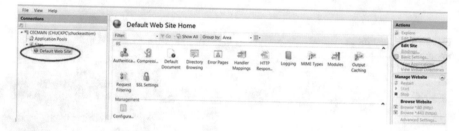

Fig. 13.16 HTTPS bindings

Finally, you have to set the bindings for this site. When you have navigated to a specific site, the binding's option is on the right-hand side of the screen, as shown in Fig. 13.16.

There you simply bind the HTTPS protocol to a specific port, usually 443, and select a certificate to use with this website. This is shown in Fig. 13.17.

While this is all done through a graphical user interface, you can readily see that the process involves many steps.

Conclusions

SSL/TLS protocols are a cornerstone of modern secure communications. These protocols in turn depend on other technologies such as X.509 certificates, digital signatures, and the symmetric and asymmetric algorithms we have discussed in previous chapters. A thorough understanding of SSL/TLS is critical to applying cryptography to real-world problems. After reading this chapter, you should be thoroughly familiar with digital certificates as well as the SSL/TLS handshake process.

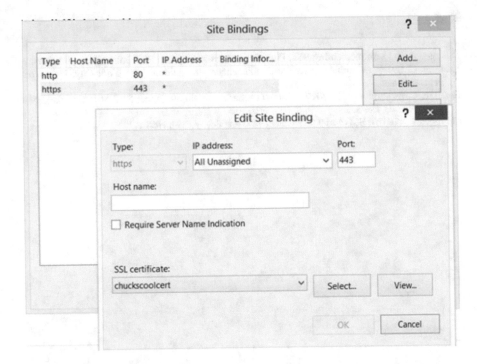

Fig. 13.17 Finalizing HTTPS bindings

You have also seen the SSL/TLS encrypted communications are important in a variety of technologies including Voice over IP, Email, and Web traffic. In this chapter, you explored the basics of setting up SSL/TLS for encrypting web traffic with both Apache and IIS.

Test Your Knowledge

1. In general terms, describe how digital signatures work.
2. Briefly describe the concept of blind signatures.
3. What does X.509 support that PGP certificates do not?
4. By whom is an X.509 certificate digitally signed?
5. What part of the SSL/TLS handshake includes the cryptographic ciphers the client can support?
6. If a certificate that is being used in SSL/TLS is on the CRL, what message will be sent?
7. What port does secure IMAP operate on by default?
8. What is an arbitrated signature?
9. Is voicemail secure?
10. Where is PGP most often used?

References

Davies, J. (2011). Implementing SSL/TLS using cryptography and PKI. John Wiley and Sons.
Easttom, C. (2019). Computer Security Fundamentals, 4th Edition. New York City, New York: Pearson Press.
Easttom, C. & Dulaney, E. (2017). CompTIA Security+ Study Guide: SY0-501. Hoboken, New Jersey: Sybex Press.
Oppliger, R. (2016). SSL and TLS: Theory and Practice. Artech House.

Chapter 14
Virtual Private Networks, Authentication, and Wireless Security

Abstract A widely used area of applied cryptography is that of virtual private networks (VPNs). VPNs allow secure access to network resources from remote locations. VPNs utilize cryptographic algorithms to secure communication. This chapter will explore a wide range of authentication protocols as well as VPNs. VPN's are one of the most widely used applications of cryptography, thus is quite important to study. The chapter will also cover wireless security. Given the ubiquitous nature of Wi-Fi, wireless security is also very critical. This chapter will also cover authentication methods. Many authentication methodologies utilize cryptography. This chapter, combined with the previous chapter, provide the reader with a working knowledge of the practical applications of cryptography.

Introduction

In the previous chapter, we explored the concepts of secure socket layer (SSL)/ transport layer security (TLS). In this chapter, we take a look at yet another common, and quite important, application of cryptography: virtual private networks (VPNs). VPNs are designed to provide a secure connection either between two sites, or between a remote worker and a site. The concept is to give the same connection one would have if one were actually physically connected to the network. This means the same level of security, and the same (or at least substantially similar) access to network resources. Thus, the name: it is a virtual connection to a private network. VPNs have evolved over time. In this chapter, we examine general concepts as well as specific technologies. We also look at related technologies that are, while not strictly speaking VPNs, secure remote connections.

In This Chapter We Will Cover

Authentication
    Point-to-point tunneling protocol (PPTP)

Layer 2 tunneling protocol (L2TP)
Internet protocol security (IPSec)
Secure socket layer (SSL)/transport layer security (TLS) VPN
Secure Communications

Concepts

The Dictionary of Computer and Internet Terms defines a VPN as "a network where data *is* transferred over the Internet using security features preventing unauthorized access." This is a fairly typical definition similar to what one would find in any computer dictionary, but not quite detailed enough for our purposes.

The idea of a VPN is to emulate an actual physical network connection. This means it must provide both the same level of access and the same level of security. In order to emulate a dedicated point-to-point link, data are encapsulated, or wrapped, in such a way that its contents are secure. Since the packet is wrapped, it must also include a header that provides routing information allowing it to transmit across the Internet to reach its destination. This creates a virtual network connection between the two points. But this connection only provides a virtual network; next, the data are also encrypted, thus making that virtual network private.

The question becomes how does one implement the authentication and encryption required to create a VPN? There are several technologies that can facilitate establishing a VPN; each works in a slightly different manner. We will examine each of these in this chapter. Despite the differences in the protocols, the end goals are the same. First authenticate the user, ensure that this user is who they claim to be. Then exchange cryptographic information such as what algorithms to use, and finally establish a symmetric key the two parties can use to secure data between the two parties.

Authentication

Authentication is merely the process of verifying that some entity (be it a user or another system or another program) is indeed, who they claim to be. The most obvious example of authentication is when an end user provides a username and password. The password presumably verifies that this is indeed that user. However, as security breaches become more widespread, a simple username and password are not adequate. Passwords can be guessed, written down, or similarly exposed. There are three main types of authentication (Easttom 2019):

Type I: This is something you know. The aforementioned password is a classic example. However, pin numbers, pass phrases, etc. all constitute type I authentication. They all have the same weakness, they can be guessed, stolen, etc.

Type II: This is something you have. A good example would be the debit card one must put into the ATM in order to extract funds. Smart cards and keys are other examples of type II authentication. These have the advantage that they cannot be guessed, but they can certainly be stolen.

Type III: Something you are. This is biometrics. Fingerprints, retinal scans, facial recognition, even handwriting all constitute Type III authentication. This has the advantage that it cannot be guessed and cannot be stolen (at least not without the user being aware of it!).

These are the three primary means of authentication, but other methods exist. As one example, there is Out Of Band Authentication. This form of authentication involves prompting the user for information from various data sources that is likely to only be known to the authentic user. When you run a credit report online, many systems will prompt you for identifying information such as a United States Social Security Number (Type I authentication) as well as ask you question such as the amount of your monthly credit card payment and current balance. These last two pieces of information are Out Of Band Authentication (OOB).

Type I authentication can be used for VPNs and often is. Type II can also be used. The best solution, often called strong authentication, is to combine Type I and Type II authentication. Using an ATM is a good example of this. You must have a debit card (Type II) and a pin number (Type I). Some VPN solutions operate in a similar fashion requiring a password and perhaps a digital certificate (see Chap. 13 for a detailed discussion of digital certificates) or even a smart card on the client machine, in order to function. Of course, as technology improves, type III is also feasible, even for the home user. Using facial recognition or fingerprint is already possible on most smart phones. If one combines a password, a smart card, and facial recognition, that will provide a reasonably secure level of secure authentication.

It should be noted that the United States Department of Defense (US DoD), uses combined authentication modalities. The Common Access Card (CAC, often called the 'kack' card). is a smart card used by the DoD. The front of the card has identification information for the holder, including a photo. There is a two-dimensional barcode in the bottom left hand corner, and an integrated circuit chip embedded in the card. The information on the chip is encrypted with strong encryption. For computer access, one needs to insert the CAC card and provide the password or pin. This provides two-factor authentication. An example of a CAC card is shown in Fig. 14.1.

In addition to considerations as to what type of authentication to utilize, the method of transmitting authentication data as well as cryptographic information must be addressed. This is handled by VPN protocols. Each protocol establishes its own authentication method as well as how to exchange cryptographic data.

Since all VPN protocols include authentication as well as encryption, it is important to have a good understanding of authentication protocols before proceeding. Throughout this book we have examined a number of cryptographic and hashing algorithms, compared to those, authentication is a relatively simple process. The purpose of authentication is to simply verify that a given entity is who they claim to be. An added feature of more modern authentication protocols is to also prevent sophisticated attacks such as session-hijacking.

Session hijacking occurs when an attacker waits until a user has authenticated to a system and then takes over that sessions. There are various methods for doing this, all of which are at least moderately sophisticated. The key to all session hijacking is that the attacker does not need to ever actually obtain the users password, they merely take over a session after the user has supplied their password to the system.

Fig. 14.1 Exemplary
CAC Card

CHAP

Challenge Handshake Authentication Protocol was specifically designed to prevent session hijacking. There are some earlier authentication protocols but none of those are used in modern VPNs. The process of CHAP is relatively straight forward (Easttom and Dulaney 2017):

1. The client sends their username and password to the server (often called the authenticator in CHAP documentation).
2. The server/authenticator then requests that the client calculate a cryptographic hash and send that to the server.
3. Periodically, at random intervals, the server/authenticator will request that the client re-present that cryptographic hash. The purpose of this is to detect if session hijacking has occurred, and then terminate that connection.

This process is depicted in Fig. 14.2.

It should be noted that Microsoft created their own variation of CHAP, designated MS-CHAP. This conforms to the general description given above, the variations are not pertinent to our discussions of VPNs. For example, MS-CHAP v2 provides for mutual authentication, wherein both sides authenticate each other.

EAP

Extensible Authentication Protocol, as the name implies is actually a framework that can be modified, or extended, for a variety of purposes. The EAP standard was

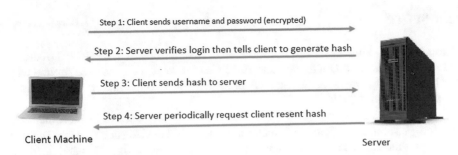

Fig. 14.2 The CHAP process

originally specified in RFC 2284, but that was supplanted by RFC 3748, then later updated by RFC 5247. The various permutations of EAP differ in how authentication material is exchanged. There are many variations of EAP, let us take a look at just a few.

LEAP

Lightweight Extensible Authentication protocol was developed by Cisco and has been used extensively in Wireless communications. LEAP is supported by many Microsoft operating systems including Windows 10. LEAP uses a modified version of MS-CHAP.

EAP-TLS

Extensible Authentication Protocol - Transport Layer Security utilizes TLS (see Chap. 13) in order to secure the authentication process. Most implementations of EAP-TLS utilize X.509 digital certificates (also discussed in Chap. 13) to authenticate the users.

PEAP

Protected Extensible Authentication Protocol encrypts the authentication process with an authenticated TLS tunnel. PEAP was developed by a consortium including Cisco, Microsoft, and RSA Security. It was first included in Microsoft Windows XP but is still in use today.

EAP-POTP

Protected One-Time Password is a variation of EAP described in RFC 4793. It uses a one-time password token to generate authentication keys. I can be used for one way or bilateral authentication (i.e., the client authenticates to the server, and the server authenticates to the client.

A One Time Password (OTP) is, as the name suggests, used only once, then it is no longer valid. It is often used in conjunction with a device such as a key fob that generates the OTP. The key fob and the server use the same algorithm and tie in the current time stamp to generate the same OTP on both ends. Often this is used in conjunction with a traditional password.

Kerberos

Kerberos is a network authentication protocol. It is designed to provide strong authentication for client/server applications by using symmetric cryptography. It was originally developed at MIT under the name Project Athena. The protocol was named after the Greek mythological character Kerberos (or Cerberus), known in Greek mythology as being the monstrous three-headed guard dog of Hades (Neuman & Ts'o, 1994).

Kerberos is widely used and has been implemented in versions from Windows 2000 to the current (Server 2019 as of this writing), as well as Red Hat Enterprise Linux, FreeBSD, IBM AIX, and other operating systems. The process itself is purposefully complex. The reason for the complexity is to prevent session hijacking attacks, as well as to ensure a more robust and reliable authentication mechanism. There are variations on the process depending on implementation, but the general process is as follows (Easttom 2019; Neuman and Ts'o 1994):

- The username is sent to the Authentication Server (AS). AS generates a secret key by retrieving the stored hash of the user password then sends two messages to client, both messages are encrypted with the key the AS generated:

  - Message A: CLIENT/Ticket Granting Session Key encrypted with secret key of client.
  - Message B: Ticket Granting Ticket (TGT) includes client ID, client network address, and validity period.

- Then the client machine attempts to decrypt message A with a secret key generated by the client hashing the users entered password. If that entered password does not match the password the AS found in the database, then the hashes won't match, and the decryption won't work. If it does work, then message A contains the Client/TGS session key that can be used for communications with the TGS. Message B is encrypted with the TGS secret key and cannot be decrypted by the client.

- When requesting services, the client sends the following messages to the TGS:
  - Message C: Composed of the TGT from message B and the ID of the requested service.
  - Message D: Authenticator (which is composed of the client ID and the timestamp), encrypted using the Client/TGS Session Key.
- Upon receiving messages C and D, the TGS retrieves message B out of message C. It decrypts message B using the TGS secret key. This gives it the "client/TGS session key." Using this key, the TGS decrypts message D (Authenticator) and sends the following two messages to the client:
  - Message E: Client-to-server ticket (which includes the client ID, client network address, validity period and Client/Server Session Key) encrypted using the service's secret key.
  - Message F: Client/Server Session Key encrypted with the Client/TGS Session Key.
- Upon receiving messages E and F from TGS, the client has enough information to authenticate itself to the Service Server (SS). The client connects to the SS and sends the following two messages:
  - Message E from the previous step (the client-to-server ticket, encrypted using service's secret key).
  - Message G: a new Authenticator, which includes the client ID, timestamp and is encrypted using Client/Server Session Key.
- The Service Server (SS) decrypts the ticket (message E) using its own secret key to retrieve the Client/Server Session Key. Using the sessions key, SS decrypts the Authenticator and sends the following message to the client to confirm its true identity and willingness to serve the client:
  - Message H: the timestamp found in client's Authenticator.
- The client decrypts the confirmation (message H) using the Client/Server Session Key and checks whether the timestamp is correct. If so, then the client can trust the server and can start issuing service requests to the server.
- The server provides the requested services to the client.

This process is intentionally complex with multiple reverification steps. This is done in order to make circumventing the system much more difficult. It is likely that you have used this process without knowing it, as it is the default authentication mechanism in Windows domains. All of the complex process described is transparent to the user. There are some basic Kerberos terms you should be familiar with:

Principal – a server or client that Kerberos can assign tickets to
Authentication Server (AS) – Server that gives authorizes the principal and connects them to the Ticket Granting Server
Ticket Granting Server (TGS) – provides tickets

Key Distribution Center (KDC) – A server that provides the initial ticket and handles TGS requests. Often it runs both AS and TGS services

Ticket Granting Ticket (TGT) – the ticket that is granted during the authentication process

Ticket – used to authenticate to the server. Contains identity of client, session key, timestamp, and checksum. Encrypted with server's key

Session key – temporary encryption key

The complexity you see in Kerberos is intentional. The constant exchange of encrypted tickets and verification of information makes session hijacking far more difficult. And the combination of the session key with the ticket granting ticket would mean an attacker would need to crack two *strong* encryption keys in order to subvert the system.

SESAME

Secure European System for Applications in a multivendor Environment (SESAME) is the European answer to Kerberos. It was developed to extend Kerberos and improve on its weaknesses. SESAME uses both symmetric and asymmetric cryptography. SESAME "Privileged Attribute Certificates" rather than tickets, PACS are digitally signed and contain the subjects identity, access capabilities for the object, access time period and lifetime of the PAC. PACS come from the Privileged Attribute Server.

NTLM

NTLM is not used in modern VPNs but is commonly used in Microsoft products to provide authentication as well other security services, and thus you should be basically familiar with it. While Microsoft recommends Kerberos as the preferred authentication protocol, many implementations still utilize NTLM. NTLM is a three-step process:

1. The first step is for the client to connect to the server then send a NEGOTIATE _MESSAGE advertising its capabilities.
2. The server then responds with CHALLENGE_MESSAGE. This is used to establish the identity of the client.
3. The process culminates with the client responding to the servers challenge with an AUTHENTICATE_MESSAGE.

Authentication in NTLM is accomplished by sending a hash of the password. NTLM version 1 uses DES-based LanMan hash. NTLM version 2 uses MD4 cryptographic hash.

NTLM is vulnerable to the pass the hash attack. Essentially the attacker acquires a hash of the user's password and sends that along. The attacker does not know the actual password but has somehow acquired the hash. This can be done if the user has the same password for a local password, thus the attacker can get that hash from the local SAM file (Security Account Manager) that is in Windows. It may also be possible to get the hash via packet sniffing. Sometimes hashes are cached on a machine, and the attacker gets the hash from that cache. The exact mechanism whereby the attacker obtains the hash is not important. Once the attacker has it, they can execute the pass the hash attack and login as that user without ever actually knowing the users password.

PPTP

Point to Point Tunneling Protocol is an older method. The standard was first published in 1999 as RFC 2637. The specification did not expressly define specific encryption or authentication modalities and left these up to the implementation of PPTP.PPTP has been supported in all versions of Windows since Windows 95. Originally the Microsoft implementation of PPTP used MS-CHAP for authentication and DES for encryption (Tiller 2017).

PPTP is based on the earlier protocol of PPP (Point to Point Protocol) and uses TCP port 1723 to establish a connection. After the initial connection is established, PPTP creates a GRE tunnel. GRE (Generic Routing Encapsulating) is a protocol developed by Cisco. Because PPTP is based on PPP, a brief description of PPP is in order. PPP was designed for moving datagrams across serial point-to-point links. It sends packets over a physical link, a serial cable set up between two computers. It is used to establish and configure the communications link and the network layer protocols, and also to encapsulate datagrams. PPP has several components and is actually made up of several protocols:

Each of these handles a different part of the process. PPP was originally developed as an encapsulation protocol for transporting IP traffic over point-to-point links. PPP also established a standard for a variety of related tasks including:

PPP supports these functions by providing an extensible Link Control Protocol (LCP) and a family of Network Control Protocols (NCPs) to negotiate optional configuration parameters and facilities. In addition to IP, PPP supports other protocols. This is no longer such an advantage as now all mainstream networks used TCP/IP. However, at the time of the creation of PPTP Novel used IPX/SPX, Apple used AppleTalk, and Unix used TCP/IP. Now all of these, as well as Windows, use TCP/IP.

PPTP supports two generic types of tunneling, voluntary and compulsory. In the case of voluntary tunneling, a remote user connects to a standard PPP session that enables the user to log on to the provider's network. The user then launches the VPN software to establish a PPTP session back to the PPTP remote-access server in the central network. This process is called voluntary tunneling because the user initiates the process and has the choice of whether to establish the VPN session. While not advisable, the user could simply use a standard PPP connection without the benefits of a VPN.

In a compulsory tunneling setup the only connection available to the host network is via a VPN. A simple PPP connection is not available, only the full PPTP connection, forcing users to use a secure connection. From a security standpoint this is the preferred option.

PPTP Authentication

When connecting to a remote system, encrypting the data transmissions is not the only facet of security. You must also authenticate the user. PPTP supports two separate technologies for accomplishing this: Extensible Authentication Protocol (EAP) and Challenge Handshake Authentication Protocol (CHAP). Both of these where described earlier in this chapter.

PPTP Encryption

The PPP payload is encrypted using Microsoft's Point-to-Point Encryption protocol (MPPE). MPPE was designed specifically for PPTP. It uses RC4 with either a 128-bit, 56-bit, or 40-bit key. Obviously, 128 bit is the strongest. One advantage to MPPE is that the encryption keys are frequently changed. During the initial negotiation and key exchange for MPPE a bit is set to indicate what strength of RC4 to use:

128-bit encryption ('S' bit set)
56-bit encryption ('M' bit set)
40-bit encryption ('L' bit set)

L2TP

L2TP sends its entire packet (both payload and header) via User Datagram Protocol (UDP). The endpoints of an L2TP tunnel are called the L2TP Access Concentrator (LAC) and the L2TP Network Server (LNS). The LAC initiates a tunnel, connecting to the LNS. After the initial connection, however, communication is bilateral. Like PPTP, L2TP can work in either voluntary or compulsory tunnel mode. However, there is a third option for an L2TP multihop connection.

The major differences between L2TP and PPTP are listed here:

L2TP provides the functionality of PPTP, but it can work over networks other than just IP networks. PPTP can only work with IP networks. L2TP can work with Asynchronous Transfer Mode (ATM), Frame Relay, and other network types.

L2TP supports a variety of remote access protocols such as TACACS+ and RADIUS, while PPTP does not.

While L2TP has more connection options, it does not define a specific encryption protocol. The L2TP payload is encrypted using the IPSec protocol. The RFC 4835 specifies either the 3DES or AES encryption. IPSec is considered far more secure that the authentication and encryption used in PPTP.

IPSEC

IPSec, short for Internet Protocol Security, is a technology used to create virtual private networks. IPSec is a security used in addition to the IP protocol that adds security and privacy to TCP/IP communication. IPSec is incorporated with Microsoft Operating Systems as well as many other operating systems. For example, the security settings in the Internet Connection Firewall that ships with Windows XP enables users to turn on IPSec for transmissions. IPSec is a set of protocols developed by the IETF (Internet Engineering Task Force, www.ietf.org) to support secure exchange of packets. IPSec has been deployed widely to implement VPNs (Davis 2001).

IPSec has two encryption modes: transport and tunnel. The transport mode works by encrypting the data in each packet but leaves the header unencrypted. This means that the source and destination address, as well as other header information, are not encrypted. The tunnel mode encrypts both the header and the data. This is more secure than transport mode but can work more slowly. At the receiving end, an IPSec-compliant device decrypts each packet. For IPSec to work, the sending and receiving devices must share a key, an indication that IPSec is a single-key encryption technology. IPSec also offers two other protocols beyond the two modes already described. Those protocols are AH (Authentication Header) and ESP (Encapsulated Security Payload).

IPSec is probably the most widely used VPN protocol today. The reason being that, unlike PPTP and L2TP, IPSEC provides a complete solution that includes built in authentication and encryption. IPSec is actually comprised of multiple protocols:

Authentication Header (AH) this is used to provide for authentication and integrity for packets. The Authentication Header contains several pieces of information including: Payload length, an Integrity Check Value (ICV), sequence number, and Security Parameters Index (SPI). The header format is shown in Fig. 14.3.

Encapsulating Security Payload (ESP) provides confidentiality as well as authentication and integrity. The Encapsulating Security Payload header has some similar items to what that Authentication Header has. Some of the responsibilities for the two headers overlap (Tiller, 2017). You can see that header in Fig. 14.4.

Offsets	Octet	0								1								2								3							
																		Authentication Header															
Octet	Bit	0	1	2	3	4	5	6	7	8	9	10	11	12	13	14	15	16	17	18	19	20	21	22	23	24	25	26	27	28	29	30	31
0	0				Next Header									Payload Len														Reserved					
4	32													Security Parameters Index (SPI)																		
8	64													Sequence Number																		
C	96													Integrity Check Value (ICV)																		

Fig. 14.3 The Authentication Header

Offsets	Octet	0								1								2								3							
																		Encapsultating Security Payload Header (ESP)															
Octet	Bit	0	1	2	3	4	5	6	7	8	9	10	11	12	13	14	15	16	17	18	19	20	21	22	23	24	25	26	27	28	29	30	31
0	0													Security Parameters Index (SPI)																		
4	32													Sequence Number																		
8	64													Payload data																		
														Padding (0-255 octets)																		
																							Pad Length						Next Header		
														Integrity Check Value (ICV)																		

Fig. 14.4 The Encapsulating Security Payload Header

The broad steps of the IPSec process are as follows

1. Some party decides to initiate an IPSec VPN. That starts the Internet Key Exchange (IKE) Process.
2. During IKE phase 1, the peers are authenticated, and security associations are negotiated.
3. During IKE phase 2 the Security Authentication parameters are negotiated and finalized.
4. Now data can be transmitted.

One concept you will see throughout IPSec is that of the Security Association, also called an SA. This is a set of parameters such as algorithms, keys, and other items necessary to encrypt and authenticate. One part of IPSec initiation is to ensure that both peers in an IPSec connection have identical Security Associations. These security associations are established using the Internet Security Association and Key Management Protocol (ISAKMP). This protocol is most often implemented with IKE, which occurs in two phases.

IKE Phase 1

The primary purpose of Internet Key Exchange phase one is to authenticate the IPSec peers and to set up a secure channel between the peers to enable IKE exchanges. In other words, first the two parties are authentication, then subsequent steps involving exchange of keys can take place. IKE phase one is subdivided into a few phases:

Authenticates the identities of the IPSec peers.
Negotiates a matching IKE SA policy between the peers.
Uses Diffie-Hellman exchange so that both peers have the same secret key.
Now using that same secret key, the IKE phase two parameters can be securely
  exchanged.

There are two methods or modes for performing IKE phase 1. Those modes are:

Main Mode

Main mode is the primary or preferred mode for IKE phase 1. In main mode, there are three two-way exchanges between the initiator and receiver. In the first exchange, the peers agree on the symmetric algorithms to secure communication and the cryptographic hashes used to protect integrity. Again, this information is being exchanged via the key established using Diffie-Hellman.

In the second exchange the two peers will generate secret key material used to generate secret keys on both ends. This is similar to the process you saw for SSL/TLS in Chap. 13. The IPSec standard allows for HMAC, SHA1, and SHA2 to be used for authentication and 3DES and AES (both with cipher block chaining) to be used for encryption. The third exchange involves each peer verifying the others identity. At the end of this exchange the two peers should have identical Security Associations (SA's). Now a secure channel exists between the two peers.

Aggressive Mode

Aggressive mode essentially condenses the process of main mode into fewer exchanges. The entire process is condensed into just three packets. All of the information required for the Security Association is passed by the initiating peer. The responding peer then sends the proposal, key material and ID, and authenticates the session in the next packet. The initiating peer then replies by authenticating the session. And at that point IKE Phase 1 is complete.

IKE Phase 2

You may think that after IKE phase 1 there is nothing left to do. However, IPSec takes additional steps to ensure that there is secure, authenticated communication. IKE phase 2 only has one mode, called quick mode (Tiller 2017). This occurs immediately after IKE phase 1 is complete. Phase 2 negotiates a shared IPSec policy

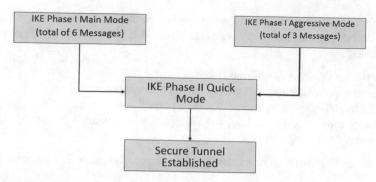

Fig. 14.5 IKE Phase I and II

and then derives secret keying material used for the IPSec security algorithms. Phase 2 also exchanges nonce's (recall these are Numbers Only used oNCE) that provide protection against a variety of attacks, including replay attacks. The nonce's are also used generate new secret key material and prevent replay attacks. IKE Phase 2 can also be used to renegotiate a new Security Associations if the IPSec SA lifetime expires but the peers have not terminated communication.

The basic flow of these phases is shown in Fig. 14.5.

During establishment of IPSec policies, there is a setting called Perfect Forward Secrecy. If this is specified in the IPSec policy, then a new Diffie-Hellman exchange is performed with each iteration of IKE phase 2 quick mode. This means new key material is used each time IKE phase 2 is executed. While IKE is the most common way to set up Security Associations, there are other methods. IPSECKEY DNS records are one method and using Kerberos in a process termed Kerberized Internet Negotiation of Keys is another.

SSL/TLS

A new type of firewall uses SSL (Secure Sockets Layer) or TLS (Transport Layer Security) to provide VPN access through a web portal. Essentially TLS and SSL are the protocols used to secure websites. If you see a website beginning with HTTPS, then traffic to and from that website is encrypted using SSL or TLS.

Now there are VPN solutions wherein the user simply logs in to a website, one that is secured with SSL or TLS, and is then given access to a virtual private network. It is important to know that simply visiting a website that uses SSL or TLS does not mean you are on a VPN. As a general rule most websites, like banking websites, only give you access to a very limited set of data, like your account balances. A VPN gives you access to the network, the same or similar access to what you would have if you were physically on that network.

The process to connect, authenticate, and establish the VPN is the same as what was described in Chap. 13. The difference is that instead of simply getting secure access to a website such as a bank or e-commerce site, the end user gets a virtual private network connection to their network.

Other Secure Communications

There are a variety of additional secure communications that are not considered virtual private networks, but they do utilize authentication and encryption to provide secure communication with some end point. SSL/TLS communication with a secure website, described in Chap. 13 would be the most common example of such a communication. In this section we will look at a few additional communication methods.

SSH

It is common for network administrators to need to utilize some secure communication channel to a server. Telnet has long been used for this purpose. However, Telnet sends the data in clear text, and is thus insecure. The only way to secure Telnet is to first establish a VPN then connect via Telnet through the VPN. Secure Shell provides an alternative that is secure, without the need to first establish a VPN.

Unix and Unix-based systems such as Linux utilize SSH to connect to a target server. The SSH standard uses asymmetric cryptography to authenticate the remote computer and, when mutual authentication is required, to authenticate the client. SSH was first released in 1995 and was developed by Tatu Ylonen, at Helsinki University of Technology. His goal was to replace insecure protocols such as Telnet, rsh, and rlogin. SSH version 1 was released as freeware. By 1999 OpenSSH had been released and is still a very popular version of SSH.

SSH version 2 has an internal architecture with specific layers responsible for particular functions.

The transport layer handles the key exchange and authentication of the server. Keys are re-exchanged usually after either 1 hour of time has passed, or 1 gigabyte of data has been transmitted. This re-negotiation of keys is a significant strength for SSH.

The user authentication layer is responsible for authenticating the client. There are a few ways this can be done, the two most common are password and public key. The password method simply checks the user's password. The public key method uses either DSA or RSA key pairs to verify the client's identity and can also support X.509 certificates.

GSSAPI (Generic Security Service Application Program Interface) authentication is variation of SSH authentication to allow for the use of either Kerberos or NTLM to authenticate. While not all versions of SSH support GSSAPI, OpenSSH does.

SSH can be used to provide secure file transfer with technologies such as SCP (Secure Copy), SFTP (SSH File Transfer Protocol), and FISH (Files transferred over SSH).

SSH can be configured to use several different symmetric algorithms including AES, Blowfish, 3DES, CAST128, and RC4. The specific algorithm is configured for each SSH implementation.

Wi-Fi Encryption

One obvious application of cryptography is encrypting wireless communications. Wi-Fi is now ubiquitous. Most homes in North America, Europe, Japan, and much of the world have wireless internet in them. Free Wi-Fi can be found at bars, coffee shops, airports, fast food restaurants, and other locations. For a fee one can even access Wi-Fi in flight on many airlines. The pervasiveness of Wi-Fi means that securing wireless communications is critical.

WEP

Wired Equivalent Privacy was released in 1999. It uses RC4 to encrypt the data and a CRC-32 checksum for error checking. Standard WEP uses a 40-bit key (known as WEP-40) with a 24-bit initialization vector, to effectively form 64-bit encryption. 128-bit WEP uses a 104-bit key with a 24 bit IV(Initialization Vector). Because RC4 is a stream cipher, the same IV must never be used twice. The purpose of an IV, which is transmitted as plain text, is to prevent any repetition, but a 24-bit IV is not long enough to ensure this on a busy network. The IV must never be repeated (recall the concept of a number only used once).

In August 2001, Scott Fluhrer, Itsik Mantin, and Adi Shamir published a crypt-analysis of WEP. They showed that due to the improper implementation of RC4 along with the re-use of initialization vectors, WEP could be cracked with relative ease. WEP should never be used today unless there is absolutely no other choice. However, the story of WEP illuminates one reason more security professionals need to understand cryptography (and thus the reason for the book you are holding in your hands right now). WEP was created by a consortium of computer companies, each contributing engineers to work on the standard. Clearly these were skilled engineers with a solid understanding of network communications. However, they lacked an appropriate knowledge of cryptography and this led to an insecure wireless encryption standard.

WPA

Wi-Fi Protected Access or WPA was released in 2003. It was meant as an intermediate step to make up for the issues with WEP while the full implementation of 802.11i (WPA2) was being finalized. One advantage of WPA over WEP is the use of the Temporal Key Integrity Protocol. TKIP is a 128-bit per-packet key that generates a new key for each packet. This regeneration of keys makes WPA much stronger than WEP.

Note: 802.11 is the IEEE standard for wireless communications. If you have an 'N' router, what you really have is a router that complies with the 802.11n wireless standard. 802.11i is the IEEE standard for wireless security.

WPA operates in one of two modes.

WPA-Personal: Also referred to as *WPA-PSK* (Pre-shared key) mode. Is designed for home and small office networks and doesn't require an authentication server. Each wireless network device authenticates with the access point using the same 256-bit key.

WPA-Enterprise: Also referred to as *WPA-802.1x* mode, is designed for enterprise networks, and requires a RADIUS authentication server. An Extensible Authentication Protocol (EAP) is used for authentication. EAP has a variety of implementation such as EAP-TLS and EAP -TTLS.

WPA-2

WPA-2 is the only Wi-Fi encryption protocol that fully implements the IEEE 802.11i standard. It uses a robust set of cryptographic algorithms. Confidentiality is maintained through the use of Advanced Encryption Standard (AES) using Cipher Block Chaining (CBC). Message integrity is protected via Message Authentication Code's. WPA-2 allows for the optional use of Pairwise Master Key (PMK) caching and opportunistic PMK caching. In PMK caching, wireless clients and wireless access points cache the results of 802.1X authentications. This improves access time.

The optional use of pre-authentication which allows a WPA2 wireless client can perform an 802.1X authentication with other wireless access points in its range even though it is still connected to the current WAP. This also speeds connectivity. In modern Wi-Fi systems you should always select WPA-2 unless you have a compelling reason to use one of the other Wi-Fi protocols. In some cases, you may need to support older systems that cannot perform WPA-2.

WPA-3

WPA-3 is the latest version of WPA and features many security enhancements. First, and perhaps most obvious, is that even on open networks, traffic between the client and the Wireless Access Point (WAP) is encrypted. WPA3 attackers to interact with your Wi-Fi for every password guess they make, making it much harder and time-consuming to crack. WPA-3 also uses mutual authentication when operating in personal mode. Specifically, it uses the Simultaneous Authentication of Equals defined in IEEE 802.11-2016 standard.

Conclusions

The point of this chapter, and the preceding chapter, is to demonstrate common applications of the cryptography you have been studying throughout this book. There are two sides to cryptography. The first is the understanding of the mathematics and algorithms being utilized to secure communications. The second is to have a working knowledge of the protocols that implement these cryptographic algorithms.

In this chapter, we explored some of the most common applications of cryptography. Virtual Private Networks (VPNs) were the focal point of this chapter. It is important that you understand the authentication methods presented as well as the VPN protocols presented. Most important are CHAP, Kerberos, L2TP, and IPSec. You were also introduced to other applications of cryptography such as Secure Shell (SSH) and Wi-Fi encryption. Both of these are common applications of the cryptographic algorithms you have learned previously in this book.

Test Your Knowledge

1. What is a Kerberos principle?
2. What attack is NTLM particularly susceptible too?
3. The IPSec standard describes three algorithms used for integrity. What are they?
4. What phase of IPSec involves the exchange of Nonces?
5. What authentication protocols does PPTP use?
6. Which authentication method is based on periodically re-requesting a hash from the client?
7. How many steps are in IKE Phase 1 in aggressive mode?
8. What encryption does PPTP use?
9. What is the major weakness in WEP?
10. Which Wi-Fi encryption protocol completely implements 802.11i?

References

Davis, C. R. (2001). IPSec: Securing VPNs. McGraw-Hill Professional New York.

Easttom, C. (2019). Computer Security Fundamentals, 4th. New York City, New York: Pearson Press.

Easttom, C. & Dulaney, E. (2017). CompTIA Security+ Study Guide: SY0-501. Hoboken, New Jersey: Sybex Press.

Neuman, B. C., & Ts'o, T. (1994). Kerberos: An authentication service for computer networks. IEEE Communications magazine, 32(9), 33-38.

Oppliger, R. (2016). SSL and TLS: Theory and Practice. Norwood, MA Artech House.

Tiller, J. S. (2017). A technical guide to IPSec virtual private networks. Boca Raton CRC Press.

Yuan, R., & Strayer, W. T. (2001). Virtual private networks: technologies and solutions. Addison-Wesley Longman Publishing Co., Inc: Boston..

Chapter 15
Military Applications

Abstract Military applications are a common application of cryptography. This chapter explores such applications, including the United States National Security Administration. We also explore cryptographic laws and regulations, as well as the application of cryptography in cyber warfare. These are common applications of cryptography and are an important part of applied cryptography. This chapter will also examine the TOR network, as that network is based on cryptographically securing all communications.

Introduction

It may seem a bit odd to some readers to have a separate chapter specifically for military applications of cryptography. After all, isn't the military the primary user of cryptography? Prior to the internet and the advent of e-commerce, that was undeniably accurate. However, for the past several decades banks have utilized cryptography, websites, individuals sending email or securing their hard drives, all use cryptography. In fact, the civilian applications for cryptography are quite numerous. And all we have discussed thus far in this book can be used for civilian purposes, though some of them (such as AES and GOST) are also used for military purposes.

In this chapter, we specifically examine the applications of cryptography that are exclusively (or nearly so) the domain of militaries, governments, and intelligence agencies. While civilian organizations certainly have a need for secure communications, the need is more pressing in military and intelligence applications for two reasons. The first reason is that the stakes are much higher. Rather than money being lost or embarrassing data being leaked, lives might be lost. In the event of armed conflict, a breach of security could provide the opponent with a tactical or even strategic advantage. The second reason the cryptographic needs of the military are different from civilians is the nature of the persons attempting to breach security. Militaries are not worried about solo hackers or similar threats so much as concerted efforts by trained intelligence personnel to breach their communication.

It should be noted that the line between military and law enforcement can sometimes become blurred. The most obvious example is the investigation of terrorist organizations. Their activities are pursued by law enforcement agencies, but also by militaries and intelligence agencies. International criminal organizations also blur the lines between intelligence agencies and law enforcement. In the latter part of this chapter, we discuss some items that fit in both criminal and terrorist categories.

In this chapter, we discuss classified equipment and algorithms. However, we only be discuss those aspects that are accessible in the public domain. So, you may find some algorithm descriptions much more vague than the algorithms you have explored earlier in this book. For what should be obvious reasons, it is not possible to give complete descriptions of classified algorithms in a published book.

In This Chapter We Will Cover

NSA And Cryptography
    U.S. Cryptography laws
    How do other nations use cryptography?
    Cryptography and Malware
    TOR

NSA and Cryptography

It would be impossible to discuss military applications of cryptography without discussing the United States National Security Agency. In Chap. 2, we briefly discussed the history of the NSA up through the cold war. In this section, we first discuss the NSA's modern cryptographic role.

Security Classifications

Throughout this chapter, you will see items designated as "secret" or "top secret." It is important to understand what these terms mean. Each nation has its own classification system, and even some agencies within the United States have their own. In this chapter, the terms used are in reference to the United States Department of Defense classifications.

The terms secret and top secret have specific meanings. The United States has a specific hierarchy of classification. The lowest being confidential. This is information that might damage national security if disclosed. Secret information is data that might cause serious damage to national security if disclosed. Top secret information is data that could be expected to cause exceptionally grave damage to national security if disclosed. There is another designation: Top Secret SCI or Sensitive Compartmented Information.

Each of these clearances requires a different level of investigation. For a secret clearance, a complete background check including criminal, work history, credit check, and check with various national agencies (Department of Homeland Security, Immigration, State Department, etc.) is required. This is referred to as a NACLC or National Agency Check with Law and Credit. The check for employment will cover the last 7 years. The secret clearance may or may not include a polygraph.

The top-secret clearance is more rigorous, as you may imagine. It uses a Single Scope Background Investigation (SSBI). This means a complete NACL for the subject and their spouse that goes back at least 10 years. It will also involve a subject interview conducted by a trained investigator. Direct verification of employment, education, birth, and citizenship are also required. At least four references are required, and at least two of those will be interviewed by investigators. A polygraph is also used. The SSBI is repeated every 5 years.

Sensitive Compartmented Information is assigned only after a complete SSBI has been completed. An SCI may have its own process for evaluating access, therefore, a standard description of what is involved is not available.

Regardless of the clearance level, in any clearance investigation, should any issues arise, the scope will be expanded to resolve those issues. For example, if a specific issue arises in regard to the applicants education, but that education occurred 20 years ago, the investigation would be expanded to address that issue.

NSA Cryptographic Standards

Since the early days of the NSA, the agency has been responsible for U.S. Government cryptography. The National Security Administration defines cryptographic algorithms in two ways. First, there are four product types labeled type 1, 2, 3, and 4. Then there are two suites of algorithms named Suite A and B.

Type 1 Products

Type 1 products are those that the NSA endorses for use on classified U.S. purposes (Bauer 2013; Davida 1981). This often includes equipment and algorithms that are classified, though in some cases, classified equipment might use an unclassified algorithm.

HAIPE-IS

HAIPE-IS: HAIPE or High Assurance Internet Protocol Encryptor is a device that can use both Suite A and Suite B algorithms. A HAIPE device is often used as a secure gateway to connect two sites. HAIPE-IS is based on IPSec with additional enhancements. The device's purpose is to encrypt IP traffic. Much of it functions in a manner transparent to the user. There are specialized Cisco routers that incorporate HAIPE into them.

HAVE QUICK

HAVE QUICK is actually a frequency hopping algorithm, originally developed for Ultra High Frequency (UHF) radios used between ground and air. Military radio traffic hops through a range of frequencies making it very difficult to jam signals. For signals to employ HAVE QUICK, it is important that both ends are initialized with an accurate time of day (often called TOD). They often also use a Word of the Day (WOD) that serves a key for encrypting transmissions. Finally, they use a NET number to select a specific network. HAVE QUICK is not itself an encryption system, however, many systems combine HAVE QUICK with encryption in order to provide confidentiality and prevent jamming.

SINCGARS

Single Channel Ground and Airborne Radio System (SINCGARS) is a VHF FM band radio using from 30 to 87.975 MHz. It is currently used by the United States Military as well as other countries. It can provide both single frequency and frequency hopping modes. Early units did not have built-in crypt and required external cryptographic units. Later versions had built-in cryptographic units.

There have been many models starting with the RT-1439 produced in 1988. More recent developments include the RT-1523G, RT-1730C, and E (for Naval applications) and RT-1702G made to be carried by an individual solider. One example of these systems is shown in Fig. 15.1.

Fig. 15.1 SINCGARS system

Type 2 Products

Type 2 products are those the NSA endorses for sensitive but unclassified purposes. The KEA asymmetric algorithm and the SKIPJACK block cipher are examples. Equipment includes Fortezza, CYPRIS, etc.

Fortezza Plus

The Fortezza Plus card, also known as the KOV-14 card, is PC card that provides cryptographic functions and keys for secure terminal equipment. The original Fortezza Crypto Card was a card that contained a security token. Each user is issued a Fortezza card that includes, among other things, private keys used to access sensitive data. The original Fortezza card was developed to use the SKIPJACK algorithm.

Fortezza Plus is an improvement that uses a classified algorithm. Fortezza Plus is used with secure terminal equipment for voice and data encryption. There is an even more modern improvement on Fortezza Plus called the KSV-21. KSV-21 is backward compatible with the Fortezza Plus card. You can see a Fortezza card shown in Fig. 15.2.

Fishbowl

Fishbowl is an entire architecture designed to provide secure voice over IP. Phones utilized two layers of encryption. When using the phone on a commercial channel, all communications must be sent through a managed server. The entire purpose is to allow for secure communications.

Fig. 15.2 Fortezza Card

Type 3 and 4 Products

Type 3 products are unclassified. Many of the algorithms we have examined in this book so far are in this category including DES, AES, SHA, etc. Type 4 products are those that the NSA has not endorsed for any government use. These are usually products that have such weak encryption as to make them of no practical value.

Suite A

Suite A algorithms are unpublished, classified algorithms. These are used for highly sensitive systems. Suite A algorithms include MEDLEY, BATON, SAVILLE, WALBURN, JOSEKI, and SHILLELAGH.

SAVILLE

The SAVILLE algorithm is often used in voice encryption. While its details are classified, some indications suggest it may have a 128-bit key. This algorithm was purportedly a joint project between NSA and British Intelligence, developed in the late 1960s.

BATON

The BATON algorithm itself is classified. However, the publicly available standard PKCS#11 has some general information about BATON. PKCS public key cryptography standards. The PKCS 11 defines an API for cryptographic tokens such as those used in smart cards. BATON is a block cipher using a 128 block with a 320-bit key. It can also be used in ECB mode with a 96-bit block.

While this was mentioned earlier in this book, it is worth mentioning again here. The NSA using classified algorithms is not in conflict with the Kerckhoff principle. To begin with the NSA is the largest employer of mathematicians and cryptographers in the world. They can subject an algorithm to internal peer review that is quite exhaustive. Furthermore, Kerckhoff teaches us that the security of a cryptographic mechanism should only depend on the secrecy of the key, not the secrecy of the algorithm. The key word being *depend*. The Suite A algorithms do not depend on the secrecy of the algorithm for security. But it does at a bit of additional security.

FIREFLY

This is a key exchange protocol (similar in purpose to Diffie-Hellman). It is used in specific cryptographic systems, including secure phones.

Suite B

Suite B algorithms are published. AES is a perfect example. Suite A includes many algorithms you have already seen. These algorithms are all publicly available and many have been fully described in previous chapters of this book:

Advanced Encryption Standard. If using a 128-bit key, AES is considered secure enough for secret information. If using a 256-bit key, AES is considered secure enough for top secret information.

– Elliptic Curve Digital Signature Algorithm.
– Elliptic Curve Diffie Hellman.
– Secure Hash Algorithm 2 (SHA2 256, 384, and 512).
– Elliptic Curve. ECC with a 384-bit key is considered secure enough for top secret information.

The Modern Role of the NSA

The NSA has been involved in encrypted communications for military purposes since its inception (Bamford 2018). This involves both developing/approving cryptography for U.S. Government use, but also attempting to compromise the cryptographic communications of other countries. In the past several years, NSA involvement in cyber espionage has increased. The Office of Tailored Access Operations is a cyber warfare and intelligence gathering unit within the NSA. One of its goals is to infiltrate foreign systems. Edward Snowden revealed that the TAO has a suite of software tools used specifically for breaking into systems.

Despite the NSA expanding into areas of cyber espionage and cyber warfare, its primary role is still secure communications. The NSA leads the United States government in creating and approving cryptography. For example, for the past few years, NSA has been recommending moving from RSA to Elliptic Curve Cryptography. And many U.S. Government agencies are now using variation of ECC, including ECC Digital Signature Algorithm and ECC Diffie Hellman. Data mining and similar roles involve heavy use of both mathematics and computer science, so it is only natural that the NSA be involved in those activities.

U.S. Cryptography Laws and Regulations

Export laws regarding cryptography were quite strict up until the year 2000 when they have relaxed. Prior to 1996, the U.S. Government regulated most cryptography exports under the auspices of the Arms Export Control Act and the International Traffic in Arms Regulations. This meant that cryptography was treated much like weapons and ammunition. The laws did not relax at one time, but it was a gradual change in the U.S. Government attitude toward exporting cryptography.

One of the first steps was made in 1992 when 40-bit RC2 and RC4 were made available to export and no longer governed by the State Department, but rather managed by the Commerce Department. When Netscape developed SSL, this required a re-examination of cryptographic export laws. The U.S. Version used RSA with key size of 1024 bits and larger, along with 3DES or 128-bit RC3. The international version used the 40-bit RC4.

There were various legal challenges to cryptography export rules, throughout the early 1990s. The expansion of e-commerce and online banking also required some re-thinking of these export laws. In 1996, President Bill Clinton signed an executive order that moved commercial cryptography off of the munitions list and treated as standard commerce. In 1999, regulations were again relaxed allowing export of 56-bit symmetric keys (DES) and 1024-bit RSA. Currently, certain technologies are exempt from any export controls:

Software or hardware specifically designed for medical use is exempted from export controls.

Cryptography that is specifically used for copyright or intellectual property protections is exempted from export controls.

In general, as of this writing, in order to export cryptographic products from the United States to other nations, one needs a license from the U.S. Commerce Department. And, for the most part, these restrictions have been relaxed. For example, McAfee's Data Protection Suite, which includes encryption, has been granted an "ENC/Unrestricted" license exception by the U.S. Department of Commerce.

When exporting any software that includes cryptographic functions, it is important to check with the appropriate government entity within your country. For example, in the United States, in 2014, the company Wind River Systems (a subsidiary of Intel) was fined $750,000 for exporting encryption to several countries, including China, Russia, and Israel. In the United States, cryptography (as well as many other items) for use in government systems is regulated by Federal Information Processing Standard. Civilian use of cryptography is not regulated (except for the export of cryptography).

How Do Other Nations Handle Cryptography?

While the National Security Administration is an obvious starting point for examining government sponsored cryptography, it is not the only place. Clearly, other nations develop their own cryptographic systems and standards. In some cases, the cryptography developed is applied to military purposes and/or classified. In this section, we will take a brief look at how other nations deal with cryptography.

This information is very important for a few reasons. The first being that you, the reader, may not reside within the United States. Furthermore, you may travel abroad and be subject to the cryptography laws of the nations you visit. And should you be involved in the creation, design, or sale of cryptographic products, a knowledge of international laws is important.

International Regulations and Agreements

It is important to note that governments have an interest in regulating import and export of cryptographic equipment and software. The export issue may be more obvious, so let us begin with that. If a given nation has cryptographic software or hardware that is highly secure, that government may wish to limit its export to other countries. It is not in any government's interest to hand other nations the means to secure their communications. Import restrictions may not be so obvious. One major reason for such restrictions is the fear of cryptographic backdoors (see Chap. 18). Essentially, there is a concern that if products are imported from foreign nations, those products could contain backdoors that subvert security.

COCOM and Wassenaar

There have been attempts to regulate and standardize the export of various products considered to be of strategic military value. Such products include both munitions and cryptographic products. One early attempt was the Coordinating Committee for Multilateral Export Controls (COCOM) which consisted of Australia, Belgium, Canada, Denmark, France, Germany, Greece, Italy, Japan, Luxemburg, The Netherlands, Norway, Portugal, Spain, Turkey, United Kingdom, and the United States (Hunt 1982). Other nations did not join the committee but cooperated with the standards it set, those countries are: Austria, Finland, Hungary, Ireland, New Zealand, Poland, Singapore, Slovakia, South Korea, Sweden, Switzerland, and Taiwan.

In 1991, COCOM decided to allow export of mass-market software that implemented cryptography. In 1995, there was a successor to COCOM named Wassenaar Arrangement on Export Controls for Conventional Arms and Dual-Use Goods and Technologies (often simply called the Wassenaar agreement).

Note: The name comes from the town where an original agreement was reached in 1995. The town is Wassenaar and is a suburb of the Hague in the Netherlands. As of this writing there are 41 countries that have signed the Wassenaar agreement.

This agreement followed most of the COCOM recommendations but provided a specific exception for the use of mass market and public domain cryptographic software. There was also a personal use exception that allowed the export of products that were for the user's personal use (i.e., encrypting their laptop hard drive).

In 1998, the Wassenaar agreements were revised to allow (Shehadeh 1999)

- Export for all symmetric cryptography products using keys up to 56 bits
- Export for all asymmetric cryptography products using keys up to 512 bits
- Export for elliptic curve-based asymmetric products using keys up to 112 bits
- Export of mass market products (software and hardware) that use up to 64-bit keys

Note: in 2000, the Wassenaar agreements lifted the 64-bit limit for mass-market cryptographic products.

Specific Governments

The European Union regulates cryptography exports under the "Council Regulation (EC) No 1334/2000 setting up a Community regime for the control of exports of dual-use items and technology." The EU essentially follows the Wassenaar agreement with a few exceptions:

- Export from one EU member nation to another is relaxed, with the notable exception of cryptanalysis products.
- Exceptions can be applied to for exporting to certain nations including Canada, Australia, Japan, and the United States.

Australia regulates the export of cryptographic products via the " Defense and Strategic Goods List." What is notable about this regulation is that it has a specific exemption for public-domain software. In 2001, Australia passed the "Cybercrime Act, No. 161, 2001" which included a clause requiring the release of encryption keys or decrypting the data if a court order was issued.

Australia passed its Defense Trade Control Act of 2012. This act describes "dual use goods" (those products with both military and civilian usage) and has segments on electronics and telecommunication. However, its requirements on encryption are not that clear and some fear it could lead to criminalizing the teaching of cryptography.

China first implemented an order "Commercial Use Password Management Regulations" in 1999 that requires a license from the government in order to import or export cryptographic functions. In 2000, the government of China clarified this indicating that the order refers to products that have cryptography as a core function. Products such as mobile phones and web browsers, for which cryptographic functions are ancillary, are exempted.

Several nations require a license in order to utilize encrypted communications. In Iran, a license is required in order to use cryptography for any communications. Israel requires a license from the government to export or import cryptographic products. Pakistan requires government approval for the sale or use of cryptography products.

India, under the Information Technology Act 2000, does not require a license, but does require that any organization or individual comply with any agency of the government to assist in decrypting data if requested.

While Russia has signed the Wassenaar agreement, its laws go a bit further. A license is required to import cryptography products manufactured outside of Russia. Exporting cryptography requires government approval.

Belarus has strict regulations regarding the import and export of cryptography. A license is required for the design, production, sale, repair, and operation of cryptography. Belarus, as well as other countries, may not recognize a personal use exception.

Many of us use encrypted hard drives on our personal computers. When travelling internationally, it is recommended you check the laws of the countries you will be travelling through. It may even be necessary for you to take a separate laptop, one without encryption.

Belgium passed a law "the Law on information-science crime" in 2000. Article 9 of that law allows a court to issue an order to someone that is reasonably suspected of having "special knowledge" of encryption services, to give information to law enforcement on how to decrypt data. This means that if a suspect has encrypted data, the court can order the vendor of the cryptography product to provide assistance in decrypting the data. Failure to comply can result in 6–12 months of incarceration.

There has been a great deal of debate regarding how to handle evidence that has been encrypted by parties suspected of criminal activity. Currently, the United States has no laws specifically compelling anyone to reveal encrypted data or to assist in decrypting the data. However, several countries do. And there has been discussion of having such laws in the United States.

Cryptography and Malware

Malware, even cryptographic malware, may seem to be an unrelated topic for this chapter. How does malware relate to military applications of cryptography? As cyber warfare becomes increasingly real, malware has become a common weapon used by governments and other entities. Malware that utilizes cryptography in one form or another has become increasingly common.

The term cryptovirology is sometimes applied to the study of how cryptography is combined with malware. One application of cryptography to malware, is the use of encryption to protect the virus from detection. This is one example of what is called an armored virus. One of the first armored viruses was named Cascade and was discovered in the late 1980s. This clearly demonstrates that using encryption to bolster viruses is not a new technique.

Malware encryption is done not only to prevent detection, but also to make it difficult to analyze the malware. To use encryption, the malware needs at least three components:

- The actual malware code (which is encrypted)
- A module to perform encryption/decryption
- A key

Cryptoviral extortion, or ransomware is another application of cryptovirology. Ransomware is malware that once it has infected the target machine, encrypts sensitive files on that machine. The machine user is then sent a ransom message, demanding money for the attacker to decrypt the files. The one-half virus was an early virus that would encrypt certain files. However, it did not demand ransom. That virus was first reported in 2007. Virus.Win32.Gpcode.ag used a 660-bit version of RSA to encrypt certain files. It then asked for a ransom. If the ransom was paid, the victim was sent a key.

One of the most widely known examples is the infamous CryptoLocker. It was first discovered in 2013. CryptoLocker utilized asymmetric encryption to lock the user's files. Several varieties of CryptoLocker have been detected.

CryptoWall is a variant of CryptoLocker first found in August of 2014. It looked and behaved much like CryptoLocker. In addition to encrypting sensitive files, it would communicate with a command and control server, and even take a screenshot of the infected machine. By March of 2015, a variation of CryptoWall had been discovered which is bundled with the spyware TSPY_FAREIT.YOI and actually steals credentials from the infected system, in addition to holding files for ransom.

The viruses mentioned so far are just a sample of viruses that use encryption. The purpose of this chapter is not to educate you on viruses, however, the line between viruses and cryptography is becoming blurred. As you have seen already, cryptography can be used with malware to create both new forms of malware (ransomware) and enhanced viruses (armored viruses).

Weaponized Malware

Most analysts believe that the Stuxnet and Flame viruses were designed for the express purpose of cyber espionage and/or sabotage of the Iranian government. Moving forward, one should expect to see more examples of malware-based attacks and state-sponsored malware espionage. A variety of nations have either already been found to have been engaged in such activities or are strongly suspected of being engaged in such activities.

As our first example of cyber warfare via malware infections, consider the Stuxnet virus. Stuxnet first spread via infected thumb drives, however, once it was on an infected machine it would spread over the entire network and even over the internet. The Stuxnet virus then searched for a connection to a specific type of Programmable Logic Controller (PLC), specifically the Siemens Step7 software. Once that particular PLC was discovered, Stuxnet would load its own copy of a specific DLL (Dynamic Linked Library) for the PLC, in order to monitor the PLC and then alter the PLC's functionality. It was meant to only target centrifuge controllers involved in Iran's uranium enrichment. Clearly, this was meant to be a targeted attack. However, it spread beyond the intended targets. While many reported no significant damage from Stuxnet, outside the Iranian reactors, it was detected on numerous machines.

While Stuxnet's targeting was clearly inadequate, its design was a classic virus design. Stuxnet has three modules: a worm that executes routines related to the attack; a link file that executes the propagated copies of the worm; and a rootkit responsible for hiding files and processes, with the goal of making it more difficult to detect the presence of Stuxnet. It is not the purpose of this current paper to explore the intricacies of Stuxnet, Rather Stuxnet is introduced as both an example of state-sponsored malware attacks, and at least an attempt to target such attacks.

Our second exemplary malware is the Flame virus. The Flame virus was first discovered in May 2012 in Iran. It was spyware that recorded keyboard activity, network traffic, took screen shots, and is even reported to record Skype conversations. It also would turn the infected computer into a Bluetooth beacon attempting to download information from nearby Bluetooth enabled devices.

Kaspersky labs reported that the Flame file contained an MD5 Hash that only appeared on machines in the Middle East. This indicates the possibility that the virus authors intended to target the malware attack to a specific geographical region. The Flame virus also appears to have had a kill function allowing someone controlling it to send a signal directing it to delete all traces of itself. These two items indicate an attempt to target the malware, though like Stuxnet the outcome of that targeting was less than optimal.

Cyber Warfare

Cyber warfare may seem like a topic far removed from cryptography. While a complete discussion of cyber warfare is beyond the scope of this text, it is appropriate to briefly introduce the topic. As you have already seen, weaponized malware is an element of cyber warfare, and cryptovirology certainly has a role to play.

Let us begin by defining what we mean by cyber warfare. According to the Rand Corporation "Cyber warfare involves the actions by a nation-state or international organization to attack and attempt to damage another nation's computers or information networks through, for example, computer viruses or denial-of-service attacks." It is important to realize that cyber warfare is not just a topic for science fiction. Here are a few real-world incidents:

- 2008 CENTCOM is infected with spyware. CENTCOM or Central Command is the United States Army entity that is responsible for command and control throughout the Middle East and Central Asia. That means that during the conflicts in Iraq and Afghanistan, spyware was infecting a critical system for at least a certain period of time.
- 2009 Drone video feed is compromised. In Afghanistan, an unknown attacker was able to tap into the video feed of a United States Drone and see the video feed.
- On December 4, 2010, a group calling itself the Pakistan Cyber Army hacked the website of India's top investigating agency, the Central Bureau of Investigation (CBI).
- In December of 2009, Hackers broke into computer systems and stole secret defense plans of the United States and South Korea. Authorities speculated that North Korea was responsible. The information stolen included a summary of plans for military operations by South Korean and U.S. troops in case of war with North Korea, though the attacks traced back to a Chinese IP address.
- The security firm, Mandiant tracked several APTs over a period of 7 years, all originating in China, specifically Shanghai and the Pudong region. These APTs were simply named APT1, APT2, etc. The attacks were linked to the UNIT 61398 of the China's Military. The Chinese government regards this unit's activities as classified, but it appears that offensive cyber warfare is one of its tasks. Just one of the APTs from this group compromised 141 companies in 20 different industries. APT1 was able to maintain access to victim networks for an average of 365 days, and in one case for 1764 days. APT1 is responsible for stealing 6.5 terabytes of information from a single organization over a 10-month time frame.
- The BlackEnergy malware explicitly affected power plants. The malware is a 32-bit Windows executable. BlackEnergy is versatile malware, capable of initiating multiple attack modalities. It was a multifaceted malware that included distributed denial-of-service (DDoS) attacks. BlackEnergy also can deliver KillDisk, an aspect that would erase every file on the target system. In December 2015, a substantial region of the Ivano-Frankivsk province in Ukraine had power shut down for approximately 6 hours due to the BlackEnergy malware. The attacks have been attributed to a Russian cyber espionage group named Sandworm.
- In 2016, the hacker group Shadow Brokers appeared. They publish what they claim are hacking tools they stole from the NSA. The most famous was the tool EternalBlue which was extremely effective at breaching Windows 7 machines.
- The Shamoon virus first hit Saudi Aramco in 2012 and made a resurgence in 2017. Shamoon functions as spyware but also deletes files following it has uploaded them to the attacker. The malware attacked Saudi Aramco computers. The group named "Cutting Sword of Justice" took credit for the attack.

Each of these incidents reveals that cyber warfare and cyber espionage are indeed realities. However, they further demonstrate the role cryptography plays in at least some of cyber warfare. For example, in the drone incident of 2008, it is likely that the encrypted communications with the drone were compromised in some fashion. At a minimum, each of these incidents reveals the need for robust, secure communications.

TOR

TOR, or The Onion Router, may not seem like a military application of cryptography, however, it is appropriate to cover this topic in this chapter for two reasons:

1. The TOR project is based on an earlier Onion Routing protocol developed by the United States Navy, specifically for military applications. So, TOR is an example of military technology being adapted to civilian purposes.
2. TOR is used by privacy advocates every day. But it is also used by terrorists groups and organized criminals.

TOR consists of thousands of volunteer relay's spread around the world. Each relay uses encryption to conceal the origin, and even final destination of the traffic passing through it. Each relay is only able to decrypt one layer of the encryption, revealing the next stop in the path. Only the final relay is aware of the destination, and only the first relay is aware of the origin. This makes tracing network traffic, practically impossible.

The basic concepts for onion routing were developed at the U.S. Naval Research Laboratory in the mid-1990s and later refined by the Defense Advanced Research Projects Agency (DARPA). The goal was to provide secure intelligence communication online. Onion routers communicate using TLS (covered in depth in Chap. 13) and ephemeral keys. Ephemeral keys are so-called because they are created for one specific use, then destroyed immediately after that use. 128-bit AES is often used as the symmetric key.

While the TOR network is a very effective tool for maintaining privacy, it has also become a way to hide criminal activity. There exist markets on the TOR network that are used expressly to sell and distribute illegal products and services. Stolen credit card numbers and other financial data are a common product on TOR markets. A screenshot of one of these markets is depicted in Fig. 15.3.

Even more serious criminal activity can be found on TOR. There are places that distribute child pornography, weapons, and others that even purport to perform murder for hire. Drug sales on the dark web are quite common and are shown in Fig. 15.4.

More closely related to military and intelligence operations, it has been reported that various terrorists groups have used TOR for communication and recruitment. TOR's anonymity makes it a perfect venue for terrorist communications. The criminal activity on TOR also provides a means for fundraising for terrorist organizations.

Fig. 15.3   Stolen Financial Data Market on TOR

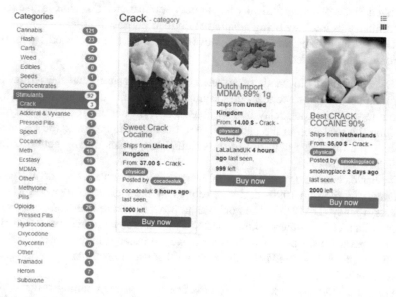

Fig. 15.4   Drug Sales on the Dark Web

In 2015, the founder of Silk Road, Ross Ulbricht, the most well known of the TOR markets, was sentenced to life in prison. Many people on various news outlets are claiming this sentence is draconian. And one can certainly argue the merits of any sentence. But allow me to give you food for thought. Silk Road was not simply a venue for privacy, or even for marijuana exchanges. It became a hub for massive drug dealing including heroin, cocaine, meth, etc. It was also used to traffic in arms, stolen credit cards, child pornography, and even murder for hire. The venue Mr. Ulbricht created was a hub for literally thousands of very serious felonies.

Conclusions

In this chapter, we have examined various military and government applications of cryptography. We have looked at the National Security Administration classifications of cryptography as well as specific military applications. We also examined various laws regarding cryptography for the United States as well as other countries. The examination of cryptography and malware covered cryptovirology and weaponized malware. As cyber warfare becomes more prevalent, it is likely that cryptovirology will play an important part as one of the weapons of choice in cyber war.

Test Your Knowledge

1. What suite of algorithms does the NSA designated as classified?
2. Which products are used for classified purposes only?
3. Which products are used for sensitive but unclassified purposes?
4. What key length of ECC is considered adequate for top secret information?
5. What key length of AES is considered adequate for top secret information?
6. What government agency currently regulates the export of cryptography from the United States?
7. What was the successor to COCOM?
8. _____ is the study of how cryptography is combined with malware.
9. Software that encrypts data and only provides the decryption key once a fee has been paid is called _____.
10. What symmetric algorithm does TOR use?

References

Bamford, J. (2018). The Puzzle Palace: a report on NSA, America's most secret agency. Houghton Mifflin Harcourt. Boston

Bauer, C. P. (2013). Secret history: The story of cryptology. CRC Press Abingdon.

Davida, G. I. (1981). Cryptographic Research and NSA: Report of the Public Cryptography Study Group. Academe: Bulletin of the AAUP, 67(6), 371-82.

Hunt, C. (1982). Multilateral cooperation in export controls-the role of CoCom. U. Tol. L. Rev., 14, 1285.

Shehadeh, K. K. (1999). The Wassenaar Arrangement and Encryption Exports: An Ineffective Export Control Regime that Compromises United States Economic Interests. Am. U. Int'l L. Rev., 15, 271.

Wagner, M. (2006). The Inside Scoop on Mathematics at the NSA. Math Horizons, 13(4), 20-23.

Chapter 16
Steganography

Abstract Steganography is not strictly speaking cryptography. It is, however, closely related. Modern steganography depends on algorithms to hide one message in another. Both steganography and cryptography are utilized to provide security for messages, though both take different approaches to that process. Steganography is used for digital rights management as well as by criminals attempting to hide evidence. Therefore, it is important to study cryptography in the context of cryptography.

Introduction

Strictly speaking, steganography is not cryptography. But the topics are often covered in the same course or textbook. The reason being that both technologies seek to prevent unwanted parties from viewing certain information. Cryptography tries to accomplish this via applying mathematics to make the message undecipherable without the key. Steganography attempts to secure data by hiding it in other, innocuous media. In this chapter, we examine how steganography is done, the history of steganography, and how to detect steganography.

In This Chapter We Cover

Historical steganography
    Steganography tools
    How steganography works

What Is Steganography?

Steganography is the art and science of writing hidden messages in such a way that no one, apart from the sender and intended recipient, suspects the existence of the message, a form of security through obscurity. Often, the message is hidden in some

Fig. 16.1 This book's
cover

other file such as a digital picture or an audio file, so as to defy detection (Easttom 2019). The advantage of steganography, over cryptography alone, is that messages do not attract attention to themselves. If no one is aware the message is even there, then they won't even try to decipher it. In many cases, messages are encrypted and hidden via steganography.

The most common implementation of steganography utilizes the least significant bits in a file in order to store data. By altering the least significant bit, one can hide additional data without altering the original file in any noticeable way. There are some basic steganography terms you should know (Easttom and Dulaney 2017).

Payload is the data to be covertly communicated. In other words, it is the message you wish to hide.

The **carrier** is the signal, stream, or data file into which the payload is hidden.

The **channel** is the type of medium used. This may be still photos, video, or sound files.

The most common way steganography is accomplished today is by manipulating the least significant bits in a graphics file. In every file, there are a certain number of bits per unit of the file. For example, an image file in Windows is 24 bits per pixel. There are 8 bits for red, 8 bits for green, and 8 bits for blue. If you change the least significant of those bits, then the change is not noticeable with the naked eye. And one can hide information in the least significant bits of an image file.

Let us walk through the basic concept of altering the least significant bit. Consider the picture of this book's author, shown in Fig. 16.1.

If one selects a single pixel, in this case denoted by the area in the lower right-hand part of the image, circled in white, then you can see the RGB (Red–Green–Blue) settings in Fig. 16.2.

The settings are Red: 174, Green: 211, and Blue: 253. Let's take one of those, red, and change just one bit, the least significant bit. Decimal value 174 when converted to binary is 1010 1110. So, let us change that last bit, resulting in 1010 1111, which would be 175 in decimal. In Fig. 16.3, you can see a comparison of what changes occur by changing 1 bit of that pixel.

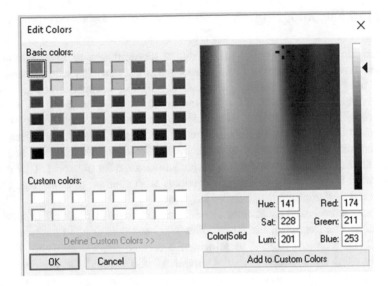

Fig. 16.2 Selecting a single pixel

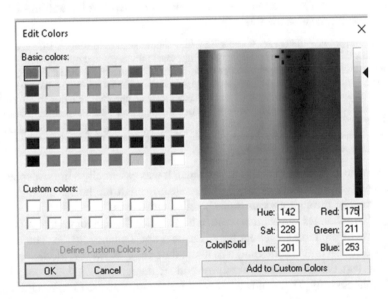

Fig. 16.3 One bit changed in a picture

As you can see, it is impossible to tell a difference. Given that the average picture is made of tens of thousands of pixels, one could change the least significant bit of thousands of these pixels and in those LSBs, store some covert message. That message would be undetectable to the human eye. This is the basic concept behind modern steganography.

Historical Steganography

In modern times, steganography means digital manipulation of files to hide messages. However, the concept of hiding messages is not new. There have been many methods used throughout history. The ancient Chinese wrapped notes in wax and swallowed them for transport. If the messenger was intercepted in transit, no matter how thoroughly he was searched, the message could not be found (Kahn 1996).

In ancient Greece a messenger's head might be shaved, a message written on his head, then his hair was allowed to grow back. Obviously, this method had some significant drawbacks. It took time to prepare a message for transport. This method was reported by the Greek historian Herodotus who claimed that this was used to warn Greece of the impending Persian invasion.

Another method used by ancient Greeks was to scrape the wax off of a wooden folding table, then write on the wood. New wax was then applied covering up the message. The recipient needed to remove the wax to see the message. This was reportedly used by a Greek named Demaratus who warned the Spartans of the impending invasion by Xerxes. Aeneas Tacitus lived in Greece in the fourth century BCE (Easttom 2019). He wrote on the art of war and is considered one of the first to provide a guide for secure military communications. Among his writings on secure military communications is this excerpt:

> Those who employ traitors must know how they should send in messages. Dispatch them, then, like this. Let a man be sent openly bearing some message about other matters. Let the letter be inserted without the knowledge of the bearer in the sole of his sandals and be sewed in, and, to guard against mud and water, have it written on beaten tin so that the writing will not be effaced by the water. And when he reaches the one intended and goes to rest for the night, this person should pull out the stitches of the sandals, take out and read the letter, and, writing another secretly, let him send the man back, having dispatched some reply and having given him something to carry openly. For in this way no one else, not even the messenger, will know the message.

Among Aeneas' innovations was the astragal. It was essentially a hollow sphere with holes representing letters. String was threaded through the holes, the order of the strings passage through various holes, spelled out words.

In 1499, Johannes Trithemius wrote a book entitled Steganographia. It is the first known use of the term steganography (Easttom and Dulaney 2017). The book was actually a three-volume series and was about the occult. However, hidden within that text, was text concerning cryptography and steganography. Aside from his work with Steganography, Trithemius was a very interesting person. He was a Benedictine abbot, but also a cryptography and involved in the occult. He wrote extensive histories; however, it was later discovered that he had inserted several fictional portions into his historical works.

Another interesting form of hiding messages was the Cardano grill, invented and named after Girolama Cardano. Essentially an innocuous message is written on paper, but when the grill is laid over the paper, it has specific holes revealing certain

This is a very simple message, nothing tantalizing at all. It can be used without anyone knowing there is another message hidden along with this message, within the letters and words, unseen.

Fig. 16.4 The Cardano grill

letters that combine to form a different message, the hidden message. You can see an example of this technique in Fig. 16.4.

During WWII, the French Resistance sent messages written on the backs of couriers using invisible ink. If the courier was intercepted, even a strip search would not reveal the message. When the courier reached his or her destination, the message was retrieved.

Microdots are images/undeveloped film the size of a typewriter period, embedded in innocuous documents. These were said to be used by spies during the Cold War. With this technique, a very close examination of a document, using a magnifying class might reveal the microdot. However, the detection process was so tedious, that detection was highly unlikely.

Also, during the cold war, the U.S. Central intelligence Agency used various devices to hide messages. For example, they developed a tobacco pipe that had a small space to hide microfilm but could still be smoked.

Methods and Tools

As was stated earlier in this chapter, using the least significant bit is the most common method for performing steganography. However, it is not the only method. As you have seen already, historically there have been a number of methods that pre-date computers. Even in this digital age, there are alternative ways to hide data and different carrier files. Most books and tutorials focus on hiding data in an image. However, one can also hide data in a wave file, video file, or in fact any sort of digital file.

Whatever technique one uses for steganography, there exists the issues of capacity and security. Capacity refers to the amount of information that can be hidden. Obviously, there is a relationship between the size of the carrier file and the size of data that can be hidden in that carrier file. Security is how well hidden the data are. How easy it is to detect the hidden message using steganalysis techniques is examined later in this chapter. As you will see, some tools do a better job of providing security than do other tools. There are a number of tools available for implementing steganography. Many are free or at least have a free trial version. In the next section, we will examine a few such tools.

Classes of Steganography

While LSB is the most common method, there are three general classes of steganographic methods. These are injection based, substitution based, and generation based.

Injection-based techniques hide data in sections of file that are not processed by the processing applications. For example, comment blocks of an HTML file. This does change file size.

Substitution-based techniques literally substitute some bits of the data to be hidden, for some bits of the carrier file. This replaces bits in the carrier file and does not increase file size. The least significant bit (LSB) method is the most obvious example of a substitution method.

There is a third technique, called the **generation** technique. Essentially the file that is to be hidden is altered to create a new file. There is no carrier file. Obviously, there are limits to what one can do with generation techniques.

In addition to classifying steganography by the techniques used to hide data, it is also possible to categorize steganographic techniques based on the medium used. As was previously stated, hiding files in images is the most common technique, but literally any medium can be used.

Discrete Cosine Transform

The discrete cosine transform or DCT is referenced throughout the literature on steganography. It has been applied to image steganography, audio steganography, and video steganography. So, it is important that you at least have some familiar with this technique. DCT takes a finite sequence of data points and expresses them in terms of the sum of cosine functions oscillating at different frequencies. DCTs express a function or a signal in terms of a sum of sinusoids with different frequencies and amplitudes. A sinusoid is a curve similar to the sine function but possibly shifted in phase, period, amplitude, or any combination thereof.

DCTs work on a function only at a finite number of discrete data points. DCTs only use cosine functions (DFTs can use cosine or sine functions). There are variations of the DCT simply termed DCT-I, DCT -II, DCT-III, DCT-IV, DCT-V, DCT-VI, DCT-VII, and DCT-VIII (i.e. 1 through 8). DCTs are a type of Fourier-related transforms that are similar to the discrete Fourier transform (DFT). DFTs convert a list of samples of a function, which are equally spaced, into a list of coefficients, ordered by their frequencies. This is a somewhat simplified definition. A full explanation is beyond the scope of this text, but you may wish to consult one of these resources.

http://www.dspguide.com/ch8.htm
http://www.robots.ox.ac.uk/~sjrob/Teaching/SP/l7.pdf
http://mathworld.wolfram.com/DiscreteFourierTransform.html

It is not imperative that you master DCTs to understand steganography. However, a general understanding of the concept is necessary, as DCTs are frequently used to implement steganography. If you aspire to develop your own steganographic tool or technique, a deeper understanding of DCTs will be required.

Steganophony

Steganophony is a term for hiding messages in sound files (Artz 2001). This can be done with the LSB method or other methods, such as echo hiding. This method adds extra sound to an echo inside an audio file, that extra sound conceals information. Audio steganography/steganophony can use the LSB method to encode hidden data. Usually audio files, such as MP3 or Wav files, have sufficient size to hide data. MP3 files such as are often used with mobile music devices, typically are 4 MB to 10 MB in size. This provides a large number of bytes wherein the least significant bit can be manipulated. If one begins with a 6 MB file and uses only 10% of the bytes in that file, storing data in the least significant bits allows for approximately 600 K bits or 75,000 bytes. To give some perspective on how much data this encompasses, a typical 20+ page Word document will occupy far less than 75,000 bytes.

Another method used with steganophony is parity coding. This approach divides the signal into separate samples and embeds the secret message into the parity bits. Phase coding can also be used to encode data. This is a bit more complex but very effective. Jayaram, Ranganatha, and Anupam describe this method as follows:

> The phase coding technique works by replacing the phase of an initial audio segment with a reference phase that represents the secret information. The remaining segments phase is adjusted in order to preserve the relative phase between segments. In terms of signal to noise ratio, Phase coding is one of the most effective coding methods. When there is a drastic change in the phase relation between each frequency component, noticeable phase dispersion will occur. However, as long as the modification of the phase is sufficiently small, an inaudible coding can be achieved.
>
> This method relies on the fact that the phase components of sound are not as perceptible to the human ear as noise is. Phase coding is explained in the following procedure:
>
> (a) Divide an original sound signal into smaller segments such that lengths are of the same size as the size of the message to be encoded.
> (b) Matrix of the phases is created by applying Discrete Fourier Transform (DFT).
> (c) Calculate the Phase differences between adjacent segments.
> (d) Phase shifts between adjacent segments are easily detectable. It means, we can change the absolute phases of the segments but the relative phase differences between adjacent segments must be preserved. So, the secret information is inserted only in the phase vector of the first signal segment as follows.
> (e) Using the new phase of the first segment a new phase matrix is created and the original phase differences.
> (f) The sound signal is reconstructed by applying the inverse Discrete Fourier Transform using the new phase matrix and original magnitude matrix and then concatenating the sound segments back together.

The receiver must know the segment length to extract the secret information from the sound file.

Then the receiver can use the DFT to get the phases and extract the secret information.

Video Steganography

Information can also be hidden in video files (Fridrich 2009). There are various methods to accomplish this. Certainly, the LSB method can be used. Discrete cosine transform is often used for video steganography. This method alters values of certain parts of the individual frames. The usual method is to round up the values.

Tools

There are a number of steganography tools available on the internet, either for free or very low cost.

Quick Stego/QuickCrypto

This tool has been available as a free download from the internet for many years. The name has now changed to QuickCrypto and new features have been added. You can download the product from http://quickcrypto.com/download.html. You can see the main screen in Fig. 16.5 with the steganography options highlighted via red rectangles.

Fig. 16.5 QuickCrypto main screen

When you click on the menu hide file or click the stego button at the bottom of the screen (both are highlighted in Fig. 16.5), you can find the original QuickStego product, shown in Fig. 16.6.

You will notice you can either type in text you wish to hide or open a text file to import that text to hide. For this demonstration, I will type in a message, then select "Open File" in the carrier file section to select a carrier file. This is shown in Fig. 16.7.

The next step is to click the "Hide Data" button. If you look in the folder where you found the carrier file, you will see a new file named "carrierfilename 2." You can open the original and the new file side by side and you won't be able to see any difference. The QuickStego tool only works with hiding text files, not hiding other images. However, due to that limitation, the ratio of hidden data to carrier file is very large, making detection more difficult.

Invisible Secrets

Invisible Secrets is a very popular tool for steganography. You can download a trial version of Invisible Secrets from http://www.invisiblesecrets.com/download.html. This tool has a number of capabilities including encryption as well as steganography. The main screen is shown in Fig. 16.8 with the steganography option highlighted with a red rectangle.

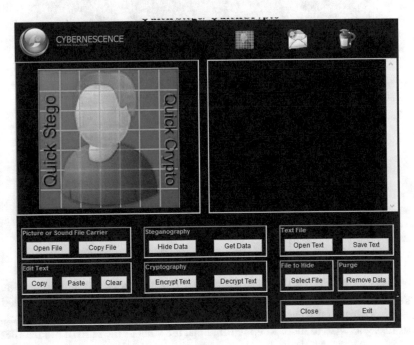

Fig. 16.6 QuickStego main screen

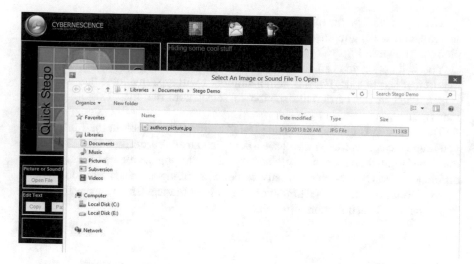

Fig. 16.7 Selecting a carrier file with QuickStego

Fig. 16.8 Invisible Secrets main screen

We will walk through the basic process of steganographically hiding data in an image using Invisible Secrets. First, simply click on the hide button highlighted in Fig. 16.8. That will take you to the screen, shown in Fig. 16.9, where you select the file or files you wish to hide. Keep in mind that the smaller the ratio of hidden files to

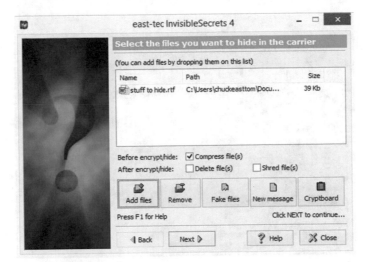

Fig. 16.9 Selecting files to hide with Invisible Secrets

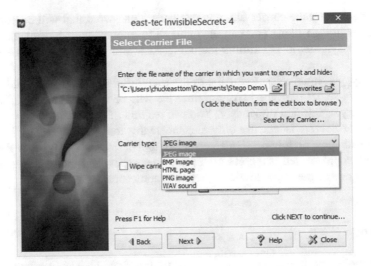

Fig. 16.10 Selecting a carrier file with Invisible Secrets

carrier file, the easier it will be to detect. So, for example, if you choose to hide five jpegs in one jpeg carrier file, it will most likely be detectable. For this demonstration, I will select only one text file.

After clicking the next button, you will be prompted to select a carrier file. One of the features that makes Invisible Secrets a preferable tool is that it gives you multiple options for carrier file, including HTML or .wav (audio) files. You can see this in Fig. 16.10, though for this demonstration, I am selecting a .jpg image file.

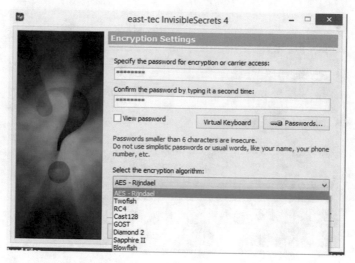

Fig. 16.11 Password and encryption with Invisible Secrets

After selecting the carrier file, you can enter a password that will be required to extract the hidden files. You can also choose to encrypt your hidden files with a number of symmetric algorithms including AES and Blowfish; this is shown in Fig. 16.11.

Finally, you must select the name of the resulting file (carrier with hidden files). You cannot select file type here; the final file will be the same type of file as the carrier file you selected previously. You can see this in Fig. 16.12.

Invisible Secrets is not the only steganography tool available. The internet is replete with free or low-cost steganography tools. However, Invisible Secrets is relatively sophisticated, supports multiple carrier types, and integrates encryption, making it one of the better low-cost steganography tools.

MP3 Stego

This is another tool; one you can download for free from http://www.petitcolas.net/fabien/steganography/mp3stego/. This tool is used to hide data into MP3 files. From the MP3 Stego readme file are these instructions on how to encode or decode data into a .wav or .mp3 file.

encode -E data.txt -P pass sound.wav sound.mp3 compresses sound.wav (the carrier file) and hides data.txt. This produces the output called sound.mp3. The text in data.txt is encrypted using the password "pass."

decode -X -P pass sound.mp3 uncompresses sound.mp3 into sound.mp3.pcm and attempts to extract hidden information. The hidden message is decrypted, uncompressed, and saved into sound.mp3.

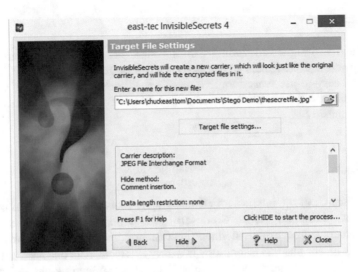

Fig. 16.12 Naming the resulting file with Invisible Secrets

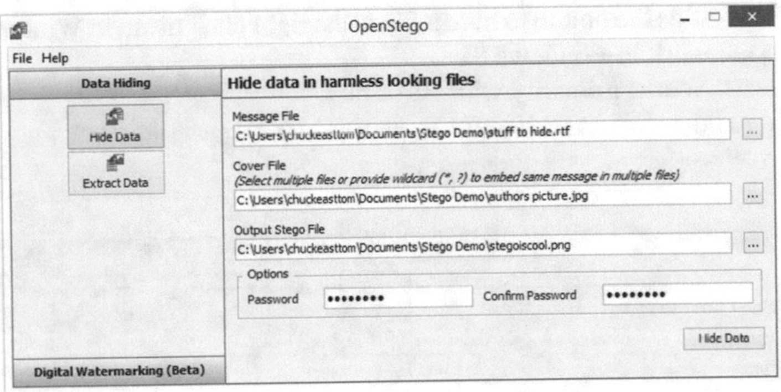

Fig. 16.13 OpenStego

You can see this is a very simple program to use, and freely available on the internet. It works entirely from the command line, and only works with sound files as the carrier file. However, given the ubiquitous nature of sound files, this tool is a good choice for hiding data in a secure manner.

OpenStego

OpenStego is a simple tool, but it is a free download (you can get it from http://openstego.sourceforge.net/) and easy to use. The main screen simply directs the user to select a) the file to hide, b) the carrier file, c) the resulting file, and a password. Then click the "Hide Data" button. This is shown in Fig. 16.13.

DeepSound

DeepSound is designed for hiding data in sound files. It is a free download from http://jpinsoft.net/. The main screen is shown in Fig. 16.14.

It is an easy-to-use, very intuitive tool. It can work with mp3, wav, wma, and other sound files. It takes just a few seconds to hide a document, image, or any other file into a music file.

Other Tools

A simple search of the internet for steganography tools will reveal a host of free or low-cost steganography tools. Some are general purposes, much like Invisible Secrets, others have narrowly defined functionality. This proliferation of steganography tools means that this technology is widely available. And one need not have an understanding of steganographic methods or of programming, in order to use steganography. A few of the widely used tools include:

Snow: This hides data in the white space of a document.
Camouflage: Adds the option to hide a file to the right click menu in Windows (note: it does not work in Windows 8).
BMP Secrets: Works primarily with BMP files.
Hide4PGP: Hides data in BMP, WAV, or VOC files.

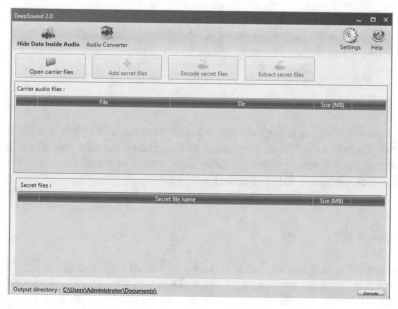

Fig. 16.14 DeepSound main screen

Current Use of Steganography

Steganography is a powerful tool for hiding data. For this reason, it is widely used today, both for innocuous purposes as well as for nefarious purposes. As early as 2001, there was speculation that terrorists were using steganography to communicate. A 2015 paper from the Sans institute has this to say about terrorists using steganography: Using image files to transfer information is the method that first comes to mind. Many newspapers have reported that according to nameless 'U.S. officials and experts' and 'U.S. and foreign officials,' terrorist groups are hiding maps and photographs of terrorist targets and posting instructions for terrorist activities on sports chat rooms, pornographic bulletin boards and other Web sites. Confessions from actual terrorists have verified that Al-Qaeda used steganography to hide both operations details, as well as training materials, in pornographic material.

Steganography is not just a tool used by international terrorists. Certain criminals find it necessary to hide their communications. In particular, more technically savvy child pornographers have been known to use steganography in order to hide their illicit images in innocuous carrier files. This poses a significant issue for forensics analysts.

Steganography has also been used in industrial espionage cases. In one case, an engineering firm suspected one of their employees of stealing intellectual property. Investigators found that this employee had sent out emails with pictures attached, which seemed innocuous, but actually had data hidden in them via steganography.

These are just three nefarious purposes for steganography. The wide proliferation of steganography tools, discussed earlier in this chapter, means that this technology is available to anyone who can use a computer, regardless of whether or not they understand the principles of steganography. Many forensics tools are now including functionality that attempts to detect steganography. One should expect to see the use of steganography increase in the coming years.

Not all uses of steganography involve illicit intent. One good example is watermarking. Watermarking embeds some identifying mark or text into a carrier file for the purpose of being able to identify copyright protected materials. For example, an artist who generates digital versions of his or her art may want to embed a watermark to identify the image should someone use that image without the artist's permission.

Steganalysis

If you can hide data in images, or other carrier files, there must be some way to detect it? Fortunately, there is. Steganalysis is the attempt to detect steganographically hidden messages/files in carrier files. It should be noted that any attempt to detect steganography is simply a best effort, and there is no guarantee of success. One of the most common methods is to analyze close color pairs. By analyzing changes in an image's close color pairs, the analyst can determine if it is likely that LSB steganography was used. Close color pairs consist of two colors whose binary values differ only in their least significant bit. Of course, you would expect a certain number of pixels to vary only in the least significant bit. But if the number of such pixels which meet this criteria is greater than one would expect, that might indicate steganography was used to hide data.

A related method is the Raw Quick Pair method. The RQP method is essentially an implementation of the close color pair concept. The Raw Quick Pair method is based on statistics of the numbers of unique colors and close color pairs in a 24-bit image. RQP analyzes the pairs of colors created by LSB embedding.

Another option uses the chi-squared method from statistics. Chi-square analysis calculates the average LSB and builds a table of frequencies and pair of values. Then it performs a chi-square test on these two tables. Essentially, it measures the theoretical vs. calculated population difference. The details of chi-square analysis are beyond the scope of this text. However, any introductory university text on statistics should provide a good description of this, and other statistical techniques.

Fredrich and Goljan give an overview of various steganalysis methods stating "Pfitzman and Westfeld introduced a powerful statistical attack that can be applied to any steganographic technique in which a fixed set of Pairs of Values (PoVs) are flipped into each other to embed message bits. For example, the PoVs can be formed by pixel values, quantized DCT coefficients, or palette indices that differ in the LSB. Before embedding, in the cover image the two values from each pair are distributed unevenly. After message embedding, the occurrences of the values in each pair will have a tendency to become equal (this depends on the message length). Since swapping one value into another does not change the sum of occurrences of both colors in the image, one can use this fact to design a statistical Chi-square test. We can test for the statistical significance of the fact that the occurrences of both values in each pair are the same. If, in addition to that, the stego-technique embeds message bits sequentially into subsequent pixels/indices/coefficients starting in the upper left corner, one will observe an abrupt change in our statistical evidence as we encounter the end of the message."

More advanced statistical methods can also be used, for example, using Markov Chain Analysis has been applied to steganalysis. A Markov chain is a collection of random variables $\{X_t\}$ that transits from one state to another with the property that looking at the present state; the future state is not dependent on the past states. This is sometimes referred to as being "memoryless." It is named after Andrey Markov. According to Sullivan et al., "In this paper, we take the logical next step toward computing a more accurate performance benchmark, modeling the cover data as a Markov chain (MC). The Markov model has the advantage of analytical tractability, in that performance benchmarks governing detection performance can be characterized and computed explicitly."

Another method is to compare similar files. If, for example, several mp3 files all came from the same CD, the analyst can look for inconsistencies in compression, statistical anomalies, and similar issues to see if one of the mp3 files was different from the others. That difference might indicate that presence of steganography.

Distributed Steganography

There have been various techniques for distributing payload across multiple carrier files. My first patent, U.S. Patent No. 8,527,779 B1 Method and apparatus of performing distributed steganography of a data message, was for just such a method, so I will describe this method for you in this section. This invention was designed specifically for covert communications for undercover law enforcement officers and intelligence agencies. Unfortunately, it can also be applied to nefarious communications. But the intent was a virtually undetectable communication channel for use with sensitive law enforcement and similar activities.

The purpose of steganography, regardless of the implementation, is to hide some underlying message so that an observer is not even aware the message is present. This is very useful in covert communications, particularly in the intelligence community. Most permutations of steganography deal with how to embed the message (text, image, video, or audio) into the carrier file. Some permutations, such as SNOW, even use blanks at the end of text files in order to hide messages. However, this invention is concerned with how to fragment the message and hide it in various carrier/cover files making the detection of the entire message extremely difficult, approaching impossibility.

With distributed steganography, as described in US Patent 8,527,779, the message is distributed across multiple carrier signals/sources in order to further hide the message. For example, a single text message would be broken into blocks, each block hidden in a different image. It should also be noted that the block size can vary, and the blocks are not necessarily stored in order. This means that the first carrier file will not necessarily hold the first segment of the hidden message/file. This is applying permutation to the blocks. Of course, the parties communicating would have to be able to re-order the blocks in their appropriate order.

Consider an example with using 8-bit blocks on a message "Steganography is cool." Each character represents 8 bits, so every 8 characters would be a separate block. Keep in mind that blanks are also represented by 8 bits, so this message would have 5 separate blocks stored in 5 separate images. This is shown in Fig. 16.15.

The obvious issue now is how to retrieve the blocks. This issue would involve knowing how many blocks in total were to be retrieved, the order of each block (i.e., block 2 of 4, 3 of 7, etc.), and knowing the carrier/cover file to retrieve the blocks from. This invention deals with all three issues.

Total Blocks and Block Order

Each block stored in an image would have an additional 2 bytes (16 bits) appended to the image. The first byte would contain information as to which block this was (i.e., block 3 of 9), and the second byte would store the total number of blocks the message contained (i.e., 9 blocks).

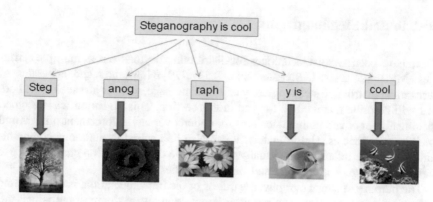

Fig. 16.15 Distributing payload across multiple carrier files

Since 8 bits can store decimal numbers between 0 and 255, this would necessitate breaking a message down into no more than 255 blocks. The size of the block would be determined by the size of the original message divided by 255.

It would also be possible to use additional bytes to store the block numbering data. For example, one could use 2 bytes (16 bits) to store the value of the current block and an additional 2 bytes (16 bits) to store the total number of blocks. This would allow a message to be broken into 65,535 total blocks. Use of up to 4 bytes (64 bits) for the value of the current block and 4 bytes (64 bits) for the total number of blocks would allow a message to be broken into 4,294,967,295 blocks. This would be appropriate for video or audio messages hidden in audio or video signals. These additional bytes indicating block number and total blocks are called *block pointers*.

The use of block numbering is similar to how TCP packets are sent over a network. Each packet has a number such as "packet 2 of 10." This same methodology is applied to hiding blocks of data in diverse images. This requires distributed steganography to have a key, much like the keys used in encryption. However, this key would contain the following information:

Block size
Size of block pointer (i.e., the bytes used to indicate block numbering)

The preferred way to find the location of the images containing secret messages would be to add that information to the key. This information could be an ip address or URL to find the image at (if images are stored at different locations), or the image name (if all images are on a single storage device). You can see this in Fig. 16.16.

Notice that it is possible to store images on web pages, file servers, or FTP servers. This means the actual message could be fragmented and stored around the internet in various locations. In some cases, it could even be stored on third-party servers without their knowledge. It is also possible, even recommended, to use different carrier file types. Parts of the hidden message could be stored in images, such as JPEG's, while other parts could be stored in audio or video files.

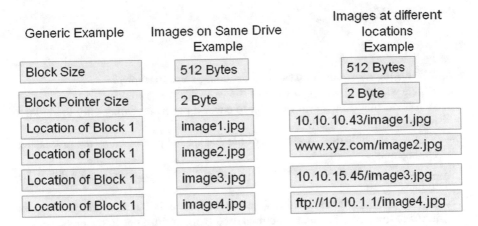

Fig. 16.16 Distributed steganography key

If desired, it would be possible to implement distributed steganography in such a manner that the locations where data would be hidden could be pre-determined. For example, messages would always be hidden in specific images at pre-determined locations. Thus, the person who needs to receive those messages would simply check those images at regular intervals. This would, obviously, be less secure.

The actual encoding of the message could be done with any standard steganography technique, such as using the least significant bits or discrete cosine transform to store the hidden message. It would also be advisable to have the message first encrypted using any preferred encryption algorithm, before hiding it using distributed steganography. It would also be advisable to at least encrypt the steganography key. It should be noted that combining encryption and steganography makes for a very powerful way to protect data. Some of the tools we examined previously in this chapter, like Invisible Secrets, provide the option to both encrypt the data and to hide it in a carrier file.

Conclusions

In this chapter, we have examined a fascinating area of data security. We have examined steganography. The most common way to perform steganography is the LSB method. However, in this chapter you also were given a brief overview of additional methods. You should be familiar with the concepts of steganography as well as at least one of the steganography tools mentioned in this chapter. You also were introduced to steganalysis and were provided with a brief description of common methods for detecting steganography. Finally, the chapter concluded with a good description of a specific, patented method for distributed steganography.

Test Your Knowledge

1. What is the most common method for doing steganography?
2. Where was the first known used of the word steganography?
3. A method that uses a grill or mask to cover a paper, revealing only the hidden message?
4. _____ hide data in sections of file that are not processed by the processing applications.
5. _____ replace some bits of the carrier file with some bits of the data to be hidden.
6. _____ takes a finite sequence of data points and expresses them in terms of the sum of cosine functions oscillating at different frequencies.
7. Parity coding is often used with what type of steganography?
8. One of the most common methods of steganalysis is _____.
9. _____ is a statistical test that can be used to detect steganography.
10. The file that data are hidden in is called what?

References

Artz, D. (2001). Digital steganography: hiding data within data. IEEE Internet computing, 5(3), 75-80.

Easttom, C. & Dulaney, E. (2017). CompTIA Security+ Study Guide: SY0-501. Hoboken, New Jersey: Sybex Press

Easttom, C. (2019). Computer Security Fundamentals, 4th. New York City, New York: Pearson Press

Fridrich, J. (2009). Steganography in digital media: principles, algorithms, and applications. Cambridge University Press.

Kahn, D. (1996, May). The history of steganography. In International Workshop on Information Hiding (pp. 1-5). Springer, Berlin/Heidelberg.

Chapter 17
Cryptanalysis

Abstract Cryptanalysis is a rather advanced cryptographic topic. Many introductory works on cryptography completely ignore this topic. In this chapter, we explore the fundamentals of cryptanalysis, sufficient to provide the reader with at least a basic knowledge of what cryptanalysis is and how it functions. General concepts as well as specific methodologies are covered. This is a topic that the reader need not master, but should develop a general familiarity with.

Introduction

What does it mean to "break" a cipher? This means finding any method to decrypt the message that is more efficient than simple brute force attempts. Brute force is simply trying every possible key. If they algorithm uses a 128-bit key that means 2^{128} possible keys. In the decimal number system that is $3.402 * 10^{38}$ possible keys. If you are able to attempt one million keys every second, it could still take as long as 10,790,283,070,806,014,188,970,529 years to break.

This brings us to cryptanalysis, what is it? Cryptanalysis is using techniques (other than brute force) to attempt to derive the key. You must keep in mind that any attempt to crack any non-trivial cryptographic algorithm is simply an "attempt." There is no guarantee of any method working. And whether it works or not, it will probably be a long and tedious process. This should make sense to you. If cracking encryption where a trivial process, then encryption would be useless.

In fact, cryptanalysis is a very tedious and at times frustrating endeavor. It is entirely possible to work for months, and your only result is to have a little more information on the key that was used for encrypting a given message. What one sees in movies wherein encryption is broken in hours or even minutes is simply not realistic. Who needs to understand cryptanalysis? Obviously, certain intelligence gathering agencies and military personnel have a need for a strong working knowledge of cryptanalysis, beyond what this chapter will provide. In some cases, law enforcement officers and forensics analysts have a need to at least understand cryptanalysis well enough to know what is feasible and what is not. It is not uncommon for suspects to use encryption to hide evidence from authorities.

Furthermore, cryptanalysis techniques are often used to test algorithms. If a new algorithm, or a variation of an old algorithm, is proposed, one way to begin testing that algorithm is to subject it to appropriate cryptanalysis techniques.

Clearly, a thorough understanding of cryptanalysis would take more than a single chapter of a book. In fact, many introductory cryptography books ignore this topic altogether. The goal of this chapter is simply to acquaint you with the fundamentals of cryptanalysis. If you wish to go on to study this topic in more depth, then this chapter should form a solid foundation. You will also find links and suggestions in this chapter for further study. If you don't wish to learn more about cryptanalysis, then this chapter will provide more than enough information for most security professionals.

In This Chapter We Will Cover

Classic techniques of cryptanalysis
Modern methods
Rainbow tables
The birthday paradox
Other methods for breaching cryptography

Classic Methods

Recall that this book began with a study of classic ciphers. This was done to help you, the reader, become comfortable with the concepts of cryptography, before delving into more modern algorithms. A similar issue exists with cryptanalysis. It is often easier for a student of cryptography to understand class methods before attempting to study modern methods. In this section, we will examine methods that are only effective against classic ciphers. These won't help you break RSA, AES, or similar ciphers, but might aid you in understanding the concepts of cryptanalysis.

Frequency Analysis

This is the basic tool for breaking most classical ciphers. In natural languages, certain letters of the alphabet appear more frequently than others. By examining those frequencies, you can derive some information about the key that was used. This method is very effective against classic ciphers like Caesar, Vigenere, etc. It is far less effective against modern methods. In fact, with modern methods, the most likely result is that you will simply get some basic information about the key, but you will not get the key. Remember in English the words' *the* and *and* are the two most common three letter words. The most common single letter words are *I* and *a*. If you see two of the same letters together in a word, it is most likely *ee* or *oo*.

A few general facts that will help you, at least with English as the plaintext:

T is the most common first letter of a word.
E is the most common last letter of a word.
"The" is the most common word.
HE, RE, AN, TH, ER, and IN are very common two-letter combinations.
ENT, ING, ION, and AND are very common three-letter combinations.
These methods are quite effective with any of the classic ciphers discussed in
 Chaps. 1 and 2 but are not effective against more modern ciphers.

When teaching introductory cryptography courses, I often conduct a lab to illustrate this. In this lab, I have each student write a brief paragraph, then select some substitution shift (i.e., +1, −2, +3, etc.) and apply that single substitution cipher to their message. Essentially, they are applying a Caesar cipher to their message. Then I ask the students to exchange the resulting ciphertext with a classmate. Each student then applies frequency analysis to attempt to break the cipher. Having conducted this lab on numerous occasions of many years, I find that it is typical that approximately ½ the class can crack the cipher within 10–15 min. These are usually students with no prior background in cryptography or cryptanalysis. This lab has multiple purposes. The first is to introduce the student to primitive cryptanalysis, and the second is to illustrate the fact that classic ciphers are no longer adequate for modern cryptographic purposes.

Kasiski

Kasiski examination was developed by Friedrich Kasiski in 1863. It is a method of attacking polyalphabetic substitution ciphers, such as the Vigenère cipher (Swenson 2008). This method can be used to deduce the length of the keyword used in the polyalphabetic substitution cipher. Once the length of the keyword is discovered, you lineup the ciphertext in n columns, where n is the length of the keyword. Then, each column can be treated as a monoalphabetic substitution cipher. Then each column can be cracked with simple frequency analysis. The method simply involves looking for repeated strings in the ciphertext. The longer the ciphertext, the more effective this method will be. This is sometimes also called Kasiski' s Test or Kasiski' s Method.

Note: Kasiski lived in the nineteenth century and was a German officer as well as a cryptographer. He published a book on cryptography entitled *Secret writing and the Art of Deciphering*

Modern Methods

Obviously cracking modern cryptographic methods is a non-trivial task. In fact, the most likely outcome to your attempt is failure. However, with enough time and resources (i.e., computational power, sample cipher/plaintexts, etc.), it is possible. Next, we will generally discuss some techniques that can be employed in this process (Swenson 2008).

Known plaintext attack: With this technique, the attacker obtains a number of plaintext-ciphertext pairs. Using the information, the attacker attempts to derive information about the key being used. This will require many thousands of plaintext/ciphertext pairs in order to have any chance of success.

Chosen plaintext attack: In this attack, the attacker obtains the ciphertexts corresponding to a set of plaintexts of his own choosing. This can allow the attacker to attempt to derive the key used and thus decrypt other messages encrypted with that key. This can be difficult but is not impossible.

Ciphertext-only: The attacker only has access to a collection of ciphertexts. This is much more likely than known plaintext, but also the most difficult. The attack is completely successful if the corresponding plaintexts can be deduced, or even better, the key. The ability to obtain any information at all about the underlying plaintext is still considered a success.

Related-key attack: Like a chosen-plaintext attack, except the attacker can obtain ciphertexts encrypted under two different keys. This is actually a very useful attack if you can obtain the plaintext and matching ciphertext.

The chosen plaintext and known plaintext attacks often puzzle students who are new to cryptanalysis. How, they ask, can you get samples of plaintext and ciphertext? You cannot simply ask the target to hand over such samples, can you? Actually, it is not that difficult. Consider, for example, that many people have signature blocks in all their emails. If you send an email to the target and get a response, that is an example of a plaintext (i.e., the signature block). If you later intercept an encrypted email from that target, you know that the matching plaintext is for the encrypted text at the end of the email. This is just one trivial example of how to get plaintext-ciphertext pairs. There are more mathematically sophisticated and difficult methods. Linear, differential, and integral cryptanalysis are three widely used methods. Some of these are applications of known plaintext or chosen plaintext attacks.

Linear Cryptanalysis

This technique was invented by Mitsuru Matsui. It is a known plaintext attack and uses a linear approximation to describe the behavior of the block cipher (Swenson 2008). Given enough pairs of plaintext and corresponding ciphertext, bits of

$$P1 \oplus P3 \oplus C1 = K2$$

Fig. 17.1 The basics of linear cryptanalysis

$$P_{i1} \oplus P_{i2} \cdots \oplus C_{j1} \oplus C_{j2} \cdots = K_{k1} \oplus K_{k2} \cdots$$

Fig. 17.2 The form of linear cryptanalysis

information about the key can be obtained. Clearly, the more pairs of plaintext and ciphertext one has, the greater the chance of success. Linear cryptanalysis is based on finding affine approximations to the action of a cipher. It is commonly used on block ciphers (Matsui 1993).

Remember cryptanalysis is an attempt to crack cryptography. For example, with the 56-bit DES key brute force could take up to 2^{56} attempts. Linear cryptanalysis will take 2^{47} known plaintexts. This is better than brute force, but still impractical for most situations. Matsui first applied this to the FEAL cipher, then later to DES. However, the DES application required 2^{47} known plaintext samples, making it impractical.

With this method, a linear equation expresses the equality of two expressions, which consist of binary variables that are XOR'd. For example, the following equation, XORs sum of the first and third plaintext bits and the first ciphertext bit is equal to the second bit of the key. This is shown in Fig. 17.1.

You can use this method to slowly recreate the key that was used. Now after doing this for each bit, you will have an equation of the form shown in Fig. 17.2.

You can then use Matsui's Algorithm 2, using known plaintext-ciphertext pairs, to guess at the values of the key bits involved in the approximation. The right-hand side of the equation is the partial key (the object is to derive some bits for part of the key). Now count how many times the approximation holds true over all the known plaintext-ciphertext pairs. This count is called T. The partial key which has a T value that has the greatest absolute difference from half the number of plaintext-ciphertext pairs is determined to be the most likely set of values for those key bits. In this way, you can derive a probable partial key.

Differential Cryptanalysis

Differential cryptanalysis is a form of cryptanalysis applicable to symmetric key algorithms. This was invented by Eli Biham and Adi Shamir. Essentially, it is the examination of differences in an input and how that affects the resultant difference in the output. It originally worked only with chosen plaintext. However, it could also work with known plaintext and ciphertext only.

The attack is based on seeing pairs of plaintext inputs that are related by some constant difference (Biham and Shamir 2012). The usual way to define the differences is via XOR operation, but other methods can be used. The attacker computes the differences in the resulting ciphertexts and is looking for some statistical pattern. The resulting differences are called the differential. Put another way "differential cryptanalysis focuses on finding a relationship between the changes that occur in the output bits as a result of changing some of the input bits."

The basic idea in differential cryptanalysis is that analyzing the changes in some chosen plaintexts, and the difference in the outputs resulting from encrypting each one, it is possible to recover some properties of the key. Differential cryptanalysis measures the XOR difference between two values. Differentials are often denoted with the symbol Ω. Thus, you might have a differential Ωa and another differential Ωb. A characteristic is composed of two differentials. For example, differential Ωa in the input produces differential Ωb in the output. These matching differentials are a characteristic. What the characteristic demonstrates is that the specified differential in the input leads to a particular differential in the output.

Differential cryptanalysis is about probabilities. So, the question being asked is: what is the probability that a given differential in the input Ωa will lead to a particular differential in the output Ωb (Swenson 2008)? In most cases, differential analysis starts with the s-box. Since most symmetric ciphers utilize s-boxes, this is a natural and convenient place to start. If you assume you an input of X1 that produces output of Y1 and an input of X2 that produces an output of Y2, this produces a differential (i.e., the difference between X1 and X2 produces the difference between Y1 and Y2). This is expressed as follows:

$\Omega i = X1 \oplus X2$ this is the input differential.
$\Omega o = Y1 \oplus Y2$ this is the output differential.

Now we need to consider the relationship between input differentials and output differentials. To do this, we have to consider all possible values of Ωi and measure how this changes the values of Ωo. For each possible value of X1, X2 and Ωi, you measure the change in Y1, Y2, and Ωo and record that information.

Even though differential cryptanalysis was publicly discovered in the 1980s, DES is resistant to differential cryptanalysis based on the structure of the DES s-box. Since the DES s-box is the portion of DES that was constructed by the NSA, it stands to reason that the NSA was aware of differential cryptanalysis in the 1970s.

Higher Order Differential Cryptanalysis

This is essentially an improvement on differential cryptanalysis that was developed by Lars Knudsen in 1994. Higher order differential cryptanalysis focuses on the differences between differences that would be found with ordinary differential cryptanalysis. This technique has been shown to be more powerful than ordinary differential cryptanalysis. Specifically, it has been applied to a symmetric algorithm known as the KN-cipher, which had previously been proven to be immune to standard differential cryptanalysis. Higher order differential cryptanalysis has also been applied to a variety of other algorithms including CAST.

Truncated Differential Cryptanalysis

Ordinary differential cryptanalysis focuses on the full difference between two texts and the resulting ciphertext, but truncated differentials cryptanalysis analyses only partial differences. Taking partial differences into account, it is possible to use two or more differences within the same plaintext/ciphertext pair to be taken into account. As the name suggests, this technique is only interested in making predictions of some of the bits, instead of the entire block. It has been applied to TwoFish, Camelia, Skipjack, IDEA, and other block ciphers.

Impossible Differential Cryptanalysis

Put another way, standard differential cryptanalysis is concerned with differences that propagate through the cipher with a greater probability than expected. Impossible differential cryptanalysis is looking for differences that have a probability of 0 at some point in the algorithm. This has been used against Camellia, ARIA, TEA, Rijndael, TwoFish, Serpent, Skipjack, and other algorithms.

Integral Cryptanalysis

Integral cryptanalysis was first described by Lars Knudsen. This attack is particularly useful against block ciphers based on substitution–permutation networks, an extension of differential cryptanalysis (Knudsen and Wagner 2002). Differential analysis looks at pairs of inputs that differ in only one-bit position, with all other bits identical. Integral analysis, for block size b, holds b-k bits constant and runs the other k through all 2 k possibilities. For $k = 1$, this is just differential cryptanalysis, but with $k > 1$, it is a new technique.

Mod-n Cryptanalysis

This can be used for either block or stream ciphers. This method was developed in 1999 by John Kelsey et al. (1999). This excerpt from the inventor's paper gives a good overview of the technique:

Nearly all modern statistical attacks on product ciphers work by learning some way to distinguish the output of all but the last rounds from a random permutation. In a linear attack, there is a slight correlation between the plaintext and the last-round input; in a differential attack, the relationship between a pair of inputs to the last round isn't quite random. Partitioning attacks, higher-order differential attacks, differential attacks, and related-key attacks all t into this pattern.

Mod *n* cryptanalysis is another attack along these lines. We show that, in some cases, the value of the last-round input modulo *n* is correlated to the value of the plaintext modulo *n*. In this case, the attacker can use this correlation to collect information about the last-round subkey. Ciphers that sufficiently attenuate statistics based on other statistical effects (linear approximations, differential characteristics, etc.) are not necessarily safe from correlations modulo *n*.

Asymmetric Cryptanalysis

So far, we have focused on symmetric ciphers, particularly block ciphers. However, there are also known weaknesses in asymmetric ciphers. Since RSA is the most widely used asymmetric cipher, we will focus our attention on it. The goal of this section is to simply introduce you to the issues with RSA. For more details, refer to any of the papers referenced in this section.

Recent studies have discovered potential flaws in RSA. Heninger and Shacham found that RSA implementations that utilized a smaller modulus were susceptible to cryptanalysis attacks. In their study, they considered RSA implementations that utilized a small exponent in the algorithm. A smaller modulus is sometimes used to increase the efficiency of the RSA algorithm. However, the size of the modulus value also could be used to reduce the set of possible factors and thus decrease the time required to factor the public key. In fact, a great many RSA implementations use $e = 2^{16} + 1 = 65{,}537$. Consequently, a cryptanalyst already has the public key and thus has e and n. And the n is relatively small, making it possible, with extensive computing power and time, to derive the private key. The authors of this study clearly showed that it is possible to derive the private RSA key, which would render that particular RSA encryption implementation useless.

In their methodology, Heninger and Shacham formulated a series of linear equations that would progressively approximate the RSA private key. The approximations were based on approximations of factoring the public key. This technique is very similar to the linear cryptanalysis method for cryptanalysis of symmetric key algorithms.

Zhao and Qi also utilized implementations that have a smaller modulus operator. The authors of this study also applied modular arithmetic, a subset of number theory, to analyzing weaknesses in RSA. Many implementations of RSA use a shorter modulus operator in order to make the algorithm execute more quickly. Like

Heninger and Shacham, Zhao and Qi showed that, based on the mathematical relationships between the elements of the RSA algorithm, increases in efficiency resulting from a smaller modulus will also render a decrease in the efficacy of that RSA implementation.

In their study, Zhao and Qi utilized a lattice matrix attack on the RSA implementation in order to attempt to factor the public key and derive the private key. The specifics of this mathematical methodology are not relevant to this paper. What is significant is that the researchers used a different approach than Heninger and Shacham and achieved the same results on RSA applications using a small modulus.

Aciicmez and Schindler examined the RSA cryptographic algorithm, as implemented in SSL. Given that SSL/TLS is used for online banking and e-commerce, the security of any implementation of the protocol is an important topic. Aciicmez and Schindler wanted to understand if there were flaws in the implementation that would allow an unintended third party to break the SSL implementation. The authors explained how a particular type of crypto-analysis can be used to break this particular specific implementation of RSA. Since RSA and SSL are both used extensively in e-commerce, exploring weaknesses in either is important. It is important to note that this analysis was dependent upon essential elements of number theory.

In their study of SSL using RSA, Aciicmez and Schindler examined the timing for modular arithmetic operations used in that specific implementation of RSA. This ultimately led to a method for factoring the public key, thus yielding the private key used in that RSA implementation. This methodology is important because normal approaches to factoring the public key are entirely too time consuming to be of practical use (Swenson 2008). It is important to derive some additional information about the implementation of RSA in order to attempt a more practical approach to factoring. By utilizing number theory, specifically in respect to the functionality of modular arithmetic, the researchers were able to significantly decrease the time required for factoring the public key.

Given that the authors of this study were able to significantly decrease the time required to factor the public key in order to derive the private key, clearly these findings are significant. The study clearly shows a problem with some implementations of RSA.

These studies mentioned are simply a sample of known attack vectors against RSA. Many of these attacks depend on a small modulus. And we have seen that many RSA implementations utilize the same small modulus ($e = 2^{16} + 1 = 65{,}537$). It is also true that increases in computing power will make these attacks, as well as brute force attempts to crack RSA, even more practical. So far, the cryptography community has reacted by simply using ever larger key sizes for RSA. It seems likely that an entirely new asymmetric algorithm may be needed. But in the meantime, when you implement RSA, make sure you not only use a large key size but be wary of using too small a modulus value.

General Rules for Cryptanalysis

Regardless of the technique used, there are three resources for cryptanalysis:

Time—the number of "primitive operations" which must be performed. This is quite loose; primitive operations could be basic computer instructions, such as addition, XOR, shift, and so forth, or entire encryption methods.

Memory—the amount of storage required to perform the attack.

Data—the *quantity* of plaintexts and ciphertexts required.

In essence, this means that with infinite time, memory, and data, that any cipher can be broken. Of course, we do not have infinite resources. In fact, resources are generally quite limited, particularly time. It would do no good to break the encryption of military communications and learn of an attack...if that break occurred several weeks after the attack. The information is only of use if it is timely.

In general, the primary advantage that a government entity has in cryptanalysis is the resource of memory. A supercomputer, or even a cluster of high-end servers, can put far more resources to breaking a cipher, than can an individual with a single computer. Even with those resources, however, breaking a modern cipher is far from trivial. It is still an onerous, resource-intensive task with no guarantee of success. Breaking a modern cipher with your individual computer is simply not feasible.

There are also varying degrees of success:

Total break—the attacker deduces the secret key.

Global deduction—the attacker discovers a functionally equivalent algorithm for encryption and decryption, but without learning the key.

Instance (local) deduction—the attacker discovers additional plaintexts (or ciphertexts) not previously known.

Information deduction—the attacker gains some Shannon information about plaintexts (or ciphertexts) not previously known.

Distinguishing algorithm—the attacker can distinguish the cipher from a random permutation.

Consider this list carefully. A total break may be the only way many readers have thought about success in cryptanalysis. However, that is only one possible definition of a successful cryptanalysis. In fact, it is the least likely outcome. In general, if your cryptanalysis produces more information about the key than was previously known, then that is considered a success.

Rainbow Tables

Many passwords are stored with a cryptographic hash. This prevents the network or database administrator from reading the password, since it is hashed. If you will recall from Chap. 9, cryptographic hashes are not reversible. That means that it is not merely computationally difficult, or impractical to "unhash" something, but that it is mathematically impossible. It would seem that breaking passwords protected by cryptographic hashes is impossible. That, however, is not the case.

A rainbow table is essentially a precomputed table of hashes. The most primitive way to accomplish this would be to simply precompute hashes of all possible passwords of a given size. Assuming a standard English keyboard, there are 26 characters in upper or lower case (i.e., 52 possibilities), 10 digits, and about 8 special characters (#, !, $, etc.) or about 70 possible values each character could take on (the value 70 is just a rough estimate to illustrate this concept). So, a single character password could have 70 possible values, whereas a two-character password could have 70^2 or 4900. Even an 8-character password could have up to 576,480,100,000,000 possible values. So, calculating tables that account for all passwords of any length from 5 characters to 10 characters would be both computationally intensive and require a great deal of storage.

The method for composing precomputed tables of hashes, described above, is the most primitive way to accomplish this task. Hash chains are used to make this process more efficient and to reduce the space needed to store the pre-computed hashes. The concept of a hash-chain is to use a reduction function, we will call R, that maps hash values back to plaintext values. This is not unhashing or reversing a hash, it is just a method to more quickly pre-compute hashes. The next, even more advanced method is to replace the reduction function with a sequence of related reduction functions $R_1 \ldots R_k$.

The issue then becomes how to implement this process. For example, Microsoft Windows stores the hash of passwords in the SAM file. One has to first obtain the SAM file for a target machine, then take the contents and search through rainbow tables for matches in order to find the passwords. The tool Ophcrack automates this process for you. It can be placed on a CD/DVD and will boot to a Live version of Linux. Then it launches a tool that copies the SAM file from the Windows machine and searches the rainbow tables on the CD/DVD for a match. You can see the interface of Ophcrack in Fig. 17.3. This screenshot comes from an actual system, so some information is redacted for security purposes.

Rainbow tables get very large. As of this writing, there are no portable Rainbow tables that can return passwords that are more than 12–14 characters in length. Therefore, longer passwords are useful in thwarting this attack.

The Birthday Paradox

There is a mathematical puzzle that can help. It is called the birthday paradox (sometimes called the birthday problem). The issue is this: how many people would you need to have in a room to have a strong likelihood that two would have the same birthday (i.e., month and day, not year)? Obviously, if you put 367 people in a room, at least two of them must have the same birthday, since there are only 365 days in a year + Feb 29 on leap year. However, we are not asking how many

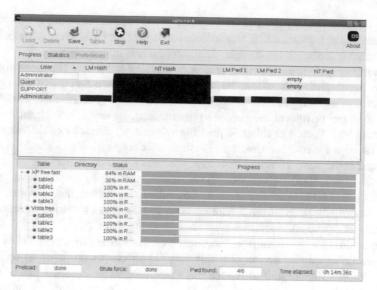

Fig. 17.3 Ophcrack

people you need to *guarantee* a match, just how many you need to have a strong probability. It just so happens that with even 23 people in the room you have a 50% chance that two have the same birthday.

How is this possible? How is it that such a low number can work? Basic probability tells us that when events are independent of each other, the probability of all of the events occurring is equal to a product of the probabilities of each of the events. Therefore, the probability that the first person does not share a birthday with any previous person is 100%, since there are no previous people in the set. That can be written as 365/365. Now for the second person, there is only one preceding person, and the odds that the second person has a different birthday than the first are 364/365. For the third person, there are two preceding people he or she might share a birthday with, so the odds of having a different birthday than either of the two preceding people are 363/365. Since each of these are independent, we can compute the probability as follows:

365/365 * 364/365 * 363/365 * 362/365 ... * 342/365 (342 is the probability of the 23 person sharing a birthday with a preceding person). Let us convert these to decimal values, which yields (truncating at the third decimal point):

1 * 0.997 * .994 * .991 * .989 * .986 * .... .936 = .49 or 49%. This 49% is the probability that they will not have any birthdays in common, thus there is a 51% (better than even odds) that two of the 23 will have a birthday in common.

Just for reference, if you have 30 people the probability that two have the same birthday is 70.6%. If you have 50 people, the probability raises to 97%, which is quite high. This does not simply apply to birthdays. The same concept can be applied to any set of data. It is often used in cryptography and cryptanalysis. The birthday paradox represents a guideline for how one might get a collision in a hashing algorithm.

In reference to cryptographic hash functions, the goal is to find two different inputs that produce the same output. When two inputs produce the same output from a cryptographic hash, this is referred to as a collision. It just so happens that the number of samples from any set of n elements, required to get a match or collision, is $1.174 \sqrt{n}$. Returning to the preceding birthday problem, $1.174 \sqrt{365} = 22.49$. For more details on this mathematical problem and how it applies to cryptography, you may wish to consult: Chap. 2 of *Modern Cryptanalysis: Techniques for Advanced Code Breaking*, which provides even more detailed coverage of the Birthday Paradox.

Other Methods

Cryptanalysis is a formal process whereby one applies specific techniques in an attempt to crack cryptography. However, as you have seen in this chapter, the ability to completely crack a cipher is quite limited. In many cases, the best one can achieve is to derive some additional information about the target key. In many cases, breaking cryptography is so time consuming, and the probability of actually deriving the key is so small, that cryptanalysis provides no practical means of breaking ciphers. Put another way: cryptanalysis is an excellent tool for testing the security of a given cipher, but it is usually not efficient enough to make it a practical means for situations that require the cipher be compromised in a short period of time. For example, a law enforcement agency that needs to break the encryption on a hard drive is unlikely to find cryptanalysis techniques of much utility. However, there are other methods for compromising ciphers that might provide more immediate results. These methods all depend on some flaw in the implementation of the cryptography.

Other Passwords

Particularly with email and hard drive encryption, there is usually some password the user must know in order to decrypt the information and access it. Many hard drive encryption tools utilize very strong encryption. For example, Microsoft Bitlocker uses AES with a 128-bit key. Several open source hard drive encryption tools use

AES with a 256-bit key. It is simply not feasible to break the key. However, it is entirely possible that the user has utilized the same password (or a substantially similar permutation) somewhere else, for example, with their email account, or with their Windows password. So, you can check these sources. For example, you may wish to use Ophcrack on their Windows computer and take those passwords (and again use permutations of those passwords as well) to try and decrypt the encrypted partition or email.

Many people are not even aware that their email password has been cracked. Several websites keep lists of email accounts that have been breached. Sites including:

haveibeenpwned.com
PwnedList.com

If, for example, you are a law enforcement officer attempting to breach an encrypted drive belonging to a suspect, you may wish to check these sites to see if that suspect's email account has been breached. You can then use the email password, and close permutations, to attempt to decrypt the hard drive.

Related Data

It is often the case that people choose passwords that have meaning for them. This makes memorization much easier. I frequently advise forensic analysts to learn all they can about a suspect. In particular, photographing or videotaping the area where the computer is seized can be quite valuable. If, for example, the suspect is an enthusiastic fan of a particular sports team, with memorabilia extensively displayed in his home or office, then this might aid in breaking encrypted drives, phones, and emails. It is at least reasonably likely that this individual's cryptography password is related to his beloved sports team.

Spyware

Of course, the easiest way to breach encryption is to see the password when the user types it in. This is often accomplished via spyware. For intelligence gathering agencies, this is a viable option. For law enforcement, this can only be done with a warrant, and for civilians this is simply not an option.

Resources

As stated at the beginning of this chapter, the goal of this chapter is just to get you a general familiarity with the concepts of cryptanalysis. If you are seeking more specifics, perhaps even tutorials on cryptanalysis, then I suggest you first master the information in this chapter, then consider the following resources:

A Tutorial on Linear and Differential Cryptanalysis by Howard M. Heys Memorial University of Newfoundland http://www.engr.mun.ca/~howard/PAPERS/ldc_tutorial.pdf

The Amazing King tutorial on Differential Cryptanalysis http://www.theamazingking.com/crypto-diff.php Note that this source is not a typical scholarly resource, but is very easy to understand.

The Amazing King tutorial on Linear Cryptanalysis http://www.theamazingking.com/crypto-linear.php Note that this source is not a typical scholarly resource but is very easy to understand.

Conclusions

This chapter has provided you with a broad general introduction to cryptanalysis. You should be very familiar with classic techniques such as frequency analysis. It is also important that you understand the concept of the birthday paradox, as well as how it is applied to hashing algorithms. For modern methods such as linear cryptanalysis and differential cryptanalysis, it is only necessary that you have a general understanding of the concepts and some grasp of the applications. It is not critical that you be able to apply these techniques. You should, however, have a working knowledge of rainbow tables.

Test Your Knowledge

1. What is brute force cracking?
2. Applying the birthday paradox, how many keys would you need to try out of n possible keys to have a 50% chance of finding a match?
3. In English __ is the most common first letter of a word.
4. In English ___ is the most common word.
5. With _____, the attacker obtains a number of plaintext–ciphertext pairs.

6. _____ is based on finding affine approximations to the action of a cipher

7. _____ the attacker discovers additional plaintexts (or ciphertexts) not previously known.

8. _____ the attacker discovers a functionally equivalent algorithm for encryption and decryption, but without learning the key.

9. A _____ is essentially a precomputed table of hashes.

10. The concept of a hash-chain is to use a _____ we will call R, that _____.

References

Biham, E., & Shamir, A. (2012). Differential cryptanalysis of the data encryption standard. Springer Science & Business Media.

Kelsey, J., Schneier, B., & Wagner, D. (1999, March). Mod n cryptanalysis, with applications against RC5P and M6. In International Workshop on Fast Software Encryption (pp. 139–155). Springer, Berlin, Heidelberg.

Knudsen, L., & Wagner, D. (2002, February). Integral cryptanalysis. In International Workshop on Fast Software Encryption (pp. 112–127). Springer, Berlin, Heidelberg.

Matsui, M. (1993, May). Linear cryptanalysis method for DES cipher. In Workshop on the Theory and Application of of Cryptographic Techniques (pp. 386–397). Springer, Berlin, Heidelberg.

Swenson, C. (2008). Modern cryptanalysis: techniques for advanced code breaking. John Wiley & Sons.

Chapter 18
Cryptographic Backdoors

Abstract Cryptographic backdoors are a special type of cryptographic attack. These are alterations in cryptographic algorithms and protocols to allow the attacker to subvert the security of the algorithm. Creating cryptographic backdoors is not particularly difficult; however, implementing them on a widespread basis requires substantial resources. In this chapter, we explore the concepts of cryptographic backdoors as well as some specific examples. Since many of the readers of this book are concerned about cybersecurity, it is important that you be familiar with this topic.

Introduction

In 2013, Edward Snowden released classified documents that demonstrated that the NSA had placed a cryptographic backdoor into a particular pseudorandom number generator, the DUAL_EC_DRBG (Landau 2014), or Dual Elliptic Curve Deterministic Random Bit Generator (some sources cite it as DUAL_ECC_DRBG emphasizing the Elliptic Curve Cryptography). This made quite a bit of news in the computer security world. Suddenly, everyone was talking about cryptographic backdoors, but few people knew what such a backdoor was or what it meant. Did it mean the NSA could read your email as easily as reading the New York Times? Did it mean you were vulnerable to hackers exploiting that backdoor? In this chapter, we address these questions.

In this chapter, you are given a general overview of what cryptographic backdoors are and how they function, and we look at specific cryptographic backdoors. While the specifics are quite interesting, the most important part of this chapter are the general concepts. You should complete this chapter with a working understanding of what a cryptographic backdoor is and how it affects security. We will also discuss a few general counter measures.

In This Chapter We Will Cover

What are cryptographic backdoors?

General concepts of cryptographic backdoors
Specific examples of cryptographic backdoors
How prevalent are cryptographic backdoors
Countermeasures

What Are Cryptographic Backdoors?

What is a cryptographic backdoor? To begin with, the concept of a backdoor is borrowed from computer security and computer programming. In computer programming, a backdoor is some way to bypass normal authentication. It is usually put in by the programmer, often for benign purposes. He or she simply wishes to be able to circumvent the normal authentication process (which may be cumbersome) for testing purposes. It is just important that these backdoors be removed from the code before distribution of the final product (Easttom 2018a). In security, a backdoor is some method that allows an attacker to circumvent the normal authentication for a target computer. For example, an attacker might send a Trojan Horse to a victim, and that Trojan Horse contains some remote desktop software, such as Timbuktu, or opens up a reverse command shell, using a tool like netcat. Now the attacker can circumvent normal authentication methods and directly access the target computer.

The concept of a cryptographic backdoor is not exactly the same, but it is related. A software backdoor gives unfettered access to the intruder. A cryptographic backdoor simply makes the encrypted message more susceptible to cryptanalysis. In essence most cryptographic backdoors simply mean that a given element if a cipher is modified to be susceptible to some attack. For example, if the pseudorandom number generator used to generate a key for a symmetric cipher, contains a backdoor, then that key is not really random.

Nick Sullivan describes the issue as follows: "In some cases, even this might not be enough. For example, TrueCrypt, like most cryptographic systems, uses the system's random number generator to create secret keys. If an attacker can control or predict the random numbers produced by a system, they can often break otherwise secure cryptographic algorithms. Any predictability in a system's random number generator can render it vulnerable to attacks. It's absolutely essential to have an unpredictable source of random numbers in secure systems that rely on them" (Sullivan 2014).

General Concepts

Kleptography is a colloquial term for creating cryptographic algorithms that resemble the original/actual algorithms but provide the creator an advantage in cracking encrypted messages. Whether you refer to cryptographic backdoors or kleptography, there are some basic properties a successful backdoor must have (Easttom 2018b).

Output Indistinguishability

This means that if one compares the output of the standard algorithm (without the backdoor) with the output of the algorithm with the backdoor, no difference should be discernible. Without output indistinguishability, anyone with even moderate cryptographic skills could simply compare the output of your algorithm implementation with the backdoor to other algorithm implementations and discover the existence of the backdoor.

Confidentiality

If a person is using a cryptographic algorithm that has a backdoor, that backdoor should not generally reduce the security of the cipher. In other words, the vulnerability should only be to the person who created the backdoor. A random attacker, unaware of the existence of the backdoor, should not have an increased chance to break the cipher.

Ability to Compromise the Backdoor

Obviously, the backdoor must provide a means whereby the attack who created the backdoor can compromise the cipher. This is, indeed, the entire purpose of having a cryptographic backdoor. As mentioned previously, however, that should not make the cipher more susceptible to general attacks.

Bruce Schneier described three other properties, he believed, a successful cryptographic backdoor should possess:

> "Low discoverability. The less the backdoor affects the normal operations of the program, the better. Ideally, it shouldn't affect functionality at all. The smaller the backdoor is, the better. Ideally, it should just look like normal functional code. As a blatant example, an email encryption backdoor that appends a plaintext copy to the encrypted copy is much less desirable than a backdoor that reuses most of the key bits in a public IV (initialization vector).
>
> High deniability. If discovered, the backdoor should look like a mistake. It could be a single opcode change. Or maybe a "mistyped" constant. Or 'accidentally' reusing a single-use key multiple times. This is the main reason I am skeptical about _NSAKEY as a deliberate backdoor, and why so many people don't believe the DUAL_EC_DRBG backdoor is real: they're both too obvious.
>
> Minimal conspiracy. The more people who know about the backdoor, the more likely the secret is to get out. So, any good backdoor should be known to very few people. That's why the recently described potential vulnerability in Intel's random number generator worries me so much; one person could make this change during mask generation, and no one else would know"

In general, a cryptographic backdoor needs to, first and foremost, allow the attacker to subvert the security of the backdoored cipher. It must also be relatively easy to use. Preferably, it would compromise a wide number of ciphers, giving the attacker the greatest range of potential access.

Specific Examples

The following examples of specific cryptographic backdoors are given to provide a practical illustration of the concepts. The RSA example given is probably the easiest to understand. The Dual_EC_DRBG backdoor has received so much attention in the media, that it would be impossible to have a discussion of backdoors, and not discuss Dual_EC_DRBG. Keep in mind that it is not critical that you memorize the backdoor techniques discussed in this section, just that you know the general concepts being used.

Dual_EC_DRBG

Dual_EC_DRBG or Dual Elliptic Curve Deterministic Random Bit Generator is a pseudorandom number generator that was promoted as a cryptographically secure pseudorandom number generator (CSPRNG) by the National Institute of Standards and Technology. This prng is based on the elliptic curve discrete logarithm problem (ECDLP) and is one of the four CSPRNGs standardized in the NIST SP 800-90A (Scott 2013).

As early as 2006, cryptography researchers suggested the algorithm had significant security issues. Other researchers in 2007 discovered issues with the efficiency of Dual_EC_DRBG and questioned why the government would endorse such an inefficient algorithm. Then in 2007, Shamow and Ferguson published an informal presentation suggesting that Dual_EC_DRBG might have a cryptographic backdoor (Easttom 2018a).

In 2013, The New York Times reported internal NSA memos leaked by Edward Snowden suggest an RNG generated by the NSA which was used in the Dual_EC_DRBG standard does indeed contain a backdoor for the NSA. Given that there had been discussions in the cryptographic community regarding the security of Dual_EC_DRBG for many years, this revelation was not a particular shock to the cryptographic community. However, it generated a great deal of surprise in the general security community.

Details

Note: This section is much easier to follow if you have a good understanding of elliptic curve cryptography in general. You may wish to review Chap. 11 before proceeding.

The algorithm specification specifies an elliptic curve, which is basically just a finite cyclic (and thus Abelian) group G. The algorithm also specifies two group elements P,Q. It doesn't say how they were chosen; all that is known is that they were chosen by the NSA. It is the choice of P and Q that form the basis for the backdoor. In the simplified algorithm, the state of the PRNG at time t is some integer s. To run the PRNG forward one step, we do the following:

You compute sP (recall we use additive group notation; this is the same as P s, if you prefer multiplicative notation), convert this to an integer, and call it r.

You compute rP, convert it to an integer, and call it s' (this will become the new state in the next step).

You compute rQ and output it as this step's output from the PRNG. (OK, technically, we convert it to a bit string in a particular way, but you can ignore that.)

Now here's the observation: it is almost certain that $P = eQ$ for some integer e. We don't know what e is, and it's hard for us to find it (that requires solving the discrete log problem on an elliptic curve, so this is presumably hard). However, since the NSA chose the values P,Q, it could have chosen them by picking Q randomly, picking e randomly, and setting $P = eQ$. In particular, the NSA could have chosen them so that they know e.

And here the number e is a backdoor that lets you break the PRNG. Suppose the NSA can observe one output from the PRNG, namely, rQ. They can multiply this by e, to get erQ. Now notice that $erQ = r(eQ) = rP = s'$. So, they can infer what the next state of the PRNG will be. This means they learn the state of your PRNG! That's really bad—after observing just one output from the PRNG, they can predict all future outputs from the PRNG with almost no work. This is just about as bad a break of the PRNG as could possibly happen.

RSA Backdoor

The RSA backdoor involves compromising the generation of keys for RSA. This example is based on an example by Yung & Young RSA labs (Young and Yung 2005). The specific steps are similar to the usual method of generating RSA keys, but slightly modified.

Choose a large value x randomly that is of the appropriate size for an RSA key (i.e., 2048 bits, 4096 bits, etc.)

Compute the cryptographic hash of this x value, denoted as H(x). We will call that hash value p.

If the cryptographic hash of x, the value p, is either a composite number or p-1 not relatively prime to e, then go to step 1, repeat until p is a prime then proceed to step 4. It should be noted that this step may need to be repeated several times to find a suitable H(x), thus this entire process is not particularly efficient.

Now choose some large random number, we will call R

Encrypt the value of x with the attackers own private key. Denote this value as c. Essentially, c is the digital signature of x, signed by the attacker. It should be noted that this is yet another flaw in this particular approach. You have to ensure you generate a public/private key pair just for this purpose. If you use a private/public key pair that you utilize for general communication, then you will have just signed your backdoor.

Now you solve for (q, r) in $(c \| R) = pq + r$.

Much like computing H(x), if it turns out that q is composite or q-1 not co-prime to e, then go to step 1.

Now you output the public key $(n = pq, e)$ and the private key p.

The victim's private key can be recovered in the following manner:

The attacker obtains the public key (n, e) of the user.
Let u be the 512 uppermost bits of n.
The attacker sets $c_1 = u$ and $c_2 = u + 1$ (c_2 accounts for a potential borrow bit having been taken from the computation

$$n = pq = (c|R) - r$$

The attacker decrypts c_1 and c_2 to get s_1 and s_2, respectively.
Either $p_1 = H(s_1)$ or $p_2 = H(s_2)$ will divide n.

Only the attacker can perform this operation since only the attacker knows the needed private decryption key corresponding to Y.

Compromising a Hashing Algorithm

Some researchers have even been exploring methods for compromising cryptographic hashing algorithms. Albertini et al. wrote a paper in which they created a cryptographic backdoor for SHA-1. In that paper, the authors state "SHA-1 as a target because it is (allegedly) the most deployed hash function and because of its background as an NSA/NIST design. We exploit the freedom of the four 32-bit round constants of SHA-1 to efficiently construct 1-block collisions such that two valid executables collide for this malicious SHA-1. Such a backdoor could be trivially added if a new constant is used in every step of the hash function. However, in SHA-1 only four different 32-bit round constants are used within its 80 steps, which significantly reduces the freedom of adding a backdoor. Actually, our attack only modifies at most 80 (or, on average, 40) of the 128 bits of the constants." (Joux and Youssef 2014).

The details of their approach are not important to our current discussion. What is important is that this is one more avenue for compromising cryptographic algorithms. Now not only symmetric ciphers and asymmetric ciphers, but cryptographic hashes as well.

The Prevalence of Backdoors

While a cryptographic backdoor can be a significant security issue, the problem is not particularly prevalent. Consider that resources are needed to implement a cryptographic backdoor. As you have seen in this chapter, the algorithms are not overly difficult, but how does one get a target organization to utilize a cipher, pseudorandom number generator, key generator, or hash that contains a backdoor. Clearly, there are going to be two primary methods for doing this. In this section, we look at both methods.

Governmental Approach

The first approach is the one that most people are familiar with, and that is government agencies either implementing backdoors in standards, as was the case with DUAL_EC_DRBG or working with a company to get a backdoor in their product. This method always runs the risk of being discovered. Even the most secure of intelligence agencies, when working with outside vendors, runs a risk of operational security being compromised.

The first such attempt was not really a backdoor, so much as it was a method whereby law enforcement, with a warrant, could gain access to symmetric keys used in secure voice communications. This was the Skipjack symmetric cipher implemented on the Clipper chip. However, negative reactions to the possibility of the federal government being able to listen in on secure communications prevented this project from moving forward.

According to Bruce Schneier "The FBI tried to get backdoor access embedded in an AT&T secure telephone system in the mid-1990s. The Clipper Chip included something called a LEAF: a Law Enforcement Access Field. It was the key used to encrypt the phone conversation, itself encrypted in a special key known to the FBI, and it was transmitted along with the phone conversation. An FBI eavesdropper could intercept the LEAF and decrypt it, then use the data to eavesdrop on the phone call." (Schneier 2013).

Edward Snowden also has claimed that Cisco cooperated with the NSA to introduce backdoors in Cisco products. Specifically, Snowden claimed that Cisco routers built for export (not for use in the United States) were equipped with surveillance tools. He further claimed that this was done without the knowledge of Cisco. According to an article in InfoWorld "Routers, switches, and servers made by Cisco are booby-trapped with surveillance equipment that intercepts traffic handled by those devices and copies it to the NSA's network." In other cases, a government agency may get the full cooperation of a technology company. Several sources claim that the computer security firm RSA took ten million dollars from the NSA, to make the Dual_EC_DRBG the default for its BSAFE encryption tool kit.

Private Citizen/Group Approach

It may seem like cryptographic backdoors are a tool that only government agencies, generally intelligence gathering agencies can use, but that is not the case. There are several methods whereby a small group, or even an individual might introduce a cryptographic backdoor. I am not aware of any real-world cases of this occurring, but it is certainly possible.

PRNG: People often select third-party products to generate random numbers for cryptographic keys. Any competent programmer with a working knowledge of cryptographic backdoors could create a product that generated random numbers/keys with a backdoor. Such a product could then be released as freeware, or via a website, guaranteeing a significant number of people might use the PRNG.

Product Tampering: Any individual working in a technology company, if that individual works in the right section of the company, could potentially introduce a backdoor without the company being aware.

Hacking: Though this is more difficult and thus less likely to occur, there is always the potential to hack into a device and compromise its cryptography. For example, a router with VPN capabilities must generate new keys for each VPN sessions. Theoretically, one could hack that router and alter its key generation. Such a task would be difficult.

These are just three ways that an individual or small group might implement a cryptographic backdoor. Clearly, this makes such backdoors a serious concern.

Note: While I indicate that I have not yet heard of individuals or small groups creating cryptographic backdoors that may be due simply to a lack of skills. The primary purpose in writing this book was to correct the problem that so few security professionals have any real understanding of cryptography. The same lack of knowledge about the details of cryptography also permeates the hacking community.

Counter Measures

Detecting cryptographic backdoors is a very difficult task. As we have seen earlier in this chapter, one of the goals of a cryptographic backdoor is to be undetectable. You could, of course, subject any random number generator or cipher to extensive cryptanalysis to determine if a backdoor is likely, but that process is very time-consuming and beyond the capabilities of most organizations. And waiting until some researcher discovers a likely backdoor, is inefficient. You could be using that backdoor cipher for quite a few years before some researcher discovers the backdoor (Easttom 2018b).

The first issue is key generation. There is always a risk when you rely on a third party for key generation. In Chap. 12, we discussed pseudorandom number generators. The best solution, assuming you have the necessary skills in programming and basic mathematics, is to write code for your own key generation. For symmetric keys, I would recommend implementing Blum, Blum, Shub, Yarrow, or Fortuna.

Crepeau and Slakmon state "This suggests that nobody should rely on RSA key generation schemes provided by a third party. This is most striking in the smartcard model, unless some guarantees are provided that all such attacks to key generation cannot have been embedded." For many, this is not a viable option. For example, many programmers who write applications for Microsoft Windows rely on the Microsoft CryptoAPI for key generation. A simple mechanism, at least for generating symmetric keys, would be to take that api (or any other) and subject it to a cryptographic hash that has the appropriately sized output, then use that output as the key. An even simpler approach would be to generate two random numbers, then XOR them together using the resulting number as your key. Either technique will essentially bury the cryptographic backdoor in another operation. The cryptographic hashing technique is the more secure and reliable of the two methods.

There have been some attempts in the industry to address this issue as well. The Open Crypto Audit Project (https://opencryptoaudit.org/) is an attempt to detect cryptographic flaws, including backdoors, in open source software. Their website lists the organizations charter:

"provide technical assistance to free open source software ('FOSS') projects in the public interest

to coordinate volunteer technical experts in security, software engineering, and cryptography,

to conduct analysis and research on FOSS and other widely used software in the public interest

contract with professional security researchers and information security firms to provide highly specialized technical assistance, analysis and research on FOSS and other widely used software in the public interest."

The group includes a number of respected cryptographers.

Conclusions

This chapter has provided you with an introduction to cryptographic backdoors. Since the Snowden revelations of 2013, this topic has become very widely discussed in the security community. Cryptographic researchers have been exploring backdoors for many years before those revelations.

You need not know the details of the specific implementations discussed in this chapter. However, the general concepts of output indistinguishability, confidentiality, etc. are critical and you should have a solid working knowledge of these topics. It is also important that you have some knowledge of counter measures discussed in this chapter.

Test Your Knowledge

1. In regard to cryptographic backdoors, what is Output Indistinguishability?
2. In regard to cryptographic backdoors, what is confidentiality?
3. What made the clipper chip relevant to backdoors?
4. _____ is a colloquial term for creating cryptographic algorithms that resemble the original/actual algorithms but provide the creator an advantage in cracking encrypted messages.
5. In simple terms, the backdoor in Dual_EC_DRBG is based on what?

References

Easttom, C. (2018a) An Overview of Cryptographic Backdoors. Journal of Information System Security, 13 (3), 177–185.

Easttom, C. (2018b). A Study of Cryptographic Backdoors in Cryptographic Primitives. In Electrical Engineering (ICEE), Iranian Conference on (pp. 1664–1669). IEEE

Hanzlik, L., Kluczniak, K., & Kutyłowski, M. (2016, December). Controlled randomness–a defense against backdoors in cryptographic devices. In International Conference on Cryptology in Malaysia (pp. 215–232). Springer, Cham.

Joux, A., & Youssef, A. (Eds.). (2014). Selected Areas in Cryptography--SAC 2014: 21st International Conference, Montreal, QC, Canada, August 14-15, 2014, Revised Selected Papers (Vol. 8781). Springer.

Landau, S. (2014). Highlights from making sense of Snowden, part II: What's significant in the NSA revelations. IEEE Security & Privacy, 12(1), 62–64.

Scott, M. (2013). Backdoors in NIST elliptic curves. https://www.certivox.com/blog/bid/344797/Backdoors-in-NIST-elliptic-curves.

Schneier, B. (2013). How to Design—And Defend Against—The Perfect Security Backdoor.

Sullivan, N. (2014). How the NSA (may have) put a backdoor in RSA's cryptography: A technical primer. https://arstechnica.com/information-technology/2014/01/how-the-nsa-may-have-put-a-backdoor-in-rsas-cryptography-a-technical-primer/

Young, A., & Yung, M. (2005, August). A space efficient backdoor in RSA and its applications. In International Workshop on Selected Areas in Cryptography (pp. 128–143). Springer, Berlin, Heidelberg.

Chapter 19
Quantum Computing and Cryptography

Abstract Many researchers believe quantum computing will be a practical reality within the next 5–10 years. There have been significant advances in the research for quantum computing in recent years that support that opinion. When quantum computing becomes a reality, current asymmetric algorithms will be rendered obsolete. This will impact e-commerce, SSL/TLS, authentication mechanisms, and many other aspects of network security. It is imperative that cyber security professionals be aware of the impact of quantum computing, and understand the state of research into quantum proof algorithms. This chapter provides a general overview of the state of quantum computing and post-quantum research as well as a discussion of the impact this research has on cyber security. Obviously, this chapter can only provide a brief, layperson's view of quantum computing and cryptography. It would take an entire book to thoroughly cover the topic. The goal of this chapter is for the reader to have a general knowledge of the impact quantum computing will have an cryptography and cybersecurity.

Introduction

Advances in quantum computing will significantly increase computing power. While quantum computing promises to provide advancements in a wide range of computer science applications there will be a deleterious effect on cyber security. It has already been demonstrated that quantum computing will render current asymmetric (i.e., public key) cryptographic protocols obsolete (Easttom 2019). Symmetric algorithms will not be affected so significantly. Algorithms such as AES, Blowfish, Serpent, and GOST will still be usable, but may need longer keys.

It is the nature of currently widely used algorithms, such as RSA, Diffie-Hellman, and Elliptic Curve that leads to the problem presented by quantum computing. Almost all facets of network security depend at least to some degree on asymmetric cryptography. This includes e-commerce, online banking, secure conferencing, and a variety of other tasks. Many of which you have studied earlier in this book. Presently, widely used algorithms are secure because they are based on mathematical problems that are difficult to solve. By difficult, it is meant that they cannot be solved in practical time using classical (i.e., non-quantum) computers. RSA is based on the difficulty of factoring integers into their prime factors. Diffie-Hellman is based on the difficulty of solving the discrete logarithm problem. The various improvements to Diffie-Hellman such as ElGamal and MQV are also predicated on the difficulty of solving the discrete logarithm problem. Even elliptic curve cryptography, which includes several algorithms, is based on the difficulty of solving discrete logarithm problems of a random elliptic curve element with respect to a publicly known base point. The dilemma for cyber security is that it has already been proven that these mathematical problems can be solved with quantum computers.

Because it has been proven that quantum algorithms can solve the problems that form the basis for current asymmetric cryptography, in a time that is practical, quantum computing will eventually render existing asymmetric or public key algorithms obsolete and ineffective. That means that the cryptography used for key exchange in VPN's, digital certificates, all e-commerce solutions, and even some network authentication protocols will no longer be secure. TLS which is widely used to secure internet traffic including web traffic, email, and even voice over IP, will no longer be secure. While these dangers to cyber security are not immediate because quantum computing is not yet a practical, usable reality, it is important for anyone in cyber security to be aware of the problem and to be familiar with the progress toward solutions.

What This Means for Cryptography

Peter Shor developed the eponymously named Shor's algorithm. He provides that a quantum computer can factor an integer N in polynomial time. The specific actual time is log N, where N is the integer being factored (Lavor et al. 2003). This is substantially faster than the most efficient known classical factoring algorithm which works in sub-exponential time. It has also been shown that quantum computers will also be able to solve discrete logarithm problems.

A brief history of quantum computing is useful to understand the current state of quantum computing, and thus the impact on cyber security. In the early 1980s Richard Feynman conceived of a "quantum mechanical computer." Richard Feynman first proposed complex calculations could be performed faster on a quantum computer. At that point, quantum computing was just a concept. In 1998, Los Alamos Laboratory and Massachusetts Institute of Technology propagated the first qubit through a solution of amino acids. The same year, the first two qubit machine was built by the University of California at Berkeley.

The first five-photon entanglement demonstrated by Jian-Wei Pan's group at the University of Science and Technology of China in 2004. Five qubits are the minimal number of qubits required for universal quantum error correction. In 2005, the Institute of Quantum Optics and Quantum Information at the University of Innsbruck in Austria developed the first qubyte (8 qubits) system. Just 1 year later, in 2006, the first 12 qubit quantum computer was announced by researchers at the Institute for Quantum Computing and the Perimeter Institute for Theoretical Physics in Waterloo, as well as MIT, Cambridge. Yale University created the first quantum processor in 2009.

In 2012, D-Wave claims a quantum computation using 84 qubits. Five years later, in 2017, IBM announced a working 50-qubit quantum computer that can maintain its quantum state for 90 microseconds. In 2018, Google announced the creation of a 72-qubit quantum chip called "Bristlecone." Note the issue with the IBM 50-qubit computer only maintaining state for 90 microseconds. One issue that is preventing quantum computing from advancing faster is the issue of decoherence. There are significant issues with maintaining the state of the qubits long enough to accomplish practical tasks.

This brief overview of the history of quantum computing demonstrates two facts. The first is that practical, usable quantum computers are not yet available. However, the second fact is that research in quantum computing is progressing at a rapid pace, with many organizations investing time, effort, and money into this research. That would tend to indicate that quantum computing will eventually be a practical, usable, reality. Many experts believe that will be the case within 10 years.

What Is a Quantum Computer?

It is beyond the scope of this book to provide a detailed discussion of quantum computing. However, we can provide a general description. A quantum computer is, simply put, a machine that performs calculations based on the laws of quantum mechanics, which is the behavior of particles at the sub-atomic level. Rather than using traditional bits, quantum computer utilizes qubits (Imre and Balazs 2013). A qubit is a two-state quantum-mechanical system. Qubits store data based on some quantum state of some particle, spin or polarization work well (Nielsen and Chuang 2002). The qubit, unlike a bit, is not in one state or the other, but is in a superposition of states, until it is measured. Once measured, it will render a discrete value (1 or 0) but until it is measured it is in a superposition of states. That is why a qubit is often represented by something called a Bloch sphere, shown in Fig. 19.1.

Fig. 19.1 Bloch Sphere

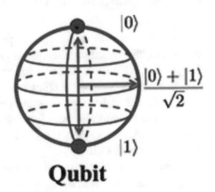

Qubit

What is occurring is that there are issues with making quantum computers continue working over a period of time that is useful for practical usage. The most prominent obstacle is controlling or removing quantum decoherence. What this means is that the state of the qubits don't keep their state for a long enough period of time. Trying to control this decoherence is the major challenge to quantum computing (Rieffel and Polak 2011). This usually means isolating the system from its environment as interactions with the external world cause the system to decohere. However, internal factors in the quantum computer itself can cause decoherence. Most quantum computing solutions are kept at supercooled temperatures, and even then the state cannot be maintained indefinitely.

The two states of a qubit are represented with quantum notation as |0>or |1>. These represent horizontal or vertical polarization. A qubit is the superposition of these two basis states. This superposition is represented as $|\psi> = \alpha|0> + \beta|1>$. The qubit will produce a single state (1 or 0) when measured but will contain any superposition of those two states (Rieffel and Polak 2011).

The issue for quantum computing is really about the qubits. The difference between a classical bit and a qubit is more than a mere naming convention change. An effective explanation for qubits is found in Neilson and Chuang:

> "What then is a qubit? Just as a classical bit has a state – either 0 or 1 – a qubit also has a state. Two possible states for a qubit are the states |0⟩ and |1⟩, which as you might guess correspond to the states 0 and 1 for a classical bit. Notation like '|⟩' is called the Dirac notation, and we'll be seeing it often, as it's the standard notation for states in quantum mechanics. The difference between bits and qubits is that a qubit can be in a state other than |0⟩ or |1⟩. It is also possible to form linear combinations of states, often called superpositions" (Nielsen and Chuang 2002)

While there are challenges to making quantum computing work practically, there are literally thousands of talented scientists and engineers working on the problem. And new advances and improvements are announced with great regularity. Most experts in the field believe these issues will be overcome in the next 5–10 years. This means that the time of current asymmetric algorithms is coming to an end.

Possible Quantum Resistant Cryptographic Algorithms

The primary reason quantum computing is a concern for cyber security professionals, is that quantum algorithms have already been developed which can solve the problems that form the basis for current asymmetric algorithms. A widely studied example is Shor's algorithm that was mentioned briefly earlier. Peter Shor developed Shor's algorithm for factoring integers. On a quantum computer it can factor an integer N in polynomial time (actual time is log N). This is substantially faster than the most efficient known classical factoring algorithm (the general number field sieve) which works in sub-exponential time.

There are several classes of algorithms that might provide a quantum Computing resistant solution. The two most widely studied are multivariate cryptography and lattice-based cryptography. Lattice-based mathematics promises to provide a range of cryptographic solutions. Lattice-based cryptography involves the construction of cryptographic primitives based on lattices. Lattices are essentially matrices, which you studied in Chap. 5 of this book. You may wish to briefly review that before continuing. A cryptographic primitive is an algorithm, such as a symmetric cipher, asymmetric cipher, cryptographic hash, or message authentication code that is part of a cryptographic application. Essentially, a complete cryptographic system must account for both confidentiality and integrity of the message. This often involves encrypting the message for confidentiality, exchanging symmetric cryptographic keys via some asymmetric algorithm, ensuring integrity with a cryptographic hash function, and digitally signing the message. Each of these aspects of security is accomplished via a different algorithm, a specific cryptographic primitive. The cryptographic primitives are combined to provide a complete cryptographic system.

Classical asymmetric cryptography is based on problems in number theory such as the factoring problem or the discrete logarithm problem. Lattice-based cryptography is simply cryptographic systems based on some problem in lattice-based mathematics. One of the most commonly used problems for lattice cryptography is the Shortest Vector Problem (SVP). Essentially, this problem is that given a particular lattice, how do you find the shortest vector within the lattice? More specifically, the SVP problem involves finding the shortest non-zero vector in the vector space V, as measured by a norm, N. A norm is a function that assigns a strictly positive length or size to each vector in a vector space. The SVP problem is a good choice for post-quantum computing (Easttom 2019).

The GGH algorithm, named after its inventors Glodreich, Goldwasser, and Halevi, is a lattice-based crypto system that was first published in 1997 and uses the closest vector problem (CVP). This problem is summarized as: given a vector space V, and a metric M for a lattice L and a vector v that is in the vector space V, but not necessarily in the lattice L, find the vector in the lattice L that is closest to the vector v. This problem is related to the previously discussed SVP problem and is also difficult to solve.

There are a variety of mathematical problems based on lattices that can form the basis for cryptographic systems, SVP and CVP are only two choices. Another such problem is the Learning With Errors (LWE) problem. This is a problem from the field of machine learning. It has been proven that this problem is as difficult to solve as several worst-case lattice problems. The LWE problem has been expanded to use algebraic rings with Ring-LWE.

Another lattice-based cryptosystem is NTRU. Many sources state that NTRU is an acronym for N'th Degree Truncated Polynomial Ring. However, most books and papers simply refer to NTRU. It was invented by Hoffstien, Pipher and Sillverman. It is the most well-known and widely studied lattice-based cryptographic system. NTRU is a cryptosystem that provides both encryption and digital signatures. It has been shown to be resistant to Shor's algorithm, unlike many other asymmetric cryptographic systems. NTRU is more secure than RSA even in a classical computing context. That makes it a viable option for classical computing (Easttom et al. 2020).

Many lattice-based cryptographic algorithms are also resistant against current cryptanalytical attacks. This means that lattice-based cryptography is a preferable solution, even when considering only classical computing. There have been advances in number sieves that are improving even classical computing's ability to break RSA. This demonstrates that there is a need for improved asymmetric cryptographic primitives, even before quantum computing becomes a reality.

The National Institute of Standards (NIST) is currently conducting a multi-year study, due to be completed in 2022, which is studying algorithms for post-quantum cryptography. The NIST study is simply allowing cryptographers from around the world to submit proposed algorithms which will then be subjected to years of cryptanalysis. This is similar to the NIST study in the late 1990s which lead to the standardization of the Advanced Encryption Standard in 2001.

Conclusions

Quantum computers are based on storing data in quantum states of particles. Such as polarization of a photon, or energy of electrons. Quantum computers are able to solve problems such as factorization of integers and discrete logarithm in polynomial time. This means that the security of RSA, Diffie-Hellman, and many other algorithms will be compromised by quantum computers. There are numerous algorithms which could potentially provide quantum computing resistant solutions.

Test Your Knowledge

1. What is the meaning of Shor's algorithm for cryptography?
2. What is a quantum computer?
3. Before measurement, what is the state of a qubit?
4. What is the closest vector problem?

References

Easttom, C. (2019). An Analysis of Leading Lattice-Based Asymmetric Cryptographic Primitives. 2019 IEEE 9th Annual Computing and Communication Conference.

Easttom, C., Ibrahim, A, Chefronov, C., Alsmadi, I., Hanson, R. (2020). Towards a deeper NTRU analysis: a multi modal analysis. International Journal on Cryptography and Information Security (IJCIS). 10(2): 11–22.

Imre, S., & Balazs, F. (2013). Quantum computing and communications: An engineering approach. Hoboken, New Jersey: John Wiley & Sons

Lavor, C., Manssur, L. R. U., & Portugal, R. (2003). Grover's Algorithm: quantum database search. arXiv preprint quant-ph/0301079.

Nielsen, M. A., & Chuang, I. (2002). Quantum computation and quantum information. Cambridge: Cambridge University Press

Rieffel, E. G., & Polak, W. (2011). A Gentle Introduction to Quantum Computing. Cambridge/London: The MIT Press

Printed in the United States
by Baker & Taylor Publisher Services